The Logic of
Quantum Mechanics

GIAN-CARLO ROTA, *Editor*
ENCYCLOPEDIA OF MATHEMATICS AND ITS APPLICATIONS

GIAN-CARLO ROTA, *Editor*
ENCYCLOPEDIA OF MATHEMATICS AND ITS APPLICATIONS

Other volumes in preparation

ENCYCLOPEDIA
OF MATHEMATICS
and Its Applications

GIAN-CARLO ROTA, Editor
Department of Mathematics
Massachusetts Institute of Technology
Cambridge, Massachusetts

Editorial Board

GIAN-CARLO ROTA, *Editor*
ENCYCLOPEDIA OF MATHEMATICS AND ITS APPLICATIONS
Volume 15

Section: Mathematics of Physics
Peter A. Carruthers, *Section Editor*

The Logic of Quantum Mechanics

Enrico G. Beltrametti and Gianni Cassinelli
University of Genoa and
Istituto Nazionale di Fisica Nucleare, Genoa

Foreword by
Peter A. Carruthers
Los Alamos National Laboratory

1981

Addison-Wesley Publishing Company
Advanced Book Program
Reading, Massachusetts

London · Amsterdam · Don Mills, Ontario · Sydney · Tokyo

Library of Congress Cataloging in Publication Data

Beltrametti, Enrico G.
  The logic of quantum mechanics.

  (Encyclopedia of mathematics and its applications;
v. 15. Section, Mathematics of physics)
  Includes bibliographical references and index.
  1. Quantum theory. I. Cassinelli, Gianni. II. Title.
III. Series: Encyclopedia of mathematics and its
applications; v. 15. IV. Series: Encyclopedia of
mathematics and its applications. Section, Mathematics of
physics.
QC174.12.B45   530.1'2   81-7904
ISBN 0-201-13514-0   AACR2

American Mathematical Society (MOS) Subject Classification Scheme (1980): 81-02, 81B10,
03G12, 03G10.

Printed in the United States of America

ABCDEFGHIJ-HA-8987654321

To our wives
Elita Beltrametti
and
Nerina Cassinelli

Contents

Editor's Statement

A large body of mathematics consists of facts that can be presented and described much like any other natural phenomenon. These facts, at times explicitly brought out as theorems, at other times concealed within a proof, make up most of the applications of mathematics, and are the most likely to survive changes of style and of interest.

This ENCYCLOPEDIA will attempt to present the factual body of all mathematics. Clarity of exposition, accessibility to the non-specialist, and a thorough bibliography are required of each author. Volumes will appear in no particular order, but will be organized into sections, each one comprising a recognizable branch of present-day mathematics. Numbers of volumes and sections will be reconsidered as times and needs change.

It is hoped that this enterprise will make mathematics more widely used where it is needed, and more accessible in fields in which it can be applied but where it has not yet penetrated because of insufficient information.

The present volume is meant to be the first of a series on quantum mechanics and its applications. It deals with the foundations as well as the fascinating logic of quantum mechanics.

This volume is more general and at places less factual than other volumes of the ENCYCLOPEDIA; nevertheless the thorough presentation will guide the reader to gain an overview of a theory that, together with relativity, is regarded as the greatest achievement in physics in this century.

GIAN-CARLO ROTA

Foreword

For many years the physical interpretation of quantum theory has been dominated by the "wave-particle duality" attitude of the "Copenhagen school." This point of view is eloquently described in Bohr's collection of essays on the subject.[1] Despite persistent concerns with apparent paradoxes and limitations to this interpretation (as exemplified by Schrödinger's cat, the paradox of Einstein, Podolsky and Rosen, among others), the Copenhagen view persists *de facto* in the daily life of the modern physicist. Traditional quantum theory has so successfully explained such a vast amount of data in atomic and molecular physics, solid state physics (and to a lesser extent, elementary particle physics) that little doubt can exist concerning its essential validity.

As a consequence of the overwhelming practical success of the "Copenhagen interpretation," the latter has acquired the status of dogma. For many years, therefore, most physicists have found it expedient to relegate the puzzling aspects of the theory to philosophers, and mathematicians. Nevertheless, a persistent interest in this subject has produced a significant and fascinating literature, reviewed by Jammer.[2] In recent years, concerns over the proper meaning to be ascribed to quantum theory have produced an increasingly deep and incisive series of investigations.

These investigations fall generally into two categories: either (1) discussion of the philosophical content of the theory, or (2) analyses of the mathematical variants of the theory and their connection with differing interpretational schemes. Physicists tend to be detached from a commitment to philosophical issues, because of their realization of the transient character of the meanings attached to theories of the day. As a simple illustration of this we can mention the profound differences between nonrelativistic and relativistic quantum mechanics. The former is the subject of most of the analysis alluded to in the foregoing. As pointed out by the authors of this treatise, profoundly different conceptual problems appear when relativity is properly taken into account. These differences are largely associated with particle creation and destruction processes and the retarded character of the sequential interactions mediated by virtual particles such as photons. Although such features are correctly accounted for by a suitable extension of quantum mechanics, the associated technical changes have a profound impact on the usual interpretative analyses. As an example we can mention the role of the position coordinate in the two theories. In nonrelativistic quantum mechanics the position variable is a full-blooded dynamical variable, while in quantum field theory it is a mere parameter (along with the time coordinate) labeling points in the space time continuum for the true dynamical variable, the quantized field itself. Associated with this situation is the conceptually distinct role played in nonrelativistic quantum mechanics by the position-momentum and energy-time uncertainty relations. Such circumstances indicate that the nonrelativistic theory is structurally incomplete and that undue focus on its detailed mathematical structure may not contribute to the mainstream of scientific progress.

Despite these reservations, the axiomatization of nonrelativistic quantum mechanics has had a decisive and creative role in the development of modern theoretical physics. Indeed the pioneering works in this field by von Neumann[3] and Dirac[4] were of such brilliance as to dominate the subject for decades, perhaps precluding heretical views. In von Neumann's book we find an almost definitive interpretation of quantum mechanics in terms of abstract relations in a Hilbert space. In addition, the essence of measurement theory is laid out in such a modern form that it is still prey to scholarly commentaries, improvement, and dispute. Dirac's incisive essay still provides the essence of the modern theorist's repertoire. Unmatched in the preeminence of physical intuition over mathematical rigor, through the stimulation of distribution theory, it has allowed modern physicists to live in a space other than Hilbert space. Indeed, nowadays one resorts to Hilbert space only to resolve mathematical niceties not clearly visible in the freewheeling world of unnormalized wave functions. This example provides foundation to the hope that future generalizations of the mathematical foundations of quantum mechanics will provide new insights, both to the conceptual and practical, by the continuing evolution of quantum field theory discoveries in elementary particle physics. In conjunction with these developments it must be mentioned that the detailed and beautiful texture of quantum mechanics can only be appreciated in the context of the group-theoretical structures dictated by both relativity theory and the "internal" symmetries discovered in weak, electromagnetic and strong interactions. These and other developments are mainly due to Wigner in a monumental sequence of works spanning the last fifty years.

At the moment of writing, quantum field theory seems to be heading in novel directions. The search for unification of the dynamics of all interactions has proceeded through promising directions through the route of non-Abelian gauge theories. The most convenient technical formulation of such theories is through the path integral formulation of field theory. Massive generalizations and possibly also modifications of our quantum-mechanical inheritance seem imminent.

In view of the illustrious history of quantum physics and the promises of contemporary problems, it is especially useful to have at hand the clear and thorough treatment of the structure of quantum theory presented in the treatise of Beltrametti and Cassinelli. Past experience shows that the refinement and analysis of the basic structural precepts of the foundations of physics can have a decisive impact on the evolution of science. Often these analyses require some time to have an impact, as exemplified by Mach's ideas on mechanics and especially his concept of the origin of inertial frames, now firmly embedded in the cosmology of modern relativity theories.

The present work is laid out in clear lines, avoiding the abyss of philosophical interpretations. The exposition adheres to a clean exposition of the mathematical content of serious formulations of rational physical alternatives of quantum theory as elaborated in recent researches, to which the authors have made noteworthy contributions. This work builds on earlier influential essays of Mackey[5], Jauch[6], Piron[7], Varadarajan[8], and others.

The treatment of the subject falls into three distinct, logical parts. In the first part, the modern version of accumulated wisdom is presented, avoiding as far as possible the traditional language of classical physics for its interpretational character. In the second part, the individual structural elements critical for the logical content of the theory are laid out, with special attention to the empirical evidence underlying each ingredient of the theory. Finally, in part three, the results of section two are used to reconstruct the usual Hilbert space formulation of quantum mechanics in a novel way. The general community of physicists and mathematicians concerned with this most decisive scientific issue will be grateful to the authors for their thoughtful and incisive analysis of this fascinating and profound problem.

PETER A. CARRUTHERS
General Editor, Section on Mathematics of Physics

1. N. Bohr, *Atomic Physics and Human Knowledge*, Wiley, New York, 1958.
2. M. Jammer, *The Philosophy of Quantum Mechanics*, Wiley-Interscience, New York, 1974.
3. J. von Neumann, *Mathematical Foundation of Quantum Mechanics*, Princeton University Press, 1955.
4. P. A. M. Dirac, *Quantum Mechanics*, 3rd ed., Oxford University Press, Oxford, 1947.
5. G. Mackey, *Mathematical Foundations of Quantum Mechanics*, W. A. Benjamin, Advanced Book Program, Reading Mass., 1963.
6. J. M. Jauch, *Foundations of Quantum Mechanics*, Addison-Wesley, Advanced Book Program, Reading, Mass., 1968.
7. C. Piron, *Foundations of Quantum Physics*, W. A. Benjamin, Advanced Book Program, Reading, Mass., 1976.
8. V. S. Varadarajan, *Geometry of Quantum Theory*, Vols. I–II., D. Van Nostrand Co., 1968.

PREFACE

"The Logic of Quantum Mechanics" appeared as the title of a scientific work in 1936, with the paper of Garrett Birkhoff and John von Neumann (9).* The use of the same title for this volume outlines the fact that a great part of the subject matter we shall deal with pertains to a research field that originated in that historic paper. This title, however, is not to be interpreted as focusing on the propositional calculus that mirrors the structure of quantum mechanics, the so-called "quantum logic" with the word "logic" used in technical sense; rather, the title should be interpreted as focusing on the mathematical foundations of quantum mechanics. The complex edifice of this theory contains simpler substructures that have direct physical bases; each of them can explain some aspects of the behavior of quantum systems. This was also the idea of the classical books by G. W. Mackey (3), J. M. Jauch (3), and V. S. Varadarajan (3), which appeared in the sixties. The present volume includes results of more than a decade of active research which followed these classical works.

The volume is divided into three parts. The first contains an exposition of the basic formalism of quantum mechanics using the theory of Hilbert spaces and of linear operators in these spaces. We shall not follow the old tradition of striving to use concepts of classical mechanics to explain quantum facts (as in the so-called principle of correspondence or the wave-particle dualism)—a tradition that reflects the unusually long delay suffered by quantum mechanics before it acquired autonomy and internal coherence, breaking with the language of the theory to be superseded. The second part follows the program of decomposing quantum theory into its conceptual constituents, singling out the basic mathematical structures, isolating what may be founded on direct empirical evidence, and controlling how single assumptions contribute to shape the theory. In the third part we face the problem of recovering the Hilbert-space formulation of quantum mechanics, starting from the simpler and more general theoretical schemes examined in the second part.

Almost every chapter contains exercises with hints. We consider them an integral part of the text. In some of them we deal with proofs of mathematical facts that are only stated in the text; in the others we provide examples that can help the comprehension of the matter.

*Papers and books mentioned in this Preface appear also in following chapters. Here we give in parentheses the number of the chapter in which the reference can be found.

This volume is primarily written for theoretical physicists and students of quantum mechanics who want to have a critical and deeper understanding of quantum theory. The knowledge of quantum mechanics is however not assumed as a prerequisite. We think that the volume can be of interest also to mathematicians. For instance, a functional analyst can find in Part I one of the most notable applications of the theory of linear operators in Hilbert space, while people dealing with lattice theory, or related fields, can find in Part II pertinent physical applications.

Throughout this volume the notion of physical system will be used. It designates the physical object to be studied, the actual object of the theory. A photon, an electron, a proton, a nucleus, a molecule, and an aggregate of molecules appear to be, in most circumstances, clear examples of physical systems. Though the notion of physical system is familiar, it contains some idealization. It is generally understood that a physical system is a portion of the physical universe, whose interaction with the rest of the universe, and with the observer in particular, can be neglected or, in any case, is such as to cause no trouble about the identity of the separated portion. The notion of physical system thus inherits, in particular, all difficulties related to the separation between the observer and what is observed. Nevertheless it works without relevant ambiguities in the cases that form the domain of elementary quantum mechanics, where destruction and creation processes are absent, or at least are not of primary importance, so that the physical system remains fixed in time. Since the passage to higher and higher energies makes destruction and creation processes more and more relevant, the expression "elementary" could here be replaced by "nonrelativistic." And, indeed, the concept of physical system becomes blurred in situations involving high-energy elementary particles, where starting with a given set of particles one can end with another set of quite different ones. Thus we shall have to do with elementary, nonrelativistic quantum mechanics.

There are some basic ingredients in the description of physical systems: we mention the concepts of "state," of "physical quantity," and of "proposition," noting that only for the first is the choice of the name universal.

By "state" is meant the result of the set of experimental procedures used to isolate and prepare the physical system. Notice that, in order to have a well-defined notion, these preparation procedures should fulfill the strong requirement of eliminating any memory of what happened to the system before the preparation procedure was started.* The notion of state collects all those attributes of the physical system that are accidental, in the sense

---

*A requirement of this sort is nonobvious in the absence of space-time confinement, as may happen in relativistic quantum field theories.

that they may be different in different situations and may change with time; the permanent attributes are instead embodied in, and form the body of, the very notion of physical system. It should be clear that the point where the definition of system ceases and the definition of state begins is, to a large extent, a matter of convention, and may depend on the historical development of scientific knowledge (as a familiar example, think of the proton and the neutron, which appeared on the scientific scene as different physical systems, with electric charge regarded as an essential attribute, but are usefully viewed as two states of one and the same system, the nucleon, when strong interactions are dominant).

By "physical quantity" is meant any quantity that can be measured: energy, position, momentum, angular momentum are familiar examples. Equally common synonyms of "physical quantity" are "observable" and "dynamical variable."

There are certain physical quantities, those that admit just two outcomes, that are of special importance and merit a name of their own. We call them "propositions." There are several alternatives to this term, each one alluding to a particular way of picturing the same idea: we may mention the terms "property," "event," "test," "question," "filter," and "yes-no experiment," all used in the literature. Characteristic of the notion here considered is a dichotomy: a proposition is either true or false, a property is either possessed or not, an event either occurs or not, a test or a filter is either passed or not, etc.

Let us anticipate that in the usual Hilbert-space formulation of quantum mechanics states are density operators, physical quantities are self-adjoint operators, and propositions are projection operators.

In some schemes of description of physical systems another basic ingredient is introduced. It may be called an "operation," though an array of somewhat different interpretations can be found: "operation" might stand for the set of instructions for performing an experiment on the physical system, or, more specifically, it might stand for the procedure of measuring a physical quantity, or it might even stand for the transformation of the state of the physical system caused by the measurement process.

The various schemes of description of physical systems that we examine in Part II make different choices about which ingredients, among states, propositions, physical quantities, and operations are taken as primitive undefined notions, and which ones are taken as derived. These approaches emphasize different mathematical structures; it will be seen in particular that what is typical of propositions is an ordered structure, which will be our main guiding theme. Relative to this ordered structure, the passage from classical to quantum systems corresponds to the abandonment of the distributive law, as already pointed out in Birkhoff and von Neumann's paper.

We favor, in this volume, the probabilistic standpoint according to which states are probability measures on the set of propositions. The nonprobabilistic standpoint, developed by C. Piron, has been eloquently exposed in his recent book (12). We only touch (in Chapter 19) on the operational approach due to G. Ludwig and B. Mielnik. We leave aside the developments of this approach due to E. B. Davies, C. M. Edwards, J. Lewis, and others (19). We also leave aside the operational statistics worked out by D. Foulis, C. Randall, and their school, which constitutes a general language promising an alternative approach to quantum theory. We hope that, within these limits, this volume contributes to the understanding of the logic of quantum mechanics.

It is a pleasure to thank Professor G. C. Rota for asking us to write this volume, and many people who helped us in various ways: R. Greechie for numerous general suggestions and for a critical analysis of the manuscript; G. W. Mackey, with whom we discussed the organization of the volume when it was in an early stage of development; M. Maczynski for many suggestions on the structure of the volume; S. S. Holland for valuable correspondence; B. Mielnik for a number of enlightening discussions; S. Bugajski for his comments and for criticisms and suggestions on Sections 3.3 and 3.4 and Chapter 18; G. T. Ruttimann for his decisive help with Chapter 19; P. Lahti for useful discussions, especially on the subjects of Sections 3.4, 3.5 and 14.7; P. Truini for his help on Section 22.3; and D. Aerts for his help on Chapter 24.

Last but not least, we are deeply grateful to Mrs. B. Basiglio for her careful and patient typing of several versions of the manuscript, and to Mrs. N. Cassinelli for her final check of the typescript and the addition of hand-drawn symbols.

ENRICO G. BELTRAMETTI

GIANNI CASSINELLI

The Logic of
Quantum Mechanics

PART I

Hilbert-Space Quantum Mechanics

This Part is a self-contained exposition of the basic structure of quantum mechanics in Hilbert space. As such it does not assume a knowledge of quantum mechanics, although some familiarity might help in understanding specific physical examples (spin, free particle, etc.). Some general acquaintance with the elementary theory of operators in Hilbert space is presupposed. When less elementary facts are needed, we refer to short appendices. We assume also some acquaintance, even if only at an intuitive level, with measure and integration theory.

CHAPTER 1

Static Description of Quantum Systems

1.1 The Hilbert-Space Description

By the static description of a physical system we mean the rules that assign specified mathematical objects to the states and to the physical quantities of the system, and the prescriptions for calculating the probability distribution of the possible values of every physical quantity when the state of the system is given.

In the usual Hilbert-space formulation of quantum mechanics, to each physical system is attached a separable Hilbert space $\mathcal{H}$ (generally infinite-dimensional) over the complex field. To every physical quantity is associated a linear, self-adjoint, not necessarily bounded operator on $\mathcal{H}$. If one deals with a strictly quantum system, then the converse is also generally assumed: every self-adjoint operator on $\mathcal{H}$ represents some physical quantity. The restriction "strictly quantum" is necessary: if the system retains some nonquantum (i.e. classical) feature, so that it requires the algorithm of so-called "superselection rules" (see Chapter 5), then there are self-adjoint operators on $\mathcal{H}$ that do not represent physical quantities. It should also be stressed that even in the strictly quantum case, most of the self-adjoint operators actually do not represent "interesting" physical quantities: only a few of them represent physical quantities that are useful and meaningful for the description of the physical system (e.g., energy, momentum, position, angular momentum). Therefore, by asserting that every self-adjoint operator on $\mathcal{H}$ corresponds to some physical quantity we mean there is no *a priori* impossibility of devising such a correspondence, but we do not claim that this correspondence is present in the real work of experimental physics.

ENCYCLOPEDIA OF MATHEMATICS and Its Applications, Gian-Carlo Rota (ed.). Vol. 15: E. G. Beltrametti and G. Cassinelli, The Logic of Quantum Mechanics

ISBN 0-201-13514-0

Another crucial ingredient in the description of a physical system is the notion of state: the Hilbert-space formulation of quantum mechanics assigns to every state a mathematical object, which is usually called a "density operator", or "statistical operator", or (especially in the physical literature) "density matrix". Technically, it is a linear, bounded, self-adjoint, positive, trace-class operator on $\mathcal{H}$ of trace one.* If we deal with a strictly quantum system, then also the converse is true: every density operator represents a state of the system, so that the set of states can be identified with the set of density operators.

1.2 Probability Distribution of a Physical Quantity

Consider now the statistics of the possible values of a physical quantity. The question we must answer is the following: if D is the density operator that represents the state of the system, and if E is a subset of the real line $\mathbb{R}$, what is the probability that the value of the physical quantity represented by the operator A lies in E? Actually, in these terms the question is not well posed, for it is impossible to construct a probability measure defined on all subsets of $\mathbb{R}$. One has to restrict oneself to a distinguished family of subsets of $\mathbb{R}$, the family of Borel subsets, to be denoted $\mathcal{B}(\mathbb{R})$, that is, to the smallest family of subsets of $\mathbb{R}$ that includes the open sets and is closed under complements and under countable intersections. Thus, our question becomes: if the system is in state D, what is the probability that the value of the physical quantity represented by A is in the Borel set E?

The answer makes use of the spectral theorem for self-adjoint operators (see Appendix B). According to it, A determines, for every Borel set E of $\mathbb{R}$, a projection operator, denoted $P_A(E)$, that satisfies:

(i) $P_A(\varnothing)=0$, $P_A(\mathbb{R})=I$,
(ii) $P_A(\cup_i E_i)=\Sigma_i P_A(E_i)$ for every disjoint sequence $\langle E_i \rangle$ in $\mathcal{B}(\mathbb{R})$.†

The function $E \mapsto P_A(E)$ is called the spectral projection-valued measure of A.

Now, it is a basic assumption of quantum mechanics that $\operatorname{tr}(DP_A(E))$ is precisely the probability that the value of the physical quantity represented by A, when the state of the system is D, lies in the Borel set E. Notice that, for any fixed A and D, the function $E \mapsto \operatorname{tr}(DP_A(E))$ has indeed the properties of a probability measure on $\mathbb{R}$, for, due to (i), (ii), and using a result of

---

*For properties of the trace and of trace-class operators we refer to Appendix A.

†In case the sum is infinite it has to be understood as a strong limit, that is, s-$\lim_{n \to \infty} \Sigma_{i=1}^{n} P_A(E_i)$.

Appendix A, we have

$$\text{tr}(DP_A(\varnothing)) = 0, \qquad \text{tr}(DP_A(\mathbb{R})) = 1,$$

$$\text{tr}\left(DP_A\left(\bigcup_i E_i\right)\right) = \sum_i \text{tr}(DP_A(E_i))$$

$$\text{if } \langle E_i \rangle \text{ is a disjoint sequence in } \mathscr{B}(\mathbb{R}).$$

The function $E \mapsto \text{tr}(DP_A(E))$, $E \in \mathscr{B}(\mathbb{R})$, is called the distribution of the physical quantity represented by A in the state D.

CHAPTER 2 _____

States

2.1 Pure and Nonpure States

As we said above, the states of a physical system are identified with the density operators in the associated Hilbert space $\mathcal{H}$. The set they form is a convex set. This means that, if D_1, D_2 are distinct density operators and w_1, w_2 are real positive numbers such that $w_1 + w_2 = 1$, then $w_1 D_1 + w_2 D_2$ is again a density operator, a fact easily checked. Thus, $w_1 D_1 + w_2 D_2$ represents a new state of the system and is called the convex combination of the states D_1 and D_2 with weights w_1 and w_2. Of course, this notion of convex combination can be extended to more than two states: if $D_1, \ldots, D_n$ are density operators and $w_1, \ldots, w_n$ are positive numbers such that $w_1 + \cdots + w_n = 1$, the operator $w_1 D_1 + \cdots + w_n D_n$ is still a density operator, thus representing a state of the system: the convex combination of $D_1, \ldots, D_n$ with weights $w_1, \ldots, w_n$.

Actually, the set of states is even more than a convex set: it is σ-convex. This means that, for every countable sequence $\langle D_i \rangle$ of density operators, and for every corresponding sequence $\langle w_i \rangle$ of positive numbers such that $\Sigma_i w_i = 1$, the uniform limit

$$\text{u-lim}_{n \to \infty} \sum_{i=1}^{n} w_i D_i$$

is still a density operator (see Exercise 1 at the end of this chapter). It is called the convex combination of the D_i's with weights w_i.

In the sequel we shall adopt the notation $\Sigma_i w_i D_i$ to denote a generic convex combination, understanding the uniform limit in case the sum is

ENCYCLOPEDIA OF MATHEMATICS and Its Applications, Gian-Carlo Rota (ed.). Vol. 15: E. G. Beltrametti and G. Cassinelli, The Logic of Quantum Mechanics

ISBN 0-201-13514-0

infinite. Moreover, when we say that $\langle w_i \rangle$ is a sequence of weights, we shall always understand that the w_i's are positive numbers and that their sum is 1.

It is important to remark that, in case D is a convex combination, say $\sum_i w_i D_i$, the probability $\mathrm{tr}(DP_A(E))$ that the physical quantity represented by A takes a value in E (see Section 1.2) has the form

$$\mathrm{tr}(DP_A(E)) = \sum_i w_i \mathrm{tr}(D_i P_A(E)). \qquad (2.1.1)$$

This is obvious when D is a finite convex combination; it is less obvious, but still true, when D is an infinite convex combination (see Exercise 2).

Notice that the decomposition of D as a convex combination need not be unique (indeed, we shall see that it is never unique in the quantum case): the equality (2.1.1) says also that the sum on the right-hand side does not depend upon the decomposition one chooses.

It may happen that a state cannot be written as a convex combination of other (distinct) states: in this case it is called a *pure state*. In the language of convex-set theory the pure states are thus the extreme points of the convex set of states. The states that are convex combinations of others are then called *nonpure*, or *mixtures*.

We list now a number of relevant facts:

(i) *Projectors onto one-dimensional subspaces are states.* This follows by checking that one-dimensional projection operators are a particular case of density operators. In the sequel, the one-dimensional subspace spanned by a vector ψ will be denoted by $\hat{\psi}$, and the projector onto $\hat{\psi}$ will be denoted by $P^{\hat{\psi}}$.

(ii) *For every state D there exist an orthonormal sequence $\langle \psi_i \rangle$ of vectors of $\mathcal{H}$ (not necessarily complete) and a corresponding sequence $\langle w_i \rangle$ of weights, such that $D = \sum w_i P^{\hat{\psi}_i}$.* This follows by remarking that D, being a trace-class operator, is compact,[1] and as such admits the canonical decomposition of the form $\sum_i w_i P^{\hat{\psi}_i}$. The positivity of D entails the positivity of the coefficients w_i, and the condition $\mathrm{tr}\, D = 1$ entails that the sum of the w_i's is 1. The reader will notice that the weights $w_1, w_2, \ldots$ can also be viewed as the eigenvalues of the operator D.

(iii) *A state is pure if and only if it is a projector onto a one-dimensional subspace.* After remarking that the canonical decomposition under (ii) is indeed a proper convex combination unless, for some index k, $w_k = 1$ (and $w_i = 0$ if $i \neq k$), it follows that if D is pure, then $D = P^{\hat{\psi}_k}$. Conversely, suppose $D = P^{\hat{\psi}}$ for some $\psi \in \mathcal{H}$. Should it be possible to write $P^{\hat{\psi}} = wD_1 + (1 - w)D_2$, with $0 < w < 1$ and D_1, D_2 distinct density operators, we would have,* for every vector φ orthogonal to ψ, $w(\varphi, D_1\varphi) + (1-w)(\varphi, D_2\varphi) = 0$,

---

*The scalar product in $\mathcal{H}$ will always be denoted by $(\cdot, \cdot)$.

and hence, from the positivity of the density operators, $(\varphi, D_1\varphi)=(\varphi, D_2\varphi)$
$=0$. Then writing $D_1=U_1^2$, $D_2=U_2^2$ with U_1, U_2 self-adjoint operators, it
would follow that $\|U_1\varphi\|^2=\|U_2\varphi\|^2=0$, whence $D_1\varphi=D_2\varphi=0$, so that both
D_1 and D_2 would be proportional to $P^{\hat{\psi}}$ and indeed would coincide with $P^{\hat{\psi}}$,
being of trace one. This would contradict the assumption $D_1 \neq D_2$.

(iv) *A state D is pure if and only if $D=D^2$*. Indeed, by using the
canonical decomposition $D=\Sigma w_i P^{\hat{\psi}_i}$ of item (ii), we have $D^2=\Sigma w_i^2 P^{\hat{\psi}_i}$, so
that the equality $D=D^2$ reads $w_i=w_i^2$ for every i. Hence, recalling that
$\Sigma_i w_i=1$, there must be some index k such that $w_k=1$, while $w_i=0$ if $i \neq k$:
thus $D=P^{\hat{\psi}_k}$. We come to the conclusion by use of item (iii).

(v) *If D is a pure state, then the probability $\mathrm{tr}(DP_A(E))$ that the value of
the physical quantity represented by A lies in E, can be computed as
$(\psi, P_A(E)\psi)$, where ψ is any unit vector in the one-dimensional subspace onto
which D projects.* This follows by remarking that it is always possible to
choose in $\mathcal{H}$ an orthonormal basis containing ψ as an element. Moreover, if
$\varphi=e^{i\lambda}\psi$, $\lambda \in \mathbb{R}$, then $(\varphi, P_A(E)\varphi)=(\psi, P_A(E)\psi)$.

In view of what said above, the pure states are in one-to-one correspondence with the one-dimensional subspaces of $\mathcal{H}$. The set of unit vectors that
differ only by a phase factor is customarily called a *ray* of $\mathcal{H}$; then it may be
said that the pure states are in one-to-one correspondence with the rays of
$\mathcal{H}$. With a slight abuse of language, it is also customary to call the unit
vectors themselves "pure states". Accordingly, the pure states are also called
vector states.

2.2 Superpositions of Pure States

The linearity of the Hilbert space provides a way of generating new pure
states out of any given pair of pure states: if ψ_1 and ψ_2 are unit vectors in $\mathcal{H}$,
then any linear combination $\lambda_1\psi_1+\lambda_2\psi_2$, $\lambda_1, \lambda_2 \in \mathbb{C}$, $|\lambda_1|^2+|\lambda_2|^2=1$, is
(representative of) another pure state which is called a *superposition* of the
pure states ψ_1 and ψ_2. This fact is often referred to as the superposition
principle of quantum mechanics. Needless to say, the superposition of pure
states has nothing to do with the convex combination of states: the
superposition of pure states is again a pure state, while the convex combination of (pure) states is, by definition, a nonpure state. The appearance of
these superpositions of pure states is a distinguishing feature of quantum
mechanics: it has no analogue in classical mechanics. True, a notion of
superposition is also present in classical mechanics (e.g., in classical wave
phenomena), but what is superposed in classical examples is quite different
from what is superposed in quantum mechanics. In the rest of this volume
we shall repeatedly return to the notion of quantum superposition of states
as a sort of guiding theme.

2.3 Nonunique Decomposability of Quantum Mixtures

In the Hilbert-space formulation of quantum mechanics the set of pure states is rich enough to generate, by convex combination, the set of all nonpure states. In other words, every nonpure state can be written as a convex combination (possibly infinite) of pure states. This fact is contained in item (ii) of the preceding section. But the question arises: is the decomposition of a nonpure state into pure states unique? In other words: does a nonpure state determine uniquely the pure states it is a convex combination of? The answer is no, as is made clear by the following remarks.

Given any nonpure state D, its decomposition into *pairwise orthogonal* pure states is unique if and only if there are no degenerate eigenvalues of D. In other words: if, writing $D = \sum_i w_i P^{\hat{\psi}_i}$, with $\langle \psi_i \rangle$ an *orthonormal* sequence of vectors, it happens that $w_i \neq w_k$ for $i \neq k$, then there is no other *orthonormal* sequence of vectors that does the trick. All this follows from the uniqueness of the spectral decomposition of self-adjoint operators.

If the density operator D has degenerate eigenvalues, then it admits infinitely many decompositions into orthonormal sequences of vectors. Any two such decompositions will read $D = \sum_i w_i P^{\hat{\psi}_i}$ and $D = \sum_i w_i P^{\hat{\varphi}_i}$, with equal weights and with the orthonormal sequence $\langle \varphi_i \rangle$ obtained from the orthonormal sequence $\langle \psi_i \rangle$ by replacing the vectors belonging to each degenerate subspace with any other orthonormal basis of that subspace.

If we give up the requirement of orthogonality of the sequence of vectors carrying the decomposition, then every mixture admits infinitely many decompositions into pure states. Let us make a few comments to justify this assertion.

Suppose that a density operator D is given that admits the canonical decomposition $D = \sum_i w_i P^{\hat{\psi}_i}$ into the orthonormal sequence $\langle \psi_i \rangle$. We are to show that we can choose, in infinitely many ways, another (nonorthogonal) sequence $\langle \varphi_i \rangle$ of unit vectors and a corresponding sequence $\langle w_i' \rangle$ of weights such that $D = \sum_i w_i' P^{\hat{\varphi}_i}$. Of course, the equivalence of the two decompositions can be expressed by the condition that

$$\mathrm{tr} \left(\sum_i w_i P^{\hat{\psi}_i} P^{\hat{\vartheta}} \right) = \mathrm{tr} \left(\sum_i w_i' P^{\hat{\varphi}_i} P^{\hat{\vartheta}} \right) \qquad \text{for all} \quad \vartheta \in \mathcal{K},$$

which is physically interpreted as the fact that the two expressions for D must assign the same probability distribution to every physical quantity. Now, expanding the φ_j's and ϑ into a basis that contains $\langle \psi_i \rangle$, and writing $b_{ij} = (\psi_i, \varphi_j)$, $a_i = (\psi_i, \vartheta)$, this condition entails, first of all, that the sequence $\langle \varphi_i \rangle$ has to span the same subspace of $\mathcal{K}$ that is spanned by $\langle \psi_i \rangle$, so that

$\Sigma_i |b_{ij}|^2 = 1$ for all i, and then it takes the form

$$\sum_i w_i |a_i|^2 = \sum_{k,i,j} w'_k a_i \bar{a}_j b_{jk} \bar{b}_{ik} \qquad \text{for every sequence } \langle a_i \rangle.$$

Therefore we have

$$w_i = \sum_k w'_k |b_{ik}|^2 \quad \text{and} \quad \sum_k w'_k b_{jk} \bar{b}_{ik} = 0 \quad \text{when} \quad i \neq j. \qquad (2.3.1)$$

To show that there are indeed infinitely many solutions to these conditions (in which the original weights w_i are given), take any two complex numbers λ_1, λ_2 such that

$$\frac{|\lambda_1|^2}{|\lambda_2|^2} < \frac{w_1}{w_2} \quad \text{and} \quad |\lambda_1|^2 + |\lambda_2|^2 = 1$$

and choose the following sequences of state vectors and weights:

$$\varphi_1 = \psi_1, \qquad\qquad\qquad w'_1 = w_1 - \frac{|\lambda_1|^2}{|\lambda_2|^2} w_2,$$

$$\varphi_2 = \lambda_1 \psi_1 + \lambda_2 \psi_2, \qquad\quad w'_2 = \frac{1}{2|\lambda_2|^2} w_2,$$

$$\varphi_3 = \lambda_1 \psi_1 - \lambda_2 \psi_2, \qquad\quad w'_3 = w'_2,$$

$$\varphi_n = \psi_{n-1}, \qquad\qquad\quad w'_n = w_{n-1} \qquad\qquad \text{when} \quad n \geq 4.$$

It is immediate that this choice satisfies (2.3.1), so that

$$D = \sum_i w_i P^{\hat{\psi}_i} = \sum_i w'_i P^{\hat{\varphi}_i}.$$

To visualize these two decompositions of D let us refer to a familiar physical example. A photon beam, travelling along the x-axis, is partially polarized in the (z, x) plane with a polarization degree of 0.6. Let ψ_1 and ψ_2 denote the states of complete polarization in the (z, x) plane and the (y, x) plane, respectively. Then we know that our beam can be described by the mixture $0.8\, P^{\hat{\psi}_1} + 0.2\, P^{\hat{\psi}_2}$. Take now $\lambda_1 = \lambda_2 = \frac{1}{2}$, so that φ_2 and φ_3 denote the states of complete polarization in the two crossed planes at $\pm 45°$ relative to (z, x) plane. According to what is said above, the beam should be also described by the mixture $0.6\, P^{\hat{\psi}_1} + 0.2\, P^{\hat{\varphi}_2} + 0.2\, P^{\hat{\varphi}_3}$, and this is indeed asserted by elementary optics.

Notice also that the sequence $\langle \varphi_i \rangle$ considered above is, in a sense, redundant, for the pair $\langle \psi_1, \psi_2 \rangle$ is replaced by the triple $\langle \varphi_1, \varphi_2, \varphi_3 \rangle$. This fact is accidental. What is not accidental is that $\langle \varphi_i \rangle$ is not orthogonal.

There is another fact that is not accidental: when we go from the orthogonal sequence $\langle \psi_i \rangle$ to the nonorthogonal sequence $\langle \varphi_i \rangle$, we make use of quantum superpositions (φ_2 and φ_3 in the construction above). Indeed, we remarked that both sequences have to span the same subspace. Therefore we come to the important conclusion that the nonunique decomposability of quantum mixtures into pure states is deeply intertwined with the fact that pure states can be superposed to get new pure states. The nonunique decomposability of mixtures is thus recognized as a genuine quantum phenomenon.

Let us add a remark. Given a nonpure state D, take any one of its decompositions into pure states, say $D = \Sigma_i w_i P^{\hat{\psi}_i}$; then the probability distribution that it assigns to the physical quantity represented by A takes the form

$$E \mapsto \sum_i w_i (\psi_i, P_A(E)\psi_i), \tag{2.3.2}$$

as follows from (2.1.1) and from item (v) of Section 2.1, and this probability distribution is independent of the particular decomposition of D we have chosen.

2.4 Mixtures and Preparation Procedures

To each decomposition of a mixture D we can, in a natural way, associate an operational procedure that leaves the physical system in that mixture. If, for definiteness, we write $D = wP^{\hat{\psi}_1} + (1-w)P^{\hat{\psi}_2}$ $(0 < w < 1)$, it is natural to think of the following preparation procedure: take a large number of replicas of the given physical system; prepare a fraction w of them in the pure state ψ_1 and the remaining fraction $1-w$ in the pure state ψ_2; then mix up the replicas and make a random choice of one of them.

For classical systems mixtures always have a unique decomposition into pure states. This justifies the "ignorance interpretation" of classical mixtures. According to this interpretation, when we write $D = wP^{\hat{\psi}_1} + (1-w)P^{\hat{\psi}_2}$ we mean that the physical system is, in reality, either in the pure state ψ_1 or in the pure state ψ_2, but our ignorance prevents our saying in which one of them the system actually is: our knowledge is limited to saying that the system is in state ψ_1 with probability w, and in state ψ_2 with probability $1-w$.

For quantum systems, on the contrary, mixtures have no unique decomposition into pure states: to each nonpure state there correspond infinitely many decompositions. Accordingly, the ignorance interpretation of mixtures becomes untenable. Given a mixture, we can never say that the system actually is in some pure state with some probability; indeed, this pure state, though present in some decomposition, might not be present in other decompositions of the same mixture. Of course, it is still obviously possible, as we said above, to associate each decomposition to a definite preparation procedure, but then the nonunique decomposability of mixtures amounts to saying that quantum mechanics associates to different experimental preparation procedures the same mathematical object, that is, the same density operator. In other words, the algorithm of quantum mechanics does not code a complete memory of the actual experimental procedure that has been used to prepare the physical system. If D admits, among others, the decompositions $wP^{\hat{\psi}_1}+(1-w)P^{\hat{\psi}_2}$ and $w'P^{\hat{\varphi}_1}+(1-w')P^{\hat{\varphi}_2}$, then quantum theory does not retain any memory of the fact that the system might have been prepared by use of experimental devices that produce and mix the pure states ψ_1,ψ_2, or by use of devices that produce and mix the pure states

In Chapter 4 we shall further discuss, on the basis of some definite physical systems, this peculiar feature of quantum mechanics.

The nonunique decomposability of quantum mixtures, and the related untenability of the ignorance interpretation of mixtures, constitute a spectacular departure of quantum phenomena from classical ones: indeed, most of the so-called "paradoxes" of quantum mechanics have their root in this nonunique decomposability of mixtures. We shall return to this point in Chapter 8.

2.5 Transition Probabilities between Pure States

Let us remark that states can be thought of as particular physical quantities, since density operators are particular examples of self-adjoint operators. Explicitly, writing the canonical decomposition $D=\Sigma w_i P^{\hat{\psi}_i}$ into the orthogonal sequence $\langle \psi_i \rangle$, the state D could be interpreted as the physical quantity whose values are the weights $w_1,w_2,\dots$. In particular, restricting D to a pure state (i.e., to a one-dimensional projector, say $P^{\hat{\psi}}$), we can interpret it as the two-valued physical quantity that has the values 1 or 0 according as the system is or is not in the pure state ψ. If φ is another pure state, then the quantity

$$\mathrm{tr}\big(P^{\hat{\varphi}}P^{\hat{\psi}}\big),$$

in which $P^{\hat{\psi}}$ is thought of as a physical quantity and $P^{\hat{\varphi}}$ as a state, gives the probability that the system is found in the state ψ if it is known to have been prepared in the state φ. This probability can be computed as

$$\mathrm{tr}\big(P^{\hat{\varphi}}P^{\hat{\psi}}\big)=\big(\varphi,(\psi,\varphi)\psi\big)=|(\varphi,\psi)|^2,$$

and it is called the *transition probability* between the pure states φ and ψ.

Exercises

1. Show that, if $\langle D_i \rangle$ is a countable sequence of density operators and $\langle w_i \rangle$ is a sequence of weights, then

$$D = \text{u-lim} \sum_{i=1}^{n} w_i D_i$$
$$\quad\;\; {}_{n \to \infty}$$

is a density operator.

[*Hint.* $\langle \Sigma_{i=1}^{n} w_i D_i \rangle$ is a Cauchy sequence in the Banach space $\mathbb{B}(\mathcal{H})$ of bounded operators on $\mathcal{H}$. Then use Theorems A.4 and A.6 of Appendix A.]

2. Show that the density operator D defined in Exercise 1 satisfies (2.1.1).
[*Hint.* Write $\text{tr}(D P_A(E)) = \text{tr}([D^{1/2}P_A(E)]^*[D^{1/2}P_A(E)]) = \|[D^{1/2}P_A(E)]^*[D^{1/2}P_A(E)]\|_1 = \|P_A(E)DP_A(E)\|_1$; then use Theorem A.7 of Appendix A.]

Reference

1. M. Reed and B. Simon, *Methods of Modern Mathematical Physics*, Vol. I, Academic Press, New York, 1972.

CHAPTER 3

Physical Quantities

3.1 Statistics of Physical Quantities

3.1.1 *Spectrum*

As has already been said, the probability distribution of values of the physical quantity represented by the self-adjoint operator A, in the state D, is given by

$$E \mapsto \mathrm{tr}(DP_A(E)), \qquad E \in \mathfrak{B}(\mathbb{R}). \tag{3.1.1}$$

The spectrum $\sigma(A)$ of A is the support of the projection-valued spectral measure $P_A(\cdot)$: it is the subset of $\mathbb{R}$ on which the probability distribution (3.1.1) is concentrated. In other words, the possible values of the physical quantity represented by A are the elements of $\sigma(A)$.

Let us consider two examples.

Example 1. A has a pure point spectrum. Denote by $\lambda_1, \lambda_2, \ldots$ the eigenvalues of A, and by $\mathfrak{M}_i$ the eigenspace belonging to the eigenvalue λ_i. Since nonpure states can always be decomposed into pure ones (see in particular the remark at the end of Section 2.3), let us restrict ourselves to the latter. If ψ is any state vector, the probability distribution

$$E \mapsto (\psi, P_A(E)\psi)$$

is an atomic measure that has the countable set $\{\lambda_1, \lambda_2, \ldots\}$ as support. The

ENCYCLOPEDIA OF MATHEMATICS and Its Applications, Gian-Carlo Rota (ed.). Vol. 15: E. G. Beltrametti and G. Cassinelli, The Logic of Quantum Mechanics

ISBN 0-201-13514-0

probability that our physical quantity has the value λ_i is

$$\left(\psi, P_A(\{\lambda_i\})\psi\right)=\left(\psi, P^{\mathfrak{M}_i}\psi\right),$$

where $P^{\mathfrak{M}_i}$ is the projector onto $\mathfrak{M}_i$. If $\{\varphi_i^k\}_{k=1}^{\dim \mathfrak{M}_i}$ is any orthonormal basis in $\mathfrak{M}_i$, then

$$P^{\mathfrak{M}_i}\psi= \sum_{k=1}^{\dim \mathfrak{M}_i} \left(\varphi_i^k,\psi\right)\varphi_i^k$$

and

$$\left(\psi, P_A\{\lambda_i\}\psi\right)= \sum_{k=1}^{\dim \mathfrak{M}_i} \left|\left(\varphi_i^k,\psi\right)\right|^2. \tag{3.1.2}$$

If λ_i is nondegenerate (that is, $\mathfrak{M}_i$ is one-dimensional, say $\mathfrak{M}_i=\hat{\varphi}_i$), then the probability in question is simply $|(\varphi_i,\psi)|^2$. In case all eigenvalues are nondegenerate, so that the spectrum of A is simple, there exists in $\mathcal{K}$ an orthonormal basis $\varphi_1, \varphi_2,\dots$, in which φ_i is a normalized eigenvector of A belonging to λ_i, and every unit vector $\psi\in\mathcal{K}$ can be written in the form

$$\psi= \sum_i c_i\varphi_i, \qquad c_i=(\varphi_i,\psi).$$

The number $|c_i|^2$ is thus recognized as the probability that the physical quantity has the value λ_i when the system is in the vector state ψ.

Example 2. A has a simple, purely continuous spectrum. According to the multiplication-operator form of the spectral theorem (see Appendix B), to each self-adjoint operator A that has a simple spectrum $\sigma(A)$ we can associate a unitary transformation U and a measure μ on $\mathbb{R}$, concentrated on $\sigma(A)$, such that U maps $\mathcal{K}$ onto the Hilbert space $L^2(\sigma(A),\mu)$ of the complex-valued functions on $\sigma(A)$ that are square-integrable with respect to the measure μ, while A is mapped into an operator, UAU^{-1}, that acts as a multiplicative operator in $L^2(\sigma(A),\mu)$. In the usual language of quantum mechanics, this isomorphic replica of $\mathcal{K}$ is called the "representation in which A is diagonal" or simply the "A-representation"; the state vector $\psi\in\mathcal{K}$ is mapped into a normalized function in $L^2(\sigma(A),\mu)$, say $\psi(\lambda)$, which is called the "wave function of the system in the A-representation". In this representation the probability distribution of the physical quantity represented by A becomes [denoting by $\chi_E(\cdot)$ the characteristic function of the Borel set E]

$$\left(\psi(\lambda), P_{UAU^{-1}}(E)\psi(\lambda)\right)=\left(\psi(\lambda), \chi_E(\lambda)\psi(\lambda)\right)$$

$$=\int_E|\psi(\lambda)|^2\mu(d\lambda).$$

The hypothesis of purely continuous spectrum makes the integral a proper one, without degeneration into a sum. Thus, $|\psi(\lambda)|^2\mu(d\lambda)$ is recognized as the probability that the physical quantity takes a value in the interval $[\lambda, \lambda + d\lambda]$. Notice that the measure μ is not uniquely determined (only its null sets are fixed), as we shall see later in explicit physical examples. Different choices of μ are referred to, in the language of quantum mechanics, as different "normalizations in the continuum". Physically relevant examples of operators having simple, purely continuous spectra are the position and the momentum of a free particle; they will be examined in Section 4.4.

3.1.2 *Mean Value*

We define the *mean value*, or expectation, of the physical quantity represented by the self-adjoint operator A in the state represented by the density operator D, as the mean value, when it exists, of the probability distribution

$$E \mapsto \mathrm{tr}(DP_A(E)), \qquad E \in \mathcal{B}(\mathbb{R}),$$

and we denote it by $\mathcal{E}(A, D)$. Explicitly,

$$\mathcal{E}(A, D) = \int_{-\infty}^{+\infty} \lambda \, \mathrm{tr}(DP_A(d\lambda)), \qquad (3.1.3)$$

provided the integral exists.

Consider first the case in which D represents a pure state, say $D = P^{\hat{\psi}}$, and write $\mathcal{E}(A, \psi)$ in place of $\mathcal{E}(A, P^{\hat{\psi}})$. We have

$$\mathcal{E}(A, \psi) = \int_{-\infty}^{+\infty} \lambda(\psi, P_A(d\lambda)\psi),$$

and this integral is known (see Exercise 2) to exist and be finite if and only if ψ is in the domain of $|A|^{1/2}$. Now, this domain contains $\mathcal{D}(A)$, so that $\mathcal{E}(A, \psi)$ is well defined, in particular, for all vector states in $\mathcal{D}(A)$, in which case (by use of the spectral theorem)

$$\mathcal{E}(A, \psi) = (\psi, A\psi), \qquad (3.1.4)$$

a fact that allows us to express the expectation directly in terms of A and ψ, without the intervention of the spectral measure of A. Of course, when A is bounded, $\mathcal{D}(A)$ is the whole $\mathcal{H}$ and (3.1.4) holds true for every ψ.

Consider now the case in which the physical system is in a nonpure state D, and let

$$D = \sum_i w_i P^{\hat{\psi}_i}$$

be one of its decompositions into pure states. By (2.3.2) we have the expression

$$\mathcal{E}(A, D) = \int_{-\infty}^{+\infty} \lambda \sum_i w_i(\psi_i, P_A(d\lambda)\psi_i),$$

and (see Exercise 3) the integral exists and is finite if and only if $\mathcal{E}(A, \psi_i)$ exists for all i and the series

$$\sum_i w_i \int_{-\infty}^{+\infty} |\lambda|(\psi_i, P_A(d\lambda)\psi_i)$$

converges. In this case

$$\mathcal{E}(A, D) = \sum_i w_i \mathcal{E}(A, \psi_i), \qquad (3.1.5)$$

a result that is physically expected on the basis of the intuitive notion of nonpure state. True, the decomposition of D into pure states is not unique, but the right-hand side of (3.1.5) does not depend upon the particular decomposition of D.

When AD is everywhere defined, bounded, and trace-class, then $\mathcal{E}(A, D)$ exists and all the pure states occurring in the decompositions of D belong to $\mathcal{D}(A)$, and we have

$$\mathcal{E}(A, D) = \text{tr}(AD) = \sum_i w_i(\psi_i, A\psi_i). \qquad (3.1.6)$$

Of course, this is always the case when A is bounded.

3.1.3 Variance

Consider now the notion of the variance of the physical quantity represented by A in the state D. By this we mean the variance $\mathcal{V}(A, D)$ of the probability distribution

$$E \mapsto \text{tr}(DP_A(E)), \qquad E \in \mathcal{B}(\mathbb{R}),$$

that is,

$$\mathcal{V}(A, D) = \int_{-\infty}^{+\infty} [\lambda - \mathcal{E}(A, D)]^2 \text{tr}(DP_A(d\lambda)),$$

or, equivalently,

$$\mathcal{V}(A, D) = \int_{-\infty}^{+\infty} \lambda^2 \text{tr}(DP_A(d\lambda)) - \mathcal{E}^2(A, D), \qquad (3.1.7)$$

provided $\mathcal{E}(A, D)$ and the integrals exist and are finite. Intuitively, the variance gives a measure of the width of the distribution of values of A around the mean value.

Let us examine first the case of a pure state, say $D = P^{\hat{\psi}}$, and write $\mathcal{V}(A, \psi)$ in place of $\mathcal{V}(A, P^{\psi})$. We have

$$\mathcal{V}(A, \psi) = \int_{-\infty}^{+\infty} \lambda^2(\psi, P_A(d\lambda)\psi) - \mathcal{E}^2(A, \psi),$$

and we know that $\mathcal{E}(A, \psi)$ is finite whenever $\psi \in \mathcal{D}(|A|^{1/2})$, while, by the spectral theorem, the integral is finite whenever $\psi \in \mathcal{D}(A)$. Since $\mathcal{D}(A) \subseteq \mathcal{D}(|A|^{1/2})$, we conclude that $\mathcal{V}(A, \psi)$ exists and is finite if and only if $\psi \in \mathcal{D}(A)$, in which case

$$\mathcal{V}(A, \psi) = (\psi, A^2\psi) - \mathcal{E}^2(A, \psi) = \|A\psi\|^2 - (\psi, A\psi)^2. \qquad (3.1.8)$$

If the physical system is in a nonpure state D, take any one of its decompositions into pure states, say

$$D = \sum_i w_i P^{\hat{\psi}_i};$$

then use (2.3.2), (3.1.5), and (3.1.7) to write

$$\mathcal{V}(A, D) = \int_{-\infty}^{+\infty} \lambda^2 \sum_i w_i(\psi_i, P_A(d\lambda)\psi_i) - \left(\sum_i w_i \mathcal{E}(A, \psi_i)\right)^2.$$

The integral exists and is finite if $\psi_i \in \mathcal{D}(A)$ for every i and if $\sum_i w_i \|A\psi_i\|^2$ converges, in which case also $\mathcal{E}(A, D)$ exists and we have

$$\mathcal{V}(A, D) = \sum_i w_i \|A\psi_i\|^2 - [\mathrm{tr}(AD)]^2. \qquad (3.1.9)$$

Notice that, as is implicit in these formulae, the variance of a physical quantity in a nonpure state D, though computed via a particular decomposition of D into pure states, does not depend upon the particular decomposition one has chosen.

3.1.4 A Remark

In quantum mechanics a physical quantity cannot, in general, be characterized by a single numerical value (given the state of the system), but only by a whole spectrum. However, there is always at least one state in which the distribution of values around the mean value is as sharp as we want. In technical terms:

> given A, for every $\varepsilon > 0$
> there exists a unit vector ψ
> such that $\mathcal{V}(A, \psi) \leqslant \varepsilon$. $\qquad$ (3.1.10)

The proof goes as follows. If A has a nonempty point spectrum, it suffices to choose for ψ any normalized eigenvector of A, thus getting $\mathcal{V}(A, \psi) = 0$. If A has no point spectrum, then take an element λ of its spectrum and choose for ψ any unit vector in the range of the projector $P_A(\lambda - \varepsilon, \lambda + \varepsilon)$, so that the measure $E \mapsto (\psi, P_A(E)\psi)$ vanishes on the complement of the interval $[\lambda - \varepsilon, \lambda + \varepsilon]$; hence $\mathcal{V}(A, \psi)$ tends to zero when ε tends to zero.

3.2 Functions of Compatible Physical Quantities; Complete Sets of Compatible Physical Quantities

Two self-adjoint operators A, B in $\mathcal{H}$ are said to *commute* whenever

$$P_A(E)P_B(F) = P_B(F)P_A(E) \qquad \text{for all} \quad E, F \in \mathcal{B}(\mathbb{R});$$

in case A and B are bounded this condition reduces[1] to the familiar one

$$AB = BA.$$

We call two physical quantities *compatible* when they are represented by commuting operators: the reason for this term will become clear in the sequel.*

If $A_1, \ldots, A_n$ are commuting self-adjoint operators and $E_1, \ldots, E_n$ are real Borel sets, then

$$P_{A_1}(E_1) \cdot P_{A_2}(E_2) \cdot \ \cdots \ \cdot P_{A_n}(E_n)$$

is still a projection operator, and the mapping

$$E_1 \times E_2 \times \cdots \times E_n \mapsto P_{A_1}(E_1) \cdot P_{A_2}(E_2) \cdot \ \cdots \ \cdot P_{A_n}(E_n)$$

has a unique extension to a projection-valued measure on the Borel sets of $\mathbb{R}^n$,[2] which we denote by

$$\Delta \mapsto P_{A_1, \ldots, A_n}(\Delta), \qquad \Delta \in \mathcal{B}(\mathbb{R}^n).$$

---

*Notice that only if two physical quantities are compatible is their "product" a physical quantity: indeed, the product of two (bounded) self-adjoint operators A, B is self-adjoint if and only if $AB = BA$.

We can now proceed with the definition of function of compatible physical quantities. Let f be a measurable complex-valued function on $\mathbb{R}^n$. Then[2] there exists a unique operator, to be denoted by $f(A_1,\ldots,A_n)$, having domain

$$\mathcal{D}(f(A_1,\ldots,A_n))$$

$$= \left\{ \psi \in \mathcal{H} : \int_{\mathbb{R}^n} |f(\lambda_1,\ldots,\lambda_n)|^2 \left(\psi, P_{A_1,\ldots,A_n}(d\lambda_1 \times \cdots \times d\lambda_n)\psi \right) < \infty, \lambda_i \in \mathbb{R} \right\}$$

and such that for every $\varphi \in \mathcal{H}$ and every $\psi \in \mathcal{D}(f(A_1,\ldots,A_n))$,

$$\left(\varphi, f(A_1,\ldots,A_n)\psi \right)$$

$$= \int_{\mathbb{R}^n} f(\lambda_1,\ldots,\lambda_n) \left(\varphi, P_{A_1,\ldots,A_n}(d\lambda_1 \times \cdots \times d\lambda_n)\psi \right), \qquad \lambda_i \in \mathbb{R}$$

An important particular case is when $n=1$, for then one recovers the simpler and more familiar notion of a function of one operator, or, in physical terms, a function of a physical quantity.

Now we list a number of relevant facts related to the notion of a function of commuting operators.

(i) If $f(\lambda_1,\ldots,\lambda_n), \lambda_i \in \mathbb{R}$, is real-valued, then $f(A_1,\ldots,A_n)$ is self-adjoint [with some restriction on the domain in case $f(\lambda_1,\ldots,\lambda_n)$ is not bounded[2]]. In this case $f(A_1,\ldots,A_n)$ is the operator characterized by the spectral measure

$$E \mapsto P_{A_1,\ldots,A_n}\left(f^{-1}(E) \right), \quad E \in \mathcal{B}(\mathbb{R}) \quad \left[f^{-1}(E) \in \mathcal{B}(\mathbb{R}^n) \right].$$

To give an intuitive meaning to this formula, suppose $E \in \mathcal{B}(\mathbb{R})$ such that $f^{-1}(E)$ has the product form $E_1 \times E_2 \times \cdots \times E_n$ $[E_i \in \mathcal{B}(\mathbb{R}), i=1,2,\ldots,n]$; then the probability that the physical quantity represented by $f(A_1,\ldots,A_n)$ takes a value in E is given by the probability that the physical quantities represented by $A_1,\ldots,A_n$ take values in $E_1,\ldots,E_n$ respectively.

(ii) Two functions of the same commuting operators, say $f(A_1,\ldots,A_n)$ and $g(A_1,\ldots,A_n)$, commute.[2]

(iii) The set of all bounded functions of a given set of commuting operators has a natural structure of an involution algebra. Specifically, the sum, the product, and the involution are

$$f(A_1,\ldots,A_n) + g(A_1,\ldots,A_n) = (f+g)(A_1,\ldots,A_n),$$

$$f(A_1,\ldots,A_n) \cdot g(A_1,\ldots,A_n) = (f \cdot g)(A_1,\ldots,A_n),$$

$$(f(A_1,\ldots,A_n))^* = \bar{f}(A_1,\ldots,A_n).$$

It can be proved[3] that this algebra is closed in the weak operator topology; hence it is an abelian von Neumann algebra, which we denote by $\mathbf{A}(A_1,\ldots,A_n)$.

(iv) $\mathbf{A}(A_1,\ldots,A_n)$ is the von Neumann algebra generated by $\{A_1,\ldots,A_n\}$. The expression "algebra generated by" means the following. Let $\mathcal{O}$ be any set of operators in $\mathcal{H}$, and let $\mathcal{O}'$ denote the commutant of $\mathcal{O}$, namely the set of all bounded operators that commute with every element of $\mathcal{O}$. Then $\mathcal{O}''$ is a von Neumann algebra,[4] which is called the von Neumann algebra generated by $\mathcal{O}$. The fact that $\mathbf{A}(A_1,\ldots,A_n)$ is precisely the von Neumann algebra generated by $\{A_1,\ldots,A_n\}$ can be rephrased by saying that a bounded operator in $\mathcal{H}$ belongs to $\{A_1,\ldots,A_n\}''$ if and only if it can be expressed as a bounded function of $A_1,\ldots,A_n$.[5]

(v) The members of a set $\{A_1,\ldots,A_n\}$ of commuting self-adjoint operators are measurable functions of the same bounded self-adjoint operator. This means that there exists $A \in \{A_1,\ldots,A_n\}''$ such that $A_i = f_i(A)$, $i = 1,2,\ldots,n$. This important theorem is due to von Neumann. Notice however that the finiteness of the set $\{A_1,\ldots,A_n\}$ is not decisive, since the theorem can be generalized to any set of commuting self-adjoint operators.[5]

In quantum mechanics an important role is played by the notion of complete set of compatible physical quantities. In terms of the associated operators, this notion amounts to the following: a (finite) set $\{A_1,\ldots,A_n\}$ of commuting self-adjoint operators is said to be complete whenever the von Neumann algebra $\{A_1,\ldots,A_n\}''$ they generate is maximal abelian. Recall that any abelian von Neumann algebra $\mathbf{A}$ satisfies $\mathbf{A} \subseteq \mathbf{A}'$, and it is called maximal when $\mathbf{A} = \mathbf{A}'$.

Noticing that the commutant $\{A_1,\ldots,A_n\}'$ is itself a von Neumann algebra because $A_1,\ldots,A_n$ are self-adjoint,[4] we get $\{A_1,\ldots,A_n\}' = \{A_1,\ldots,A_n\}'''$, so that we can rephrase the definition of a complete set of commuting operators by saying that $\{A_1,\ldots,A_n\}$ is complete whenever $\{A_1,\ldots,A_n\}'' = \{A_1,\ldots,A_n\}'$. In view of item (iv) above, we can also say that the commuting self-adjoint operators $A_1,\ldots,A_n$ form a complete set if and only if $\mathbf{A}(A_1,\ldots,A_n) = \{A_1,\ldots,A_n\}'$, i.e., if and only if every bounded operator that commutes with each of them is a bounded function of them.

It may happen that a single self-adjoint operator A forms a complete set by itself: this occurs if and only if every bounded operator that commutes with A is a bounded function of A. If we think of operators with pure point spectrum, then this corresponds to the fact that none of the eigenvalues of A is degenerate. This explains why an operator that constitutes a complete set by itself is also called an operator with simple spectrum.

We have seen in item (v) above that if $A_1,\ldots,A_n$ are commuting self-adjoint operators, then there exists a bounded self-adjoint operator A such that $A_i = f_i(A)$, $i = 1,\ldots,n$. We have now at hand a new characterization of a complete set: $\{A_1,\ldots,A_n\}$ is complete if and only if A has a simple spectrum.[6]

Another useful characterization of a complete set of commuting operators can be given in terms of cyclic vectors.[7] Recalling that a von Neumann algebra **A** is maximal abelian if and only if it has a cyclic vector (namely a vector $\varphi \in \mathcal{H}$ such that $\mathbf{A}\varphi$ is dense in $\mathcal{H}$), we have that $A_1, \ldots, A_n$ form a complete set if and only if there exists a cyclic vector of $\{A_1, \ldots, A_n\}''$, namely a vector φ such that applying to it every bounded function of $A_1, \ldots, A_n$ we get a set dense in $\mathcal{H}$. Notice also that[7] φ is a cyclic vector of $\{A_1, \ldots, A_n\}''$ if and only if the finite linear combinations of the form

$$c_1 P_{A_1, \ldots, A_n}(\Delta_1)\varphi + \cdots + c_k P_{A_1, \ldots, A_n}(\Delta_k)\varphi,$$

$$c_1, \ldots, c_k \in \mathbb{C}, \quad \Delta_1, \ldots, \Delta_k \in \mathcal{B}(\mathbb{R}^n),$$

are dense in $\mathcal{H}$.

Let us add one more way of viewing a complete set.[2] $A_1, \ldots, A_n$ form a complete set of commuting operators when they are unitarily equivalent to (nondegenerate) multiplicative operators in some functional space. More precisely, $\{A_1, \ldots, A_n\}$ is a complete set if and only if there exist a measure μ on $\mathcal{B}(\mathbb{R}^n)$, concentrated on $\sigma(A_1) \times \cdots \times \sigma(A_n)$, and a unitary operator U from $\mathcal{H}$ to $L^2(\sigma(A_1) \times \cdots \times \sigma(A_n), \mu)$, such that in this functional space the operators $A_1', \ldots, A_n'$ defined by $A_i' = U A_i U^{-1}$ ($i = 1, \ldots, n$) have the multiplicative form

$$A_i' f(\lambda_1, \ldots, \lambda_n) = \lambda_i f(\lambda_1, \ldots, \lambda_n), \qquad \lambda_i \in \sigma(A_i),$$

in the domain

$$\mathcal{D}(A_i') = \left\{ f(\lambda_1, \ldots, \lambda_n) \in L^2(\sigma(A_1) \times \cdots \times \sigma(A_n), \mu) \right.$$

$$\left. : \int_{\mathbb{R}^n} |\lambda_i f(\lambda_1, \ldots, \lambda_n)|^2 \mu(d\lambda_1 \times \cdots \times d\lambda_n) < \infty \right\}.$$

The function space $L^2(\sigma(A_1) \times \cdots \times \sigma(A_n), \mu)$ is called, in the usual language of quantum mechanics, the representation in which $A_1, \ldots, A_n$ are simultaneously diagonal. The measure μ is, in general, nonuniquely determined; it is generated by any cyclic vector φ of the algebra $\{A_1, \ldots, A_n\}''$ in the sense that $\mu(\Delta) = 0$, $\Delta \in \mathcal{B}(\mathbb{R}^n)$, if and only if the (finite) measure

$$\Delta \mapsto \left(\varphi, P_{A_1, \ldots, A_n}(\Delta)\varphi \right)$$

is zero.[7] Different choices of μ define different representations of $\mathcal{H}$ that are unitarily equivalent.

3.3 Joint Probabilities

In the rest of this chapter we shall focus attention on some facts that characterize the different behavior of compatible and noncompatible physical quantities.

A nontrivial issue that outlines the distinction between compatible and noncompatible physical quantities is that of the *joint probability distributions* of physical quantities. If A and B represent two compatible physical quantities, and D is a state of the system, consider the function $\mu_{D; A, B}$ on the Borel rectangles of $\mathbb{R}^2$ [the subsets of $\mathbb{R}^2$ of the form $E \times F$, with $E, F \in \mathcal{B}(\mathbb{R})$] defined by

$$\mu_{D; A, B}(E \times F) = \operatorname{tr}(DP_A(E)P_B(F)),$$

and notice that, writing $A = f(C)$, $B = g(C)$, it takes the form

$$\mu_{D; A, B}(E \times F) = \operatorname{tr}\left(DP_C\left(f^{-1}(E) \cap g^{-1}(F)\right)\right)$$

This function is σ-additive on disjoint unions of Borel rectangles, so that, by a classical theorem of measure theory, it extends to a unique probability measure on $\mathcal{B}(\mathbb{R}^2)$, which we still denote by $\mu_{D; A, B}$.

It is natural to interpret $\mu_{D; A, B}(E \times F)$ as the probability that, in state D, the value of the physical quantity represented by A lies in the Borel set E, while the value of the physical quantity represented by B lies in the Borel set F. Moreover the probability measure $\mu_{D; A, B}$ on $\mathcal{B}(\mathbb{R}^2)$ behaves, with respect to the probability distributions

$$E \mapsto \operatorname{tr}(DP_A(E)) \quad \text{and} \quad F \mapsto \operatorname{tr}(DP_B(F)), \qquad E, F \in \mathcal{B}(\mathbb{R}),$$

as the joint probability distribution, in the usual sense this concept has in probability theory.[8] For these reasons we call $\mu_{D; A, B}$ the joint probability distribution of the physical quantities represented by A and B in the state D.

Notice that the two distributions $E \mapsto \operatorname{tr}(DP_A(E))$, $F \mapsto \operatorname{tr}(DP_B(F))$ are the marginal distributions of $\mu_{D; A, B}$; that is,

$$\mu_{D; A, B}(E \times \mathbb{R}) = \operatorname{tr}(DP_A(E)), \qquad E \in \mathcal{B}(\mathbb{R})$$

$$\mu_{D; A, B}(\mathbb{R} \times F) = \operatorname{tr}(DP_B(F)), \qquad F \in \mathcal{B}(\mathbb{R}).$$

When dealing with noncompatible physical quantities, it is not possible to define joint probability distributions in the way we have sketched above. Since the beginning of quantum mechanics the problem of defining some weakening of the usual notion of joint probability distributions for noncompatible physical quantities has been considered. Among the various proposals and contributions we recall those of von Neumann,[9] Wigner,[10] Mojal,[11] Varadarajan,[12] Urbanik,[13] Gudder,[14] and Jauch;[15] for an extensive review of the problem and for further references we refer to Bugajski.[16] In spite of these efforts, no meaningful entity physically interpretable as the joint probability distribution of two noncompatible physical quantities has ever been defined. But we shall not insist on this point here; rather we want to stress the positive assertion that two compatible physical quantities have a joint probability distribution in the usual sense, a fact that further characterizes the notion of compatibility.

3.4 Heisenberg Inequalities

Another issue that emphasizes the difference between the behavior of compatible and of noncompatible physical quantities deals with the study of the product

$$\mathcal{V}(A, \psi)\mathcal{V}(B, \psi), \qquad \psi \in \mathcal{D}(A) \cap \mathcal{D}(B), \tag{3.4.1}$$

of the variances of A and B in the same (pure) state ψ of the physical system. This leads to the celebrated Heisenberg inequalities. We have seen [in (3.1.10)] that the variance of a physical quantity can always be made arbitrarily small by a proper choice (i.e. by a choice depending on the physical quantity) of the state. We shall now examine how the product (3.4.1) varies with ψ. Obviously, we expect that, in general, it will not become arbitrarily small, since we require the same choice of the state in the two factors.

There are two alternatives: either

(I) for every $\varepsilon > 0$ there exists a vector state ψ such that $\mathcal{V}(A, \psi)\mathcal{V}(B, \psi) < \varepsilon$, or

(II) there exists $\varepsilon > 0$ such that for every vector state ψ, $\mathcal{V}(A, \psi)\mathcal{V}(B, \psi) > \varepsilon$.

Clearly, (II) is the negation of (I), so that they exhaust all possibilities. The occurrence of (I) or (II) will depend on the choice of the physical quantities represented by A and B, as we are going to show.

If A has a nonempty point spectrum and one if its eigenvectors belongs to $\mathcal{D}(B)$, or vice versa, then alternative (I) occurs. Indeed, choosing for ψ that eigenvector, we have $\mathcal{V}(A, \psi) = 0$ with $\mathcal{V}(B, \psi)$ finite.

If A or B is bounded (or both are bounded), then again (I) occurs. Indeed, if A is bounded, then $\mathcal{D}(A)$ is the whole $\mathcal{H}$, $\mathcal{V}(A, \psi)$ is finite (and positive) for every ψ, and [see (3.1.8)] $\mathcal{V}(A, \psi) \leq \|A^2\|$, so that there exists a positive number M such that $\mathcal{V}(A, \psi) \leq M$ for every ψ. Now, for any given $\varepsilon > 0$, we know from the statement (3.1.10) that there exists ψ such that $\mathcal{V}(B, \psi) \leq \varepsilon/M$; hence this ψ does the trick.

The situations just mentioned, all leading to alternative (I), do so irrespective of whether A and B commute or not (some examples of these situations will be encountered in next chapter).

In any case, it holds true that if A and B commute, then alternative (I) necessarily occurs. The proof goes as follows. From item (v) of Section 3.2 it follows that A and B can be written as measurable functions of a self-adjoint operator C, say $A = f(C)$ and $B = g(C)$. Out of $\sigma(C)$ choose a number $\bar{\lambda}$, and observe that $f(\bar{\lambda}) \in \sigma(A)$, $g(\bar{\lambda}) \in \sigma(B)$, so that for every $\varepsilon > 0$, if we denote by $\mathcal{I}_1, \mathcal{I}_2$ two segments of length ε centered respectively on $f(\bar{\lambda})$ and $g(\bar{\lambda})$, then we have $P_A(\mathcal{I}_1) \neq 0$, $P_B(\mathcal{I}_2) \neq 0$. Moreover, $P_A(\mathcal{I}_1) \cdot P_B(\mathcal{I}_2) = P_C(f^{-1}\mathcal{I}_1) \cdot P_C(g^{-1}\mathcal{I}_2) = P_C(f^{-1}\mathcal{I}_1 \cap g^{-1}\mathcal{I}_2)$, and this projector is not zero, for $\bar{\lambda}$ belongs to $f^{-1}\mathcal{I}_1 \cap g^{-1}\mathcal{I}_2$. Now, it suffices to take any unit vector ψ in the range of $P_A(\mathcal{I}_1) \cdot P_B(\mathcal{I}_2)$ to have that both $\mathcal{V}(A, \psi)$ and $\mathcal{V}(B, \psi)$ tend to zero when ε tends to zero.

This argument shows that alternative (II) is indeed open only to quantum systems, though it is not a necessity. More precisely, we have seen that (II) can occur only if three facts are concomitant:

(1) A and B do not commute,
(2) neither A nor B is bounded,
(3) A and B have no point spectrum, or, if either of them has a nonempty point spectrum, its eigenvectors lie outside the domain of the other.

There are many quantum instances that actually fall under the alternative (II), e.g., the position and momentum operators for a free particle in a infinitely extended space. Thus, the occurrence of (II) constitutes a factual departure from the behavior of classical systems (it is indeed a tenet of quantum mechanics). In emphatic terminology, (II) is often referred to as "Heisenberg's uncertainty principle" or "Heisenberg's uncertainty relation". Since the early days of quantum mechanics, the occurrence of (II) has been the source of innumerable discussions, which have spread out into the domains of epistemology, philosophy, and logic, sometimes going beyond the actual meaning of this relation.

There is a general theorem,[17] holding for any pair of self-adjoint operators A, B (bounded or not), which states that, for every $\psi \in \mathcal{D}(A) \cap \mathcal{D}(B)$,

$$\mathcal{V}(A, \psi)\mathcal{V}(B, \psi) \geq \tfrac{1}{4}|(A\psi, B\psi) - (B\psi, A\psi)|^2. \qquad (3.4.2)$$

Since the vectors in $\mathcal{D}(A) \cap \mathcal{D}(B)$ are precisely those for which the variances of both A and B exist and are finite, the two members of (3.4.2) have identical domains of definition. This inequality can provide, when the above conditions (1), (2), (3) are all fulfilled, a more explicit version of alternative (II), in the sense that for certain pairs of physical quantities the right-hand side of (3.4.2) is directly seen to majorize some positive number ε, independently of the choice of ψ. Of course, when conditions (1), (2), (3) are not all fulfilled, it is always possible to exhibit a state ψ that makes the right-hand side of (3.4.2) arbitrarily small (or even null), in accordance with the fact that alternative (I) then occurs. We shall refer to (3.4.2) as *Heisenberg's inequality*.

In many textbooks on quantum mechanics Heisenberg's inequality is written in the form

$$\mathcal{V}(A, \psi)\mathcal{V}(B, \psi) \geqslant \tfrac{1}{4}|(\psi,(AB-BA)\psi)|^2, \qquad (3.4.3)$$

which has, in general, the unsatisfactory feature that the right-hand side need not be defined when the left-hand side is. Paradoxically, this more popular version of Heisenberg's inequality has a clear meaning only when A and B are bounded, a case in which it is nearly meaningless, for the left-hand side can be made as small as we want.

If we consider a nonpure state with density operator $D=\Sigma_i w_i P^{\hat{\varphi}_i}$, $\langle\varphi_i\rangle$ orthonormal, then Heisenberg's inequality (3.4.2) takes the form

$$\mathcal{V}(A, D)\mathcal{V}(B, D) \geqslant \tfrac{1}{4}\left|\sum_i w_i[(A\varphi_i, B\varphi_i)-(B\varphi_i, A\varphi_i)]\right|^2,$$

which is well defined whenever $\varphi_i \in \mathcal{D}(A)\cap\mathcal{D}(B)$ for all i.

Let us stress that Heisenberg's inequalities, as we have introduced them, have only a statistical nature. Though we deal with the variances of A and B in the same state, we do not imply any idea of "simultaneous measurement" on one and the same physical system. We can think of measuring the two variances on identically prepared replicas of a physical system, with one variance measured in a certain laboratory at a certain time, and the other variance measured in another laboratory at another time, without any correlation. True, one can find in the physical literature, especially in the pioneering writings on quantum mechanics, an effort to give to Heisenberg's inequalities more than a merely statistical meaning; in our opinion, however, this effort did not attain a clear formalization.

For an analysis of the role, the use, and some abuses of Heisenberg's inequalities, we refer to a paper by J. M. Levy-Leblond.[18]

3.5 Complementary Physical Quantities

"Complementarity" was a key word of quantum mechanics in the early days of the theory and for several decades after. Despite this, the meaning of this word often remained rather vague and strongly dependent upon the taste of the user. Later, it became apparent that, to a certain extent, the insistence on 'complementarity" reflected efforts to speak about quantum systems in the frame of classical concepts.

Various aspects of "complementarity" have been considered: complementarity between space-time description and causal description, complementarity between particle picture and wave picture, complementarity in the sense that the knowledge of a precise value of a physical quantity might exclude a similar knowledge for another physical quantity.* Here, we are not concerned with a discussion of the connections among these aspects: we shall only sketch, following the approach of Reference 19, how the last aspect, that of mutual exclusion between two physical quantities, can be formulated.

When we deal with two physical quantities, represented by operators, say A and B, that have a pure point spectrum (see the Example 1 of Section 3.1), we can say that they are complementary if A and B have no common eigenvectors. In this case, whenever the physical quantity represented by A has a sharp value, thus implying that the state of the system is an eigenvector of A, the physical quantity represented by B cannot have a sharp value (unless B is a constant), for the state of the system is not an eigenvector of B. In other words, if the state vector ψ makes

$$\left(\psi, P_A\{\lambda_i\}\psi\right)=1$$

for some eigenvalue λ_i of A, then

$$\left(\psi, P_B\{\lambda_i'\}\psi\right)<1$$

for every eigenvalue λ_i' of B (unless the spectrum of B reduces to a single point). Clearly, if the two physical quantities are complementary, then they are noncompatible, i.e., A and B do not commute, for we know that, should A and B commute, they would have a common basis of eigenvectors. The converse implication, however, does not hold. In fact, if A and B do not commute, we cannot exclude that they have *some* eigenvector in common[†] (we can only exclude that they have *all* eigenvectors in common). Thus we see that complementarity and noncompatibility are indeed distinct concepts.

When we give up the hypothesis of a pure point spectrum, the notion of complementarity can be expressed as follows: we say that two physical quantities, represented by the operators A and B, are complementary if, for every bounded Borel set E strictly contained in $\sigma(A)$ and every bounded

---

*For an overview of this issue and for a rich bibliography, see Reference 19.

†Consider the operators L_x and L_y representing the x- and the y-component of the orbital angular momentum: they do not commute, but they do have a common eigenvector, the state of zero orbital angular momentum.

Borel set F strictly contained in $\sigma(B)$, the range of $P_A(E)$ is (set-theoretically) disjoint from the range of $P_B(F)$. This means that if the value of one physical quantity is in E, then it becomes impossible to assert that the value of the other physical quantity is in F, provided E and F are bounded and strictly contained in the corresponding spectra. In other words, if, for some $\psi \in \mathcal{H}$ and some bounded $E \subset \sigma(A)$, we have $(\psi, P_A(E)\psi) = 1$, then we must have $(\psi, P_B(F)\psi) < 1$ for all bounded $F \subset \sigma(B)$.

Even in these more general terms, it holds true that two complementary physical quantities are always noncompatible (i.e. their operators do not commute), but, as already remarked, noncompatibility does not entail complementarity. In Section 14-6 this issue will be considered in more detail.

The most popular example of complementary physical quantities is the momentum and position of a free particle; it will be considered in the course of next chapter. Here, we just add a remark. In the pedagogical tradition of quantum mechanics, the distinction between the notion of complementarity and that of noncompatibility is often quite blurred, and there is an old argument, called "disturbance theory", which is used to justify both. According to this argument, the measurement of a physical quantity, say A, with value in some small interval, introduces an unavoidable and uncontrollable "disturbance" of the state of the system; this disturbance may affect, and spread out, the possible values of a subsequent measurement of another physical quantity, say B, in which case the two physical quantities behave as noncompatible or complementary. The disturbance, making the value of B unpredictable after A has been measured on the same physical system, gives rise, loosely speaking, to the qualitative pattern of complementarity and of Heisenberg's inequalities, for it is assumed that the more precisely the value of A is determined, the larger the disturbance becomes, and the larger is the spreading out of the values of B. Though the "disturbance theory" may have had pedagogical merits, it should be clear that it has some naive aspects and that it is unnecessary for the coherence of quantum mechanics. It constitutes a typical effort at talking about quantum systems with classical concepts: in fact, it tacitly assumes that a state (we mean a pure state) has to assign a value to a physical quantity. This is the case with classical systems but not with quantum ones: with quantum systems, a state assigns a probability distribution of values, not just a value. Only exceptionally is this probability distribution sharply concentrated around a value: the "disturbance theory" refers to this exceptional case as if it were the rule.

Exercises

1. Show that the right-hand side of (3.1.2) does not depend on the choice of the basis $\{\varphi_i^k\}$ in $\mathfrak{M}_i$.

2. Show that $\int_{-\infty}^{+\infty} \lambda(\psi, P_A(d\lambda)\psi)$ exists and is finite iff $\psi \in \mathcal{D}(|A^{1/2}|)$.

[*Hint.* From measure theory this integral exists iff $\int_{-\infty}^{+\infty} |\lambda| (\psi, P_A(d\lambda)\psi)$ exists. Then write $|\lambda|$ as $(\sqrt{|\lambda|})^2$ and use the spectral theorem.]

3. Show that if $D = \sum_i w_i P^{\hat{\psi}_i}$, the integral $\int_{-\infty}^{+\infty} \lambda \sum_i w_i(\psi_i, P_A(d\lambda)\psi_i)$ exists and is finite iff $\mathscr{E}(A, \psi_i)$ exists for all i and $\sum_i w_i \int_{-\infty}^{+\infty} |\lambda| (\psi_i, P_A(d\lambda)\psi_i)$ converges.

[*Hint.* Use Exercise 2, and Theorem 13.15.9 of Reference 20.]

4. Show that two self-adjoint operators A_1, A_2 with pure point spectra form a complete set if: (i) for every $\lambda_1 \in \sigma(A_1)$ and $\lambda_2 \in \sigma(A_2)$ there exists $\psi_{\lambda_1, \lambda_2}$ in $\mathscr{D}(A_1) \cap \mathscr{D}(A_2)$ such that $A_k \psi_{\lambda_1, \lambda_2} = \lambda_k \psi_{\lambda_1, \lambda_2}$ $(k = 1, 2)$; (ii) when λ_1 and λ_2 range over $\sigma(A_1)$ and $\sigma(A_2)$, respectively, the vector $\psi_{\lambda_1, \lambda_2}$ ranges over a basis of $\mathscr{H}$.

[*Hint.* Use the representation in which A_1 and A_2 are diagonal.]

5. Show that a self-adjoint operator with pure point spectrum is simple when its eigenvalues are nondegenerate.

References

1. M. Reed and B. Simon, *Methods of Modern Mathematical Physics*, Vol. I, Academic Press, New York, 1972.
2. E. Prugovecki, *Quantum Mechanics in Hilbert Space*, Academic Press, New York and London, 1971.
3. M. A. Naimark, *Normed Rings*, P. Noordhoff, Groningen, 1960.
4. J. Dixmier, *Les Alègbres d'Opérateurs dans l'Espace Hilbertien*, Gauthiers-Villars, Paris, 1957.
5. F. Riesz and B. Sz-Nagy, *Functional Analysis*, F. Ungar, New York, 1955.
6. J. M. Jauch, *Foundations of Quantum Mechanics*, Addison-Wesley, Advanced Book Program, Reading, Mass., 1968.
7. G. W. Mackey, *Mathematical Foundations of Quantum Mechanics*, Benjamin, Advanced Book Program, Reading, Mass., 1963.
8. L. Breiman, *Probability*, Addison-Wesley, Reading, Mass., 1968.
9. J. von Neumann, *Mathematical Foundations of Quantum Mechanics*, Princeton University Press, Princeton N. J., 1955.
10. E. P. Wigner, *Phys. Rev.* 40 (1932) 749.
11. J. Mojal, *Proc. Cambridge Philos. Soc.* 45 (1949) 99.
12. V. S. Varadarajan, *Geometry of Quantum Theory*, Vol. I, Van Nostrand, Princeton, N. J., 1968.
13. K. Urbanik, *Studia Math.* 21 (1961) 117.
14. S. Gudder, *J. Math. Mech.* 18 (1968) 325.
15. J. M. Jauch, in *Quantum Mechanics, a Half Century Later*, J. Leite Lopes and M. Paty, eds., Reidel, Dordrecht, 1977.
16. S. Bugajski, *Joint Probabilities in Quantum Mechanics*, preprint, Silesian University, Katowice, Poland, 1977.
17. K. Kraus and J. Schröter, *Int. J. Theor. Phys.* 8 (1973) 431.
18. J. M. Levy-Leblond, *Bull. Soc. Fr. Phys.* 14 (Encarte Pédagogique 1), 1973.
19. P. Lahti, "Uncertainty and Complementarity in Axiomatic Quantum Mechanics", Thesis, Univ. of Turku, Series N.D2, 1979.
20. J. Dieudonné, *Treatise on Analysis* Vol. 2, Academic Press, New York and London, 1970.

CHAPTER 4

Spin and Motion

4.1 Spin without Motion

A physical system can have an angular momentum independent of its motion: it is called the intrinsic angular momentum, or spin, of the system. The reader with a background in quantum mechanics has learned from textbooks how the notion of spin historically emerged, and how it accounts for properties of the physical system under rotations. For an exhaustive treatment of angular momentum in quantum mechanics we refer to Biedenharn and Louck's volume in this series.[1]

Spin is a nonclassical degree of freedom. It is characterized by a number j which can take only the values 0, $\frac{1}{2}$, 1, $\frac{3}{2}$,.... For elementary enough physical systems one is often faced with situations in which all the states of the system that come into play correspond to the same j; the system is then called a spin-j system. For instance, pions are spin-0 systems; electrons, protons, neutrons, and muons are spin-$\frac{1}{2}$ systems. For more complex systems, like an atomic nucleus or an atom, one is often concerned with situations in which not every state coming into play has the same j (e.g. the excited states of these systems), so that j becomes a label for the state rather than a label for the physical system. Spin is a vector quantity: if we deal with a spin-j system, the component of spin along any axis in space can take only the values

$$-j, -j+1, \ldots, j-1, j,$$

a total of $2j+1$ values.

When, given a physical system, we want to describe only its spin, disregarding its translational degrees of freedom as well as any other degrees of freedom it might have, the Hilbert space to be associated to it becomes

ENCYCLOPEDIA OF MATHEMATICS and Its Applications, Gian-Carlo Rota (ed.). Vol. 15: E. G. Beltrametti and G. Cassinelli, The Logic of Quantum Mechanics

ISBN 0-201-13514-0

$(2j+1)$-dimensional, namely $\mathcal{H}=\mathbb{C}^{2j+1}$. In this space, the components of spin along three coordinate axes x_1, x_2, x_3 are represented by three self-adjoint operators J_1, J_2, J_3 which, in the representation in which J_3 is diagonal, are matrices with entries

$$(J_1)_{m,m'} = \tfrac{1}{2}\left[j(j+1)-m(m-1)\right]^{1/2}\delta_{m,m'+1}$$
$$+ \tfrac{1}{2}\left[j(j+1)-m(m+1)\right]^{1/2}\delta_{m,m'-1},$$

$$(J_2)_{m,m'} = \frac{i}{2}\left[j(j+1)-m(m-1)\right]^{1/2}\delta_{m,m'+1}$$
$$- \frac{i}{2}\left[j(j+1)-m(m+1)\right]^{1/2}\delta_{m,m'-1}, \qquad (4.1.1)$$

$$(J_3)_{m,m'} = m\delta_{m,m'},$$

where the row and column indices m, m' run from $+j$ to $-j$. These operators satisfy the typical commutation relation

$$J_1 J_2 - J_2 J_1 = iJ_3, \qquad (4.1.2)$$

and the ones obtained therefrom by cyclical permutation of indices. Notice that the operator $J^2 = J_1^2 + J_2^2 + J_3^2$ is the constant operator $j(j+1)I$, where I stands for the identity in $\mathbb{C}^{2j+1}$.

We leave as an exercise to show that each among J_1, J_2, J_3 forms a complete set by itself (see Section 3.2), and that any two of them are complementary (see Section 3.5). Notice also that, though these operators do not commute, there is no significant Heisenberg inequality (see Section 3.4) for them, because they are bounded operators: thus the product $\mathcal{V}(J_1, \psi)\mathcal{V}(J_2, \psi)$ can be made arbitrarily small by proper choice of the state vector ψ [in case j is integer, we need only take for ψ the eigenvector of J_3 belonging to the zero eigenvalue, as is apparent from (3.4.3) and (4.1.2)].

4.2 The Convex Set of States of Spin-$\frac{1}{2}$ Systems

As far as spin is concerned, the simplest nontrivial case is the spin-$\frac{1}{2}$ system. It is a system with a dichotomic degree of freedom: the only values the component of spin along any axis can take are $-\frac{1}{2}$ and $+\frac{1}{2}$ (the number $\frac{1}{2}$ is not important at this stage, for it depends on the choice of units; what matters is that we have just two possible values). The Hilbert space associated with this system is thus $\mathbb{C}^2$.

It is customary to write J_1, J_2, J_3 as

$$J_1 = \tfrac{1}{2}\sigma_1, \qquad J_2 = \tfrac{1}{2}\sigma_2, \qquad J_3 = \tfrac{1}{2}\sigma_3,$$

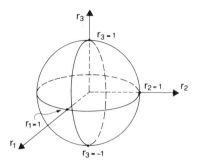

Figure 4.1. The set of states of a spin-$\frac{1}{2}$ system.

where the σ's are the *Pauli matrices*

$$\sigma_1 = \begin{pmatrix} 0 & 1 \\ 1 & 0 \end{pmatrix}, \qquad \sigma_2 = \begin{pmatrix} 0 & -i \\ i & 0 \end{pmatrix}, \qquad \sigma_3 = \begin{pmatrix} 1 & 0 \\ 0 & -1 \end{pmatrix}.$$

The self-adjoint operators on $\mathbb{C}^2$ are the hermitian 2×2 matrices: if A denotes any such matrix, we can write

$$A = \rho_1 \sigma_1 + \rho_2 \sigma_2 + \rho_3 \sigma_3 + \rho_4 I, \qquad (4.2.1)$$

where $\rho_1, \rho_2, \rho_3, \rho_4$ are real numbers and I is the unit 2×2 matrix.

Let us now proceed with the classification of states of the spin-$\frac{1}{2}$ system.

The pure states (that is, the one-dimensional projectors) take the form (see Exercise 1)

$$\tfrac{1}{2}(r_1\sigma_1 + r_2\sigma_2 + r_3\sigma_3 + I) \qquad \text{with} \quad r_1^2 + r_2^2 + r_3^2 = 1, \quad r_i \in \mathbb{R}. \quad (4.2.2)$$

Therefore, they are in one-to-one correspondence with the points on the surface of a sphere of radius 1 (Figure 4.1). The "north pole" ($r_3 = 1$) is the eigenstate of J_3 belonging to the eigenvalue $+\frac{1}{2}$, and is called the *spin-up* state relative to the x_3-axis; the "south pole" ($r_3 = -1$) is the eigenstate of J_3 with eigenvalue $-\frac{1}{2}$, and is called the *spin-down* state relative to the x_3-axis. Points that are diametrically opposite are easily seen to represent orthogonal states, i.e. orthogonal projectors.

The nonpure states take the form (see Exercise 2)*

$$\tfrac{1}{2}(r_1\sigma_1 + r_2\sigma_2 + r_3\sigma_3 + I) \qquad \text{with} \quad r_1^2 + r_2^2 + r_3^2 < 1, \quad r_i \in \mathbb{R}. \quad (4.2.3)$$

Thus they are in one-to-one correspondence with the inner points of the

---

*By use of (4.2.1)–(4.2.3) the probability distributions of physical quantities can be computed explicitly: we refer to Exercises 3 and 4.

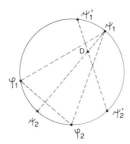

Figure 4.2. Decomposition of a non-pure state.

sphere of Figure 4.1, a fact to be expected, since nonpure states are convex combinations of pure states.

The nonunique decomposability of mixtures (see Section 2.3) is now evident. Consider, for simplicity, the section of the sphere of Figure 4.1 by a plane through the center, and take a nonpure state D as in Figure 4.2. After remarking that the convex combinations of two pure states are represented by the points on the segment they determine, we see that D admits infinitely many decompositions into pure states, with weights determined by a sort of barycentric calculus. The two orthogonal (because diametrically opposed) pure states denoted by ψ_1, ψ_2 in Figure 4.2 carry one decomposition (the unique decomposition into orthogonal states); the two nonorthogonal pure states ψ_1', ψ_2' carry another decomposition. By iterated convex combinations we can even decompose D into more than two pure states; for instance, the triple $\psi_1, \varphi_1, \varphi_2$ does the trick (and it would correspond to the explicit construction examined in Section 2.3). More generally we can say that D can be decomposed into a family of pure states if it lies inside the polygon (technically the convex hull) determined by that family.

Notice that the center of the sphere has a unique status: it can be decomposed into infinitely many pairs of orthogonal states, for there are infinitely many diameters through it. This corresponds to the fact that the center, representing the matrix $\frac{1}{2}I$, has a (doubly) degenerate eigenvalue.

The pure states (that is, the points on the surface of the sphere of Figure 4.1) can also be viewed as the states of "complete polarization" of the system. In fact, for every pure state there exists a direction z in space such that this state assigns probability 1 to the value $+\frac{1}{2}$ of the spin component along z: in other words, every pure state is the spin-up state relative to some z-axis. The nonpure states, on the contrary, can be viewed as states of partial, or incomplete, polarization: a nonpure state always assigns a nonzero probability to both values $+\frac{1}{2}$ and $-\frac{1}{2}$ of the spin component along any axis. The center of the sphere represents the completely un-polarized state, the state that assigns equal probability to the two values $+\frac{1}{2}$ and $-\frac{1}{2}$ of the spin component along any axis.

Let us notice that the convex set of states of the spin-$\frac{1}{2}$ system closely resembles the convex set of polarization states of a photon, the so-called Poincaré sphere.[2] A photon is not a spin-$\frac{1}{2}$ system: it is a spin-1 system, but of a special kind in that, due to the transversality of electromagnetic waves, the spin component along the propagation direction behaves as a two-valued (instead of three-valued) degree of freedom. It is precisely this two-valuedness that determines a convex set of polarization states (or spin states) having the same geometrical structure as the spin-$\frac{1}{2}$ system. We have again that the points on the surface of the sphere represent states of complete polarization, and the points within the sphere represent states of partial polarization. Two poles on the Poincaré sphere represent the right and the left circular polarizations, the points on the equatorial line represent the linear polarizations, and the remaining points on the surface represent the elliptical polarizations. The center of the sphere is again the completely unpolarized state, and it can be thought of as a mixture of any two diametrically opposed pure states. The nonunique decomposability of mixtures is, as in the case of the spin-$\frac{1}{2}$ system, immediately readable from the geometry of the convex set of states.

4.3 Empirical Bases of Nonunique Decomposability of Mixtures

In Section 2.4 we outlined that the nonunique decomposability of quantum mixtures makes it impossible to deduce, from the knowledge of a nonpure state, the preparation procedure actually used to prepare that state. Here we shall discuss this fact further with regard to the particular example offered by the completely unpolarized state of a spin-$\frac{1}{2}$ system and of a photon.

Consider a spin-$\frac{1}{2}$ system, and write P_z^{up} and P_z^{down} for the spin-up and spin-down states relative to some axis z (by this notation we refer to the one-dimensional projectors rather than to the vector states). From what was said in the preceding section we know that the mixture

$$\tfrac{1}{2}P_z^{\text{up}} + \tfrac{1}{2}P_z^{\text{down}}, \qquad (4.3.1)$$

which represents the completely unpolarized state (the center of the sphere of Figure 4.1), is independent of the choice of the z-axis. It has spherical symmetry. Notice that this spherical symmetry is, in a sense, highly nontrivial: in fact, if we imagine a preparation of the state (4.3.1) consisting of a random choice out of an ensemble of identical replicas of the physical system where half the replicas pass the polarizer that filters out P_z^{up} and half passed the polarizer that filters out P_z^{down}, then we shall expect, by inspection of the instruments used, only a cylindrical symmetry around z. Only our trust in quantum mechanics guarantees that (4.3.1) is indeed spherically symmetric.

Consider now a photon traveling in the x-direction. Let z be an axis perpendicular to x, and write $P_z^{\parallel}$ and $P_z^{\perp}$ for the states of complete linear polarization along z and perpendicular to the (z, x) plane. These two states are diametrically opposed in the equatorial plane of the Poincaré sphere. Again we have that the mixture

$$\tfrac{1}{2}P_z^{\parallel} + \tfrac{1}{2}P_z^{\perp}, \tag{4.3.2}$$

which represents an unpolarized photon, must be independent of z. It has cylindrical symmetry (around the propagation axis x), and again we might notice that this is a nontrivial fact, for, if one prepares the state (4.3.2) by mixing a beam filtered by a polarizer oriented along z to a beam filtered by a crossed polarizer, then the experimental setup has not the specified cylindrical symmetry.

Now, the question we want to consider is where to find direct empirical evidence in favor of the z-independence mentioned above. In practice, the preparation of a nonpure state, like (4.3.1) or (4.3.2), is seldom under experimental control: it does not consist in the actual mixing of given pure states. Beams of unpolarized particles are not produced, in laboratories, by mixing polarized beams: the simplest sources of beams produce unpolarized particles directly. Thus, instead of looking for an experimental comparison of different preparations of the nonpure states (4.3.1) and (4.3.2), it is more appropriate to shift our question a bit and ask which empirical facts relative to the polarized (hence pure) states P_z^{up} and $P_z^{\parallel}$ do motivate the z-independence of the nonpure states (4.3.1) and (4.3.2), respectively. In other words, it seems more appropriate to look for the phenomenology of polarized particles that is responsible for the quantum symmetries of unpolarized particles.

We shall not attempt any systematic analysis of the problem just posed. We only mention some very simple situations.

4.3.1 Decay of a Spin-$\tfrac{1}{2}$ Particle

Consider an unstable spin-$\tfrac{1}{2}$ system in the polarized state P_z^{up}. Let X be one of its decay products, and denote by ϑ the angle between the z-axis and the emission direction of X. Then the angular distribution of X, relative to the z-axis, has to be represented by a function $f(\vartheta)$ such that

$$f(\vartheta) + f(\pi - \vartheta) \text{ is independent of } \vartheta. \tag{4.3.3}$$

In fact, only in case this condition is met will the mixture $\tfrac{1}{2}P_z^{\text{up}} + \tfrac{1}{2}P_z^{\text{down}}$ be indifferent to the choice of the z-axis (in the sense that it will give rise to a spherically symmetric emission of X).

As an example, consider the electron-neutrino decay of a polarized muon. The electron angular distribution has the form $f(\vartheta) = k_1 + k_2 \cos \vartheta$, k_1 and k_2 being two coefficients that depend on the electron energy. This form is a typical solution of the condition (4.3.3), and it has been accurately tested by experiments.[3]

In this context, it is perhaps not superfluous to remark that if we imagine a hypothetical spin-$\frac{1}{2}$ particle that, if polarized, emits its decay products preferentially in the equatorial plane ($\vartheta = \pi/2$), then what we are imagining is nothing less than the death of quantum theory.

4.3.2 *Azimuthal Distributions in Reactions Caused by Transversally Polarized Spin-$\frac{1}{2}$ Particles*

Suppose a spin-$\frac{1}{2}$ particle, in the polarized state P_z^{up} and with momentum p perpendicular to z, impinges on a target, causing the emission of some particle X. If the emission direction of X forms an angle ϑ with p and an azimuth χ relative to the (z, p) plane, then the cross section for emission of X has to be a function $f(\vartheta, \chi)$ such that

$$f(\vartheta, \chi) + f(\vartheta, \pi + \chi) \text{ is independent of } \chi. \qquad (4.3.4)$$

This is the condition that ensures that, should the impinging particle be in the mixture $\frac{1}{2} P_z^{\mathrm{up}} + \frac{1}{2} P_z^{\mathrm{down}}$, it would give rise to a cross section independent of the choice of the z-axis.

High-energy beams of transversally polarized protons are experimentally available; however, it seems that existing experimental data are not detailed enough to provide direct, accurate tests of the condition (4.3.4).

4.3.3 *Transmission of Linearly Polarized Photons*

Suppose that a linearly polarized photon, in the state $P_z^{\parallel}$, goes through a transmitter, whose nature is, at this stage, quite arbitrary. Denote by χ the angle between the plane of photon polarization and any axis, perpendicular to the photon propagation direction, that the transmitter might determine by its own nature. Then, if we write $f(\chi)$ for the probability of transmission of the photon, we have to require that

$$f(\chi) + f\left(\left(\frac{\pi}{2} + \chi\right) \mathrm{mod}\, \pi\right) \text{ is independent of } \chi. \qquad (4.3.5)$$

Indeed, if not, the transmitter would distinguish a photon prepared by mixing (with equal weights) the states $P_z^{\parallel}$ and $P_z^{\perp}$ from a photon prepared by mixing (with equal weights) analogous states with a different choice of z.

A typical function satisfying (4.3.5) is $f(\chi) = \cos^2 \chi$. It is known to occur in case the transmitter is a polarizer and corresponds to the well-known "Malus law" of optics.

4.3.4 *Azimuthal Distributions in Photoreactions of Linearly Polarized Photons*

Consider now the situation in which a linearly polarized photon impinges on a target, causing the emission of a particle X. Let ϑ, χ be the angles defining the direction of emission of X, ϑ being the angle formed with the momentum of the impinging photon and χ the azimuth relative to the plane of photon polarization.

Then the cross section must be a function $f(\vartheta, \chi)$ such that

$$f(\vartheta, \chi) + f\left(\vartheta, \left(\frac{\pi}{2} + \chi\right) \bmod \pi\right) \text{ is independent of } \chi. \qquad (4.3.6)$$

The reason for this condition is the same as before: it reflects the required cylindrical symmetry of the mixture (4.3.2).

As an example one might think of the photoeffect due to polarized photons, in which case the calculated[4] cross section takes the typical form $g_1(\vartheta) + g_2(\vartheta) \cos^2 \chi$ and, of course, satisfies the condition (4.3.6). However, an experimental check of this result is a difficult task, for the measurement of the angular distribution of low-energy photoelectrons is a nonstandard technique. In recent years, polarized beams of high-energy gamma rays have become available, but it seems that so far there are no experimental data rich enough to constitute a direct check of (4.3.6). Clearly, this condition is a stringent one. Should one find a photoreaction exhibiting a preferred emission of particles in the azimuthal planes at $\pm 45°$ with respect to the plane of photon polarization, one would get a definite conflict with the quantum-mechanical description of mixtures.

4.4 Motion without Spin

The physical system here considered might be called an elementary particle, moving in ordinary three-dimensional space. "Elementary" is not meant in an absolute sense. It is not excluded that the system might be composed of other particles; we only mean that its internal structure is not relevant to what we are interested in. Nor do we claim that it is a point particle; we only mean that its size is not relevant to the motion under discussion.

We confine ourselves to the case of a spinless particle. Of this particle, only translational degrees of freedom come into play. Of course, the scheme to be developed might also apply to a particle with spin, provided the spin does not influence the motion and so can be neglected.

The physical system referred to is a basic example in all textbooks on quantum mechanics, for it epitomizes the problem of motion and displays many quantum features. A key notion for such a system is localizability:

roughly, the possibility of giving meaning to the assertion that the particle is in this or that volume.

For simplicity, we shall restrict ourselves to the case of a particle moving in one dimension, say along the x-axis. The generalization to the more realistic three-dimensional case is quite simple.

The Hilbert space that we associate to such a particle (see also next section) is $L^2(\mathbb{R}, dx)$, the infinite-dimensional Hilbert space of complex-valued functions on $\mathbb{R}$ that are square-integrable with respect to the ordinary (Lebesgue) measure dx on $\mathbb{R}$. For the operator representing the position of the particle we take the operator Q having domain

$$\mathcal{D}(Q) = \{\psi(x) \in L^2(\mathbb{R}, dx) : x\psi(x) \in L^2(\mathbb{R}, dx)\} \qquad (4.4.1)$$

and acting on $\psi \in \mathcal{D}(Q)$ according to

$$Q\psi(x) = x\psi(x). \qquad (4.4.2)$$

$\mathcal{D}(Q)$ is dense in $L^2(\mathbb{R}, dx)$, and Q is self-adjoint on it. By inspection of (4.4.2), Q has a simple, purely continuous spectrum extending over the whole real line (the particle can be found everywhere); $L^2(\mathbb{R}, dx)$ provides the representation in which Q is diagonal. The position constitutes, by itself, a complete set of compatible physical quantities. The projection-valued measure of Q is $P_Q(E)\psi(x) = \chi_E(x)\psi(x)$, where χ_E is the characteristic function of $E \in \mathcal{B}(\mathbb{R})$; accordingly, the probability distribution of the position values in the vector state $\psi(x) \in \mathcal{D}(Q)$ takes the form

$$E \mapsto (\psi(x), \chi_E(x)\psi(x)) = \int_E |\psi(x)|^2 \, dx, \qquad (4.4.3)$$

so that $|\psi(x)|^2$ is recognized as probability density of finding the particle at position x. As remarked in Section 3.2, the occurrence of the measure dx, the Lebesgue measure, corresponds to the choice of a cyclic vector $\varphi(x)$ of Q: in our case we have that the Lebesgue measure of E is equivalent to the measure $E \mapsto \int_E |\varphi(x)|^2 \, dx$, in the sense that the former vanishes if and only if the latter does (see Exercise 5).

Another physical quantity of interest for the system under discussion is its linear momentum. We denote by P the corresponding operator and define (in units which make Planck's constant equal to 2π)

$$\mathcal{D}(P) = F^{-1}\mathcal{D}(Q),$$
$$P\psi(x) = F^{-1}QF\psi(x) \qquad \text{if} \quad \psi(x) \in \mathcal{D}(P), \qquad (4.4.4)$$

where F stands for the Fourier-Plancherel operator in $L^2(\mathbb{R}, dx)$. Notice that on the functions [in $\mathcal{D}(P)$] that are differentiable,* the action of P takes the familiar form (see Exercise 6)

$$P\psi(x) = -i\frac{d}{dx}\psi(x). \tag{4.4.5}$$

We shall write $\tilde{\psi}(k)$ for $F\psi(x)$, stressing that the replacement of x by k is purely formal, but it can make some formulas more readable. The probability distribution of the momentum in the vector state $\psi(x) \in \mathcal{D}(P)$ follows from (4.4.4) and (4.4.3): it is

$$E \mapsto \left(\psi(x), F^{-1}P_Q(E)F\psi(x)\right) = \left(\tilde{\psi}(k), \chi_E(k)\tilde{\psi}(k)\right) = \int_E \left|\tilde{\psi}(k)\right|^2 dk,$$

where we have used the fact that F is a unitary operator (see also Exercise 7). Thus, $|\tilde{\psi}(k)|^2$ is recognized as the probability density for finding the particle with momentum k.

The operator F is precisely the one that maps the representation in which Q is diagonal into the representation in which P is diagonal. In fact, after noticing that $\psi(x) \in \mathcal{D}(P)$ is mapped by F into $\tilde{\psi}(k) \in \mathcal{D}(Q)$, we get

$$FPF^{-1}\tilde{\psi}(k) = Q\tilde{\psi}(k) = k\tilde{\psi}(k). \tag{4.4.6}$$

This makes it apparent that P too has a simple, purely continuous spectrum extending over the whole real line (the particle is allowed to have any momentum). The momentum constitutes, by itself, a complete set of compatible physical quantities. Let us also remark that in the P-representation the position operator takes the form FQF^{-1}, and on a differentiable function, $FQF^{-1}\tilde{\psi}(k) = i(d/dk)\tilde{\psi}(k)$.

There are other physical quantities that are interesting for the physical system under discussion. First of all, the energy of the particle. Here we refer only to the case of a free particle, so that the only energy that comes into play is the kinetic energy. The operator H_0 representing this energy is often called "free Hamiltonian" and is defined in terms of P according to

$$H_0 = \frac{P^2}{2m}, \tag{4.4.7}$$

where m is the mass of the particle. In the Q-representation the domain of H_0 is formed by those functions $\psi(x)$ such that $k^2\tilde{\psi}(k)$ is square-integrable. On (twice) differentiable functions of its domain, the action of H_0 becomes

---

*The set $C_b^1(\mathbb{R})$ of differentiable functions with compact support is contained in $\mathcal{D}(P)$ as well as in $\mathcal{D}(Q)$, and is dense in $L^2(\mathbb{R}, dx)$, and hence also in $\mathcal{D}(P)$ and $\mathcal{D}(Q)$.

[see (4.4.5)]

$$H_0\psi(x) = -\frac{1}{2m}\frac{d^2}{dx^2}\psi(x).$$

Going to the P-representation, H_0 becomes FH_0F^{-1} and acts as the multiplicative operator

$$FH_0F^{-1}\tilde{\psi}(k) = \frac{k^2}{2m}\tilde{\psi}(k);$$

this shows that H_0 has a purely continuous spectrum, but it has not a simple spectrum, or, in other words, it does not form a complete set by its own (see Exercise 8). From the physical point of view this is not-surprising, for the knowledge of the kinetic energy does not specify whether the particle is traveling from left to right or from right to left; by adding to the kinetic energy a new physical quantity which carries this further information, we can get a complete set of physical quantities (see Exercise 9).

The quantum behavior of the free-particle system considered in this section is emphasized by the fact that position and momentum do not commute and that they give rise to a nontrivial Heisenberg inequality. Notice that $\mathcal{D}(Q) \cap \mathcal{D}(P)$ is dense in $L^2(\mathbb{R}, dx)$, and on this set

$$QP - PQ = iI.$$

Therefore, according to what was said in Section 3.4, the Heisenberg inequality takes the particularly simple form

$$\mathcal{V}(Q, D)\mathcal{V}(P, D) \geqslant \tfrac{1}{4} \qquad (4.4.8)$$

for any (not necessarily pure) state D such that both variances exist. We stress once again that a result of this kind is possible because the noncommuting operators Q and P have a purely continuous spectrum and are both unbounded [should one consider a particle in a box, the product of the variances of position and momentum, in the same state, could be made arbitrarily small (see Exercise 10)].

The fact that, for the free particle here considered, the product of the variances of position and momentum exceeds a fixed positive number, no matter what the state of the particle is, makes (4.4.8) the exemplar of Heisenberg inequalities, and attaches to it a particularly attractive, widely celebrated physical insight.

Another typical quantum feature of the position and momentum of the free particle is that they are complementary physical quantities. According to the definition of complementarity given in Section 3.5, we have in fact that, for every bounded $E, F \in \mathcal{B}(\mathbb{R})$, the range of $P_Q(E)$ is disjoint from the range of $P_P(F)$. It must be emphasized that the restriction to *bounded* Borel sets is not superfluous, for Jauch[5] has given a counterexample of unbounded sets $E, F \in \mathcal{B}(\mathbb{R})$ such that the ranges of $P_Q(E)$ and $P_P(F)$ have a nonempty intersection.

4.5 Euclidean Invariance and One-Particle Systems

In the preceding sections of this chapter we have examined the quantum description of a free particle in the special case where it is spinless, and the quantum description of a particle with spin in the special case where it has no motion in space. In both cases the quantum description is achieved by specifying the concrete form of the Hilbert space and of the self-adjoint operators representing the physical quantities of interest.

Traditionally, this specification was based on physical intuition (for instance, in the introduction of the concept of spin by W. Pauli) and on rather vague arguments such as the so-called correspondence principle; in any case it had its final justification in the fact that it worked. The quantum description of a concrete physical system has been put on a clear and rigorous basis by the work of Mackey on infinite-dimensional unitary representations of (noncompact) groups and on imprimitivity systems.[6, 7] Loosely speaking, Mackey's idea is to start from the structure of the physical space where the system resides and from the group of symmetries of this space, deducing (*not* assuming) the concrete realization of the Hilbert space $\mathcal{H}$ and the explicit form of the operators that represent the physical quantities of interest. The choice of the Hilbert space and of the explicit form of the operators is unique up to a unitary equivalence.

In case we want to give the quantum description of a free particle in ordinary Euclidean three-dimensional space, the symmetry group of the space is the group of rigid motions in this space, the so-called Euclidean group, which is the semidirect product of the translations and rotations in three-dimensional space. We briefly sketch in the following the main conclusions of Mackey's approach.

(1) For every irreducible projective representation D^j of the three-dimensional rotation group ($2j+1$ is the dimension of the representation, $j = 0, \frac{1}{2}, 1, \dots$), let $H(D^j)$ be the complex Hilbert space in which D^j operates. Then consider the Hilbert space $\mathcal{H}(\mathbb{R}^3, d^3x, H(D^j))$ formed by the functions $f(x_1, x_2, x_3)$ defined in $\mathbb{R}^3$, taking values in $H(D^j)$, and satisfying the condition $\int_{\mathbb{R}^3} (f(x_1, x_2, x_3), f(x_1, x_2, x_3)) d^3x < +\infty$, where $(\cdot, \cdot)$ is the scalar product in $H(D^j)$ and d^3x is the ordinary three-dimensional Lebesgue measure. $\mathcal{H}(\mathbb{R}^3, d^3x, H(D^j))$ is precisely the Hilbert space of states of the particle. Thus, there is an array of possibilities labeled by the choice of j. Note that $\mathcal{H}(\mathbb{R}^3, d^3x, H(D^j))$ is unitarily equivalent to $\mathcal{H}(\mathbb{R}^3, d^3x) \otimes H(D^j)$ (see Chapter 7 for a brief description of the tensor

product of Hilbert spaces). Introducing a basis in $H(D^j)$, the elements of $\mathcal{K}(\mathbb{R}^3, d^3x) \otimes H(D^j)$ are represented by $(2j+1)$-dimensional column vectors whose elements are square-integrable complex functions defined in $\mathbb{R}^3$. This description is familiar to physicists.

(2) The operators that represent physical quantities are deduced from the geometrical transformations of the physical space they are related to. Without worring about domains of definition, we mention the following results. The x_1-component of position is represented by multiplication by x_1, and similarly for x_2 and x_3. The linear momenta along the coordinate axes are represented by $-i\partial/\partial x_1, -i\partial/\partial x_2, -i\partial/\partial x_3$. These operators are independent of the representation D^j one has chosen. The operators that represent the angular momenta along the coordinate axes are sums of two terms. For definiteness consider the x_3-component of the angular momentum: the operator that represents this quantity is the sum of the operator

$$L_3 = i\left(x_2\frac{\partial}{\partial x_1} - x_1\frac{\partial}{\partial x_2}\right)$$

and an operator J_3 in $H(D^j)$, which is explicitly given by (4.1.1). Thus, the component of angular momentum appears naturally decomposed into two terms: the first one, L_3, depends on position coordinates, but is independent of D^j; the second one, J_3, does not depend on the position coordinates (so that it commutes with position, linear momentum, and orbital angular momentum). The angular momentum, which characterizes the behavior of the system under rotations of the coordinate axes, is split into two parts: the orbital angular momentum, which depends on the motion of the particle, and the intrinsic angular momentum, or spin, which does not.

(3) The Hamiltonian operator, representing the kinetic energy of the particle, is, at this stage, not completely defined, being an arbitrary function of the Laplacian $\partial^2/\partial x_1^2 + \partial^2/\partial x_2^2 + \partial^2/\partial x_3^2$. In order to remove this arbitrariness, a richer symmetry group is required: if the Galilei group is assumed, then the Hamiltonian becomes exactly the Laplacian, up to a multiplicative factor that depends only on the choice of units.

Let us add a few remarks emerging from Mackey's constructive procedure.

(a) The usual wave mechanics of a particle (with or without spin), based on the usual representation in which the position is diagonal, provides a canonical version of quantum mechanics.

(b) The notion of intrinsic spin finds a natural place without any need of relativistic invariance. This refutes an old claim that spin is ultimately a relativistic effect: we refer to J. M. Levy-Leblond[8] for a deepening of this point.

(c) The celebrated "principle of correspondence", the historical recipe according to which the operators corresponding to physical quantities have to be deduced from the correpsonding classical expressions via the so-called "canonical quantization rules", is overcome. This gives internal coherence and autonomy to quantum mechanics, eliminating the logical vicious circle inherent in the principle of correspondence, with its claim that the formulation of quantum mechanics presupposes classical mechanics, while, at the same time, classical mechanics is required to be a limiting case of quantum mechanics.

Exercises

1. Show that, in $\mathbb{C}^2$, one-dimensional projectors have the form (4.2.2).
[*Hint.* Use (4.2.1) and impose idempotence and trace equal to one; check positivity of (4.2.2).]

2. Show that, in $\mathbb{C}^2$, the nonpure states have the form (4.2.3).
[*Hint.* Use (4.2.1) and impose the condition of unit trace and positivity.]

3. Show that the spectral decomposition of the physical quantity (4.2.1) is

$$A = \lambda^+ P^+ + \lambda^- P^-$$

with $\lambda^\pm = \rho_4 \pm (\rho_1^2 + \rho_2^2 + \rho_3^2)^{1/2}$ and

$$P^\pm = \tfrac{1}{2}\left[I \pm \left(\rho_1^2 + \rho_2^2 + \rho_3^2 \right)^{-1/2} (\rho_1\sigma_1 + \rho_2\sigma_2 + \rho_3\sigma_3) \right].$$

4. Show that the state (4.2.3) assigns to the eigenvalues $\lambda^\pm$ of Exercise 3 the probabilities

$$\tfrac{1}{2}\left[1 \pm \left(\rho_1^2 + \rho_2^2 + \rho_3^2 \right)^{-1/2} (\rho_1 r_1 + \rho_2 r_2 + \rho_3 r_3) \right].$$

5. Show that $\varphi(x) = e^{-x^2}$ is a cyclic vector for Q and that the spectral measure associated with this cyclic vector is equivalent to the Lebesgue measure dx.

6. Prove (4.4.6).
[*Hint.* Note that, if $\psi(x) \in C_b^1(\mathbb{R})$, F takes the form $F\psi(x) = (2\pi)^{-1/2}\int e^{-ikx}\psi(x)dx$.]

7. Show that if E is the finite interval $[\lambda_1, \lambda_2]$, then

$$F^{-1}P_Q([\lambda_1, \lambda_2])F\psi(x)$$

$$= (2\pi)^{-1/2} \int_{[\lambda_1, \lambda_2]} \frac{\exp(i\lambda_2(x-y)) - \exp(i\lambda_1(x-y))}{i(x-y)} dy.$$

[*Hint.* On a finite interval F has the form recalled in the hint of Exercise 6.]

8. Show that in $L^2(\mathbb{R}, dx)$, H_0 has not a simple spectrum.

[*Hint.* Refer to Section 3.2 and show that the finite linear combinations $P_{H_0}(E_1)\varphi + \cdots + P_{H_0}(E_n)\varphi$ are not dense in $L^2(\mathbb{R}, dx)$.]

9. Show that if we define the operator B as $B\psi(x) = \psi(-x)$, the pair formed by H_0 and B is a complete set of commuting operators.

[*Hint.* Build explicitly the representation in which H_0 and B become diagonal.]

10. Consider the Hilbert space $L^2([0,1], dx)$, define Q and P by (4.4.2) and (4.4.5) with domain $\mathcal{D}(P) = \{\psi \in L^2: \psi \text{ absolutely continuous}; d/dx\,\psi \in L^2; \psi(1) = \psi(0)\exp(i\vartheta), 0 \leqslant \vartheta \leqslant 2\pi\}$. Show that for the vector states of the form $\psi_n(x) = \exp[(2\pi in + i\vartheta)x]$ the variance of P is zero, while the variance of Q is obviously finite.

References

1. L. C. Biedenharn and J. D. Louck, *Angular Momentum in Quantum Physics*, Vol. 8 of Encyclopedia of Mathematics and Its Applications, Addison-Wesley, Advanced Book Program, Reading, Mass., 1981.
2. See, e.g.: H. Poincaré, *Théorie Mathematique de la Lumière* Vol. II, G. Carré, Paris, 1892 (Chapter 12); M. Born and E. Wolf, *Principles of Optics*, Pergamon Press, Oxford, 1970 (p. 30); M. S. Walker, *Am. J. Phys.* 22 (1954) 170.
3. See, e.g., J. D. Bjorken and S. D. Drell, *Relativistic Quantum Mechanics*, McGraw-Hill, New York, 1964 (Section 10.13).
4. See, e.g., L. Landau and E. Lifshitz, *Relativistic Quantum Theory*, Pergamon Press, Oxford, 1971 (Sections 56, 57).
5. J. M. Jauch, in *Quantum Mechanics, a Half Century Later*, J. Leite Lopes and M. Paty, eds., D. Reidel, Dordrecht, 1977.
6. G. W. Mackey, *Induced Representations of Groups and Quantum Mechanics*, Benjamin, Advanced Book Program, Reading, Mass., 1968.
7. V. S. Varadarajan, *Geometry of Quantum Theory*, Vol. II, Van Nostrand, Princeton, N.J., 1970.
8. J. M. Levy-Leblond, in *Group Theory and its Applications*, Vol. II, E. Loebl, ed., Academic Press, New York, 1971; *Riv. Nuovo Cimento* 4 (1974) 99.

CHAPTER 5

Superselection Rules

5.1 Physical Systems with Limited Quantum Behavior

The theoretical scheme outlined in the preceding chapters contains, as basic ingredients, the following statements:

(a) Physical quantities are in one-to-one correspondence with self-adjoint operators on $\mathcal{H}$.

(b) Pure states are in one-to-one correspondence with one-dimensional subspaces of $\mathcal{H}$.

That scheme fits in with the generality of physical systems that exhibit genuine, unlimited quantum behavior. Nevertheless, there are situations in which it is useful to imagine physical systems that do not fulfill a) and b), but do fulfill only the following weaker statements:

(a') Every physical quantity is represented by a self-adjoint operator on $\mathcal{H}$, but not every self-adjoint operator on $\mathcal{H}$ represents a physical quantity.

(b') Every pure state is represented by a one-dimensional subspace of $\mathcal{H}$, but not every one-dimensional subspace of $\mathcal{H}$ represents a pure state.

A familiar example of such a situation is encountered in nuclear physics. Protons and neutrons, the fundamental constituents of atomic nuclei, entered the scene of physics at different times, playing the role of different physical systems, as the names imply. However, insofar as one is interested in processes dominated by strong (i.e., nuclear) interactions, thus being allowed to neglect electromagnetic and weak interactions, it is useful to think of proton and neutron as different states of one and the same physical system: the *nucleon*. This is because strong interactions do not distinguish

ENCYCLOPEDIA OF MATHEMATICS and Its Applications, Gian-Carlo Rota (ed.). Vol. 15: E. G. Beltrametti and G. Cassinelli, The Logic of Quantum Mechanics

ISBN 0-201-13514-0

between protons and neutrons, so that the invention of the nucleon greatly simplifies calculations, via the so-called isotopic-spin formalism. Of course, such a simplification is achieved at the cost of limitations on the possibility of superposing pure states to get new pure states: in fact, nobody has ever succeeded in preparing a nucleon in a quantum superposition of proton and neutron states. Thus, not every unit vector in the Hilbert space of the nucleon represents a state. We are faced with case (b′). Now consider a self-adjoint operator having nonvanishing matrix elements between proton and neutron states: its diagonalization (or its spectral decomposition) necessarily calls into play vectors that are superpositions of proton and neutron states, and hence do not represent states. Thus such an operator cannot represent a physical quantity, and we are faced with case (a′). According to a well-established terminology, we say that a *superselection rule* is operating between proton and neutron states of the nucleon; it keeps them separated, preventing superpositions. Intuitively, the root of this superselection rule is the conservation of electric charge. Much in the same way, the conservation of other "charges" (for instance, the baryonic charge) generates similar superselection rules.

As another example, one might imagine a hypothetical system admitting states with integer and half-integer spin. One would then be forced to acknowledge, as an empirical fact, that superpositions of integer with half-integer spin states have never been observed. Again we say that a superselection rule is operating between these two kinds of states. We can remark that a superposition of a half-integer spin state, represented by the vector state ψ, with an integer spin state, represented by the vector state φ, would not be invariant under a 360° rotation of the coordinate axes, for ψ would undergo a change of sign, while φ would not be affected. Therefore, the superselection rule that operates between integer and half-integer spin states has its roots in the fact that nature is invariant under rotations of 360°. We see from this remark that the existence of superselection rules can, in certain instances, reflect the existence of empirically well-established symmetry properties.

Consider, as a further example, a measuring instrument, and label its states by the positions of the pointer. Then, pushing this system into the framework of quantum mechanics, one should admit that it is never possible to superpose states corresponding to different positions of the pointer, for this would contradict the very assumption that one has to do with a measuring instrument. Again, we say that states of such a system are separated by a superselection rule. Of course, when a superselection rule operates between any two states, thus dismissing entirely superpositions of states, one loses quantum behavior completely and reverts to classical behavior.

It should be clear from these examples that superselection rules are not a strict or logical necessity of quantum theory, for their occurrence depends on what is taken to be the "physical system", and this is a matter of convention. Superselection rules could be removed by a sharper definition of what to consider as physical system: in the "nucleon" example it would suffice to take proton and neutron as different physical systems; in the example of integer versus half-integer spins it would suffice to agree that the two cases refer to different physical systems; in the "measuring instrument" example it would suffice to agree that different positions of the pointer correspond to different physical systems. But it is clear that the removal of all superselection rules is too restrictive a choice, for it would make quantum theory impotent in many physically relevant situations: those in which one recognizes neither strictly quantum nor strictly classical behavior. The acceptance of superselection rules makes quantum mechanics much more flexible, providing a bridge between the strictly quantum and the strictly classical limits.

5.2 Superselection Operators

We proceed to sketch, briefly, how the formalism of Hilbert-space quantum mechanics is modified in the presence of superselection rules.

Once one assumes that not every self-adjoint operator on $\mathcal{H}$ represents a physical quantity, one must agree that the set $\{A_i\}$ of self-adjoint operators associated to the physical quantities generates a von Neumann algebra $\mathbf{A} = \{A_i\}''$ that is a proper subalgebra of $\mathbf{B}(\mathcal{H})$, the algebra of all bounded operators on $\mathcal{H}$. Recalling that the only operators that commute with every element of $\mathbf{B}(\mathcal{H})$ are the multiples of the identity, we see at once that the commutant $\mathbf{A}'$ of $\mathbf{A}$ must contain nontrivial operators, that is, operators that are not multiples of the identity. In fact, should that not be the case, we would have $\mathbf{A}'' = \mathbf{B}(\mathcal{H})$, contrary to the assumption that $\mathbf{A}$ is a von Neumann algebra (hence $\mathbf{A} = \mathbf{A}''$) strictly contained in $\mathbf{B}(\mathcal{H})$. The nontrivial elements of $\mathbf{A}'$, namely the nontrivial operators that commute with every element of $\mathbf{A}$, are called the "superselection operators."

An important fact is that all superselection operators belong to $\mathbf{A}$; in other words, they are bounded functions of the physical quantities. This is not obvious, for in principle one might imagine that $\mathbf{A}'$ has only a partial overlapping with $\mathbf{A}$. However, quantum mechanics demands $\mathbf{A}' \subseteq \mathbf{A}$, as the following argument shows. From the existence of complete sets of commuting physical quantities (see Section 3.2) follows the existence of maximal abelian subalgebras of $\mathbf{A}$. Let $\mathbf{B}$ be one of them; since $\mathbf{B}' = \mathbf{B}$ and $\mathbf{B} \subseteq \mathbf{A}$, we have

$$\mathbf{A}' \subseteq \mathbf{B}' = \mathbf{B} \subseteq \mathbf{A}.$$

Thus the superselection operators, which, by definition, are elements of $\mathbf{A}'$, belong to $\mathbf{A}$.

We have also that the center $\mathbb{Z}$ of $\mathbf{A}$, which is generally defined as $\mathbb{Z} = \mathbf{A} \cap \mathbf{A}'$, now reduces to

$$\mathbb{Z} = \mathbf{A}'.$$

Thus, the existence of superselection rules makes the center of $\mathbf{A}$ nontrivial, its nontrivial elements being precisely the superselection operators.

We may also remark that, since the set $\{A_i\}$ of the physical quantities is a set of self-adjoint operators, the commutant $\{A_i\}'$ is a von Neumann algebra, and hence $\{A_i\}' = \{A_i\}''' = \mathbf{A}' = \mathbb{Z}$. We interpret this result by saying that the superselection operators can also be characterized as the nontrivial bounded operators that commute with every physical quantity.

Notice that the superselection operators are normal operators, for they are pairwise commuting and form an abelian von Neumann algebra; hence the spectral theorem applies to them.

5.3 Coherent Subspaces

To simplify the discussion, let us make the assumption that the superselection operators have pure point spectra, so that there are pairwise orthogonal projectors P_k, k in some countable index set $\mathcal{I}$, with $\sum_{k \in \mathcal{I}} P_k = I$, such that every superselection operator is of the form $\sum_k \lambda_k P_k$ with complex eigenvalues λ_k. This assumption corresponds to the so-called *discrete superselection rules*; in its absence one would be faced with *continuous superselection rules*. The restriction to the discrete case does not remove any of the essential features of superselection rules, but avoids some technical complications.[1]

Clearly, every P_k belongs to $\mathbb{Z}$ and commutes with every element of $\mathbf{A}$. Conversely, if a bounded operator A commutes with all the P_k's, then it commutes with every central element, and hence $A \in \mathbb{Z}' = \mathbf{A}'' = \mathbf{A}$. Thus $\mathbf{A}$ consists precisely of the bounded operators that commute with all the P_k's.

Let $\mathfrak{M}_k$ be the closed subspace of $\mathcal{H}$ onto which P_k projects. Any two distinct subspaces $\mathfrak{M}_k, \mathfrak{M}_{k'}$ $(k, k' \in \mathcal{I})$ are orthogonal, and the direct sum $\oplus_{k \in \mathcal{I}} \mathfrak{M}_k$ is the whole $\mathcal{H}$. Every element of $\mathbf{A}$ leaves each $\mathfrak{M}_k$ invariant, that is, it maps the vectors of $\mathfrak{M}_k$ into vectors of $\mathfrak{M}_k$; in fact, for every $\psi \in \mathfrak{M}_k$,

$$P_k A \psi = A P_k \psi = A \psi.$$

It is customary to express this fact by saying that the subspaces $\mathfrak{M}_k$ are the coherent subspaces of $\mathcal{H}$.

By the very definition of $\mathbf{A}$ (as the algebra generated by the physical quantities) it follows that a self-adjoint operator A on $\mathcal{H}$ represents a physical quantity only if every projector occurring in its spectral decomposition belongs to $\mathbf{A}$, thus being a projector onto a subspace of some $\mathfrak{M}_k$; of course, in case A is bounded, all that amounts to saying $A \in \mathbf{A}$. With the terminology used in the theory of von Neumann algebras, we say that A represents a physical quantity when it is affiliated with $\mathbf{A}$.[2]

5.4 Correspondence between Density Operators and States

Consider now the characterization of states in terms of the algebra $\mathbf{A}$, or, equivalently, in terms of the coherent subspaces. It is still true that every density operator on $\mathcal{H}$ assigns a probability distribution to each physical quantity, but the set of all density operators on $\mathcal{H}$ is, in a sense, redundant, for there are density operators that are different as operators on $\mathcal{H}$ but are indistinguishable with regard to the probability distributions they assign to the physical quantities. In other words, when superselection rules are present, the physical quantities are not numerous enough to separate the set of all density operators on $\mathcal{H}$. Since states are characterized by probability distributions of physical quantities, the correspondence between states and density operators is no longer one-to-one.

As an example of this situation, take a unit vector φ_k in $\mathfrak{M}_k$ and a unit vector $\varphi_{k'}$ in $\mathfrak{M}_{k'}$ ($k \neq k'$), and form the superposition $\psi = \lambda \varphi_k + \lambda' \varphi_{k'}$ with $|\lambda|^2 + |\lambda'|^2 = 1$; then consider the density operators

$$D_1 = P^{\hat{\psi}}, \qquad D_2 = |\lambda|^2 P^{\hat{\varphi}_k} + |\lambda'|^2 P^{\hat{\varphi}_{k'}},$$

where $P^{\hat{\psi}}, P^{\hat{\varphi}_k}, P^{\hat{\varphi}_{k'}}$ are the projectors onto the one-dimensional subspaces spanned by $\psi, \varphi_k, \varphi_{k'}$ respectively. Choosing in $\mathcal{H}$ a basis of which φ_k and $\varphi_{k'}$ are members, we get, for every projection operator P in $\mathbf{A}$,

$$\operatorname{tr}(PD_1) = \left(\varphi_k, PP^{\hat{\psi}}\varphi_k\right) + \left(\varphi_{k'}, PP^{\hat{\psi}}\varphi_{k'}\right)$$
$$= |\lambda|^2 (\varphi_k, P\varphi_k) + |\lambda'|^2 (\varphi_{k'}, P\varphi_{k'})$$
$$= |\lambda|^2 \operatorname{tr}(PP^{\hat{\varphi}_k}) + |\lambda'|^2 \operatorname{tr}(PP^{\hat{\varphi}_{k'}}) = \operatorname{tr}(PD_2),$$

so that D_1 and D_2 are indeed indistinguishable as far as probability distributions of the physical quantities are concerned.

From this example we see that, whenever one considers a unit vector having components in different coherent subspaces, the corresponding one-dimensional projector (D_1 in the example) cannot be interpreted as a pure state, for it is fully equivalent, as long as we deal with the physical quantities, to a nonpure state (D_2 in the example). We see also that there is an infinite family of one-dimensional projectors fully equivalent to the same nonpure state: in the example, this family is obtained by arbitrarily varying the relative phase of the coefficients λ, λ'. In other words, the relative phases of vectors belonging to different coherent subspaces cannot be distinguished by the physical quantities: they are physically undetectable.[3]

What is said above makes clear that the correspondence between the set of states of the physical system and the set of all density operators on $\mathcal{H}$ cannot be one-to-one. Nevertheless, states do correspond one-to-one to the density operators that are members of the von Neumann algebra $\mathbf{A}$ generated by the physical quantities (in the foregoing example D_2 is member of $\mathbf{A}$, while D_1 is not). The proof of this fact goes as follows. Let D be any density operator on $\mathcal{H}$, and define

$$D_\mathbf{A} = \sum_{k \in \mathcal{G}} P_k D P_k, \qquad (5.4.1)$$

where the sum runs over all indices of the coherent subspaces (in case the sum is infinite, it has to be understood as a uniform limit). $D_\mathbf{A}$ is still a density operator (see Exercise 1); it belongs to $\mathbf{A}$, and assigns to every physical quantity the same probability distribution that is assigned by D. In fact (see Exercise 2)

$$\operatorname{tr}(DP) = \operatorname{tr}(D_\mathbf{A} P) \qquad \text{for every projector} \quad P \in \mathbf{A}. \qquad (5.4.2)$$

Now, if D' is another density operator on $\mathcal{H}$ that assigns to every physical quantity the same probability distribution that is assigned by D, we have

$$D_\mathbf{A} = D'_\mathbf{A}.$$

In fact, $P_k D P_k = P_k D' P_k$ for all $k \in \mathcal{G}$, as is seen by the following argument: for every $\psi \in \mathcal{H}$ we have, writing ψ_k as an abbreviation for $P_k \psi$,

$$(\psi, P_k D P_k \psi) = (\psi_k, D\psi_k) = \|\psi_k\|^2 \operatorname{tr}(DP^{\hat{\psi}_k}),$$

$$(\psi, P_k D' P_k \psi) = (\psi_k, D'\psi_k) = \|\psi_k\|^2 \operatorname{tr}(D'P^{\hat{\psi}_k}),$$

and

$$\operatorname{tr}(DP^{\hat{\psi}_k}) = \operatorname{tr}(D'P^{\hat{\psi}_k})$$

because $P^{\hat{\psi}_k}$ belongs to $\mathbf{A}$ (thus representing a particular physical quantity).

Summing up, we have seen that if the choice of density operators is restricted to $\mathbf{A}$, then the one-to-one correspondence between states and density operators is restored. Pure states are associated with unit vectors that belong to a coherent subspace $\mathfrak{M}_k$ for some k.

The statement, made at the end of Section 5.1, that superselection rules are not a strict necessity of quantum theory and could be removed by proper redefinition of the physical system now becomes more definite. If, starting from some physical system, a pattern of superselection rules emerges, so that the Hilbert space $\mathcal{H}$ can be thought of as the sum of its coherent subspaces $\mathfrak{M}_k$, then it would suffice to agree that different coherent subspaces represent different physical systems in order to prevent, for each of the new physical systems, the appearance of superselection rules. Such a redefinition procedure would reduce to a limiting case when each of the coherent subspaces is one-dimensional, that is, when the initial physical system is classical, so that the superselection rules play the role of preventing all superposition of states. But, as already remarked, the inclusion of the superselection rules in the formalism of quantum mechanics constitutes a useful enrichment of the theory, making it able to encompass physical systems that exhibit partial quantum behavior.

Exercises

1. Show that if D is a density operator, then (5.4.1) defines another density operator and we have

$$\text{tr}\left(\sum_k P_k D P_k \right) = \sum_k \text{tr}(P_k D P_k)$$

[*Hint.* See Exercises 1 and 2 of chapter 2.]

2. Prove (5.4.2).
[*Hint.* Use the equalities $\text{tr}(P_k D P_k P) = \text{tr}(D P_k P P_k) = \text{tr}(D P P_k)$.]

References

1. See, also for further bibliography, R. Cirelli, F. Gallone, and B. Gubbay, *J. Math. Phys.* 16 (1975) 201.
2. J. Dixmier, *Les Algèbres d'Opérateurs dans l'Espace Hilbertien*, Gauthiers-Villars, Paris, 1957.
3. G. C. Wick, A. S. Wightman, and E. P. Wigner, *Phys. Rev.* 88 (1952) 101.

CHAPTER 6

Dynamical Evolution

6.1 The Unitary Dynamical Group in Schrödinger's Picture

Up to now we have been concerned with the description of a physical system at a fixed instant of time. The time evolution of quantum systems constitutes a many-faced, wide issue which, however, is somewhat tangential to the main aim of this volume. We shall touch upon it only briefly, restricting ourselves to the more familiar aspects of the problem.

We assume, to begin with, that the physical system has no superselection rules, so that states and physical quantities correspond one-to-one to density operators and self-adjoint operators, respectively, on the Hilbert space $\mathcal{H}$. To describe the dynamical evolution of the system one has to specify the way the representatives of states and physical quantities evolve in $\mathcal{H}$, which is fixed in time, since the physical system is supposed to preserve its identity.

There is an old tenet according to which there are two kinds of time evolution: the discontinuous, nondeterministic evolution undergone by the quantum system when a measurement is made on it, and the continuous, deterministic evolution caused by the interaction with external forces or with other quantum systems. The first kind of evolution will be dealt with in Chapter 8, where we shall find it less alarming than is often claimed. In the present chapter we are concerned only with the second kind of evolution, and we consider it in the popular framework based on the existence of a one-parameter ($\tau \in \mathbb{R}$) group $\tau \mapsto U_\tau$ of unitary operators on $\mathcal{H}$ such that

$$U_\tau U_{\tau'} = U_{\tau+\tau'}, \qquad (6.1.1)$$

$$U_\tau^{-1} = U_{-\tau}, \qquad (6.1.2)$$

ENCYCLOPEDIA OF MATHEMATICS and Its Applications, Gian-Carlo Rota (ed.).
Vol. 15: E. G. Beltrametti and G. Cassinelli, The Logic of Quantum Mechanics

ISBN 0-201-13514-0

$$\tau \mapsto (\varphi, U_\tau \varphi') \text{ is a continuous function of } \tau \text{ for all } \varphi, \varphi' \in \mathcal{H}. \quad (6.1.3)$$

In this framework we must now specify how states and physical quantities do evolve. The more familiar prescription, called the *Schrödinger picture*, consists of the following:

(a) density operators representing states evolve from time t_1 to time t_2 according to

$$D(t_2) = U_{t_2 - t_1} D(t_1) U_{t_2 - t_1}^{-1}, \quad (6.1.4)$$

(b) self-adjoint operators representing physical quantities stay unchanged in time.

We leave it to the reader to show that the rule (6.1.4) maps density operators into density operators (see Exercise 1). The group $\tau \mapsto U_\tau$ is generally called the *dynamical group* of the physical system, while U_τ is referred to as the *evolution operator*.

It follows from the above rules that the probability distribution of a physical quantity represented by A evolves with time according to the rule: if at time t_1 the probability distribution is

$$E \mapsto \mathrm{tr}(D(t_1) P_A(E)), \qquad E \in \mathcal{B}(\mathbb{R}), \quad (6.1.5)$$

then at time t_2 it is

$$E \mapsto \mathrm{tr}\left(U_{t_2 - t_1} D(t_1) U_{t_2 - t_1}^{-1} P_A(E)\right), \qquad E \in \mathcal{B}(\mathbb{R}). \quad (6.1.6)$$

To help in understanding this sort of dynamical evolution and its limitations, we proceed with a number of remarks.

(i) In principle one might expect that the state of the system at a given instant of time depends on the whole history of the system, but we see that the evolution expressed by (6.1.4) is not of such a general kind: the state at time t_2 is uniquely determined by the knowledge of the state at any previous time t_1 and of the interactions the system has suffered from t_1 to t_2. If we know the state at time t_1, then we can forget about what happened before; the system has only a partial memory of its past. In the terminology of probability theory, we can say that we are dealing with a Markov-like evolution. Clearly, (6.1.4) expresses a deterministic evolution of states, and we can expect that the infinitesimal form of time evolution will be described by a differential equation of first order in time (see next section).

(ii) The evolution operator U_τ is a one-parameter operator, with the parameter representing a time duration, that is, a relative time. This fact, which is crucial for the semigroup property (6.1.1), reflects the assumption that the sources of external forces do not vary with time, so that there is homogeneity in time: we have not to do with absolute time; only relative time enters. Accordingly, we might say that we are dealing with a "stationary" Markovian evolution.

(iii) The requirement that the operators U_τ have an inverse, given by (6.1.2), so that they form not only a semigroup but in fact a group, corresponds to restricting the dynamical evolution to the class of reversible dynamics. This restriction excludes important classes of physical phenomena. We might mention, for instance, the case in which a quantum system, interacting with a macroscopic system, suffers an influence of a statistical or thermal nature, so that its evolution to an equilibrium state is typically irreversible: the relevance and complexity of such a class of irreversible evolutions makes their study a theory in itself.[1]

(iv) The dynamical evolution expressed by (6.1.4) preserves the convex structure of states: if $D(t_1)$ is a mixture, say $D(t_1) = w_1 D'(t_1) + w_2 D''(t_1)$, then $D(t_2) = w_1 D'(t_2) + w_2 D''(t_2)$. In case $D(t_1)$ is an infinite convex combination, an analogous result holds true (see Exercise 2). The preservation of the convex structure of states is, from the physical point of view, a rather general requirement; nevertheless, there are some concrete situations in which dynamical evolutions can occur that do not preserve convexity. As a first example, consider a polarized particle that, propagating in some medium, undergoes a depolarization process. As a second example, consider a physical system that is part of a bigger compound system: as we shall see in Chapter 7, a convexity-preserving evolution of the compound system can induce in the subsystem an evolution that does not preserve convexity. Another instance in which convexity-violating evolutions can have a physical interest is provided by physical systems endowed with superselection rules, as we shall see in Section 6.4. A deep theorem of Kadison[2] has revealed that the requirement of convexity preservation is responsible for a good deal of the dynamical evolution specified by (6.1.4): in fact, he discovered that every one-to-one map of the set of all density operators onto itself that preserves the convex combinations must take the form $D \mapsto UDU^{-1}$, where U is either a unitary or an antiunitary operator. The exclusion of the antiunitary case then follows by arguments of continuity, as Mackey[3] has shown. We refer to Section 23.2 for further details, and to an excellent overview of the subject by B. Simon.[4]

(v) Since the dynamical evolution (6.1.4) preserves convex combinations, pure states evolve into pure states. Explicitly we have (see Exercise 3) that if at time t_1 the system is in the pure state represented by the vector $\psi(t_1)$, then at time t_2 it will be in the pure state represented by the vector

$$\psi(t_2) = U_{t_2 - t_1} \psi(t_1). \tag{6.1.7}$$

From this rule we see that state vectors are transformed linearly by the time evolution, so that superpositions of pure states are preserved by the dynamics. This fact emphasizes the crucial role of superpositions in quantum mechanics.

(vi) The unitarity of U_τ can be viewed as an expression of the fact that time evolution preserves scalar products. If φ and ψ are state vectors evolving under the operator U_τ then clearly the equality $(\varphi, \psi) = (U_\tau \varphi, U_\tau \psi)$ holds if and only if U_τ is unitary. It is an important fact that the weaker (and physically natural) condition $|(\varphi, \psi)|^2 = |(U_\tau \varphi, U_\tau \psi)|^2$ (i.e., the preservation of transition probabilities), together with the condition of continuity in time of the dynamical evolution, is sufficient to imply the unitarity and the group property of U_τ. We refer again to Section 23.2 and to Simon's review paper.[4]

(vii) The continuity assumption (6.1.3)—technically, the weak continuity of U_τ—entails (see Exercise 4) that the probability distribution $\operatorname{tr}(U_\tau D U_\tau^{-1} P_A(E))$ of the (arbitrary) physical quantity represented by A [see (6.1.6)] is a continuous function of τ. Thus, (6.1.3) roughly means that probability distributions of physical quantities undergo small changes in short time intervals. This is what one expects on the basis of empirical evidence. The so-called state collapse associated with the act of measurement, which should be of a discontinuous nature, will be discussed in Chapter 8.

(viii) The evolution operator U_τ is determined up to the multiplicative function $e^{-i\lambda\tau}$, where λ is an arbitrary real number. In fact, if U_τ satisfies the conditions (6.1.1)–(6.1.3), then so does $U_\tau' = e^{-i\lambda\tau} U_\tau$, and moreover U_τ' induces the same evolution of states, since $U_\tau' D U_\tau'^{-1} = U_\tau D U_\tau^{-1}$ for all density operators.

6.2 Equation of Motion in Infinitesimal Form

Since $\tau \mapsto U_\tau$ is a weakly continuous one-parameter group of unitary operators, it follows, by Stone's theorem,[5] that U_τ is of the form $e^{-iH\tau}$ for some self-adjoint operator H. Since U_τ is defined up to the multiplicative function $e^{-i\lambda\tau}$, the operator H is correspondingly defined up to an additive real constant. H is called the Hamiltonian of the system and corresponds to the physical quantity called the energy of the system.

When the evolution of pure states is considered, one gets a particularly attractive equation of motion. In fact, as is seen in item (v) of last section, if the state at time 0 is represented by the unit vector $\psi(0)$, then at time t the state is represented by the unit vector $\psi(t) = e^{-iHt}\psi(0)$; hence, if $\psi(t)$ is in the domain of H, then as a function of time it satisfies the differential equation

$$i\frac{d}{dt}\psi(t) = H\psi(t),\tag{6.2.1}$$

which specifies the dynamical evolution in infinitesimal form and, as expected, is of first order in time. This is the *Schrödinger equation*; it specifies, in infinitesimal terms, the dynamical evolution of state vectors. Formally, we can write a differential equation of motion even in case the state is nonpure. If $D(0)$ stands for the state at time 0 and $D(t) = e^{-iHt}D(0)e^{-iHt}$ for the state at time t, we get immediately

$$i\frac{d}{dt}D(t) = [H, D(t)].\qquad(6.2.2)$$

In case H is an unbounded operator, care must be taken of its domain in order to give precise meaning to (6.2.2).

Notice that when the Schrödinger equation is deduced from the unitary dynamical group, the Hamiltonian H is necessarily time-independent, for if not, the semigroup property (6.1.1) would not be satisfied. However, the differential equation (6.2.1) admits a natural generalization to the case in which the Hamiltonian H depends on time. This generalized version of the Schrödinger equation corresponds to those situations in which the external forces explicitly vary with time.

6.3 Heisenberg's Picture

The scheme discussed above, the Schrödinger picture, attributes a time evolution only to states, leaving physical quantities fixed in time. We can ask whether there are other equivalent pictures of the dynamical evolution that make the physical quantities vary with time. In particular we can ask whether there is a picture that attributes the time evolution to physical quantities only, leaving the states fixed in time. Such a picture, if it exists, appears as diametrically opposite to Schrödinger's, and is called Heisenberg's picture.

Of course, two pictures of a dynamical evolution are physically equivalent provided they yield the same time dependence of the probability distributions of physical quantities. Now, we can easily see that a dynamics coming from a unitary dynamical group, in the sense discussed in Section 6.1, does admit Heisenberg's picture. In fact, in Schrödinger's picture, the probability distribution of the physical quantity represented by A in the state represented at time t_1 by $D(t_1)$ evolves in time according to (6.1.6). Hence, making use of the cyclic property of the trace, we can rewrite it as

$$E \mapsto \mathrm{tr}\Big(D(t_1)U_{t_2-t_1}^{-1}AU_{t_2-t_1}\Big), \qquad E \in \mathcal{B}(\mathbb{R}).$$

Thus, the Schrödinger picture expressed by (a) and (b) of Section 6.1 is fully equivalent to the one expressed by the following prescriptions:

(a') the density operators representing states stay unchanged in time,
(b') the self-adjoint operators representing physical quantities evolve, from time t_1 to time t_2, according to $A(t_2) = U_{t_2 - t_1}^{-1} A(t_1) U_{t_2 - t_1}$.

These prescriptions formalize Heisenberg's picture.

It should be clear from the deduction above that the possibility of going from Schrödinger's to Heisenberg's picture is not generally ensured: we have made essential use of the fact that our dynamics comes from a unitary evolution operator on $\mathcal{H}$.

Notice that, writing $U_\tau = e^{-iH\tau}$ as in the previous section, the infinitesimal form of the time evolution now reads, formally,

$$i\frac{d}{dt}A(t) = [A(t), H],$$

and is the counterpart, in Heisenberg's picture, of (6.2.2). When $A(t)$, or H, or both, are unbounded operators, care must be taken of their domains, to give meaning to this equation.

6.4 On the Dynamical Evolution in the Presence of Superselection Rules

We consider for simplicity only the case of discrete superselection rules and refer to Schrödinger's picture. According to Chapter 5, the Hilbert space $\mathcal{H}$ of the physical system splits into the direct sum of its coherent subspaces, and the states no longer correspond one-to-one with all density operators in $\mathcal{H}$, but they do with those density operators that commute with all the projectors onto the coherent subspaces. In particular, the pure states are represented by state vectors belonging to some coherent subspace.

Let us elaborate a little bit more on this pattern. The set of all density operators on $\mathcal{H}$ is partitioned into equivalence classes; the elements of a class assign identical probability distributions to the physical quantities. The states of the physical system can be identified with these equivalence classes. Each equivalence class contains one and only one density operator that commutes with all the projectors onto the coherent subspaces; it can be taken as canonical representative of the state. Each equivalence class contains one one-dimensional projector: if it is of the form P^{ψ_k} with ψ_k in some coherent subspace, then it represents a pure state; if it is of the form P^{ψ} with $\psi = \Sigma_k \lambda_k \varphi_k$ (φ_k in the kth coherent subspace), then the element of the equivalence class that canonically represents the state is the mixture $D = \Sigma_k |\lambda_k|^2 P^{\varphi_k}$.

Consider now a dynamical evolution described as in Section 6.1 and corresponding to a unitary evolution operator U_τ. Though U_τ certainly maps one-dimensional subspaces into one-dimensional subspaces [see item (v) of Section 6.1], we cannot conclude that pure states of the physical system are mapped into pure states: in fact, if U_τ maps the vector ψ_k of the kth coherent subspace into $\psi = \Sigma_j \lambda_j \varphi_j$, then, as remarked above, it maps a pure state into a mixture. We refer to Exercise 7 at the end of this chapter for an explicit illustration of this feature.

To get a more familiar dynamical evolution we must add a further requirement to the ones described in Section 6.1. We have to require that the dynamical evolution leave each coherent subspace invariant. Therefore, the restriction of such a dynamics to a single coherent subspace is identical to the dynamics of a system without superselection rules. Let H_k be the Hamiltonian operator in the coherent subspace $\mathfrak{M}_k$, so that $e^{-iH_k\tau}$ gives the evolution operator in $\mathfrak{M}_k$. Then the global evolution operator of the physical system takes the explicit form

$$U_\tau = \sum_k e^{-iH_k\tau} P_k,$$

where P_k is the projection operator onto $\mathfrak{M}_k$. $\tau \mapsto U_\tau$ is a weakly continuous, one-parameter group of unitary operators;* hence, by Stone's theorem, U_τ is of the form

$$U_\tau = e^{-iH\tau},$$

where H is a self-adjoint operator taking the role of a global Hamiltonian of the system. Explicitly,

$$H = \sum_k P_k H P_k = \sum_k H_k.$$

Since each of the partial Hamiltonians H_k is determined up to an additive real constant, H is determined only up to these additive constants.[6]

Exercises

1. Show that if D is a density operator, then $U_\tau D U_\tau^{-1}$ is also a density operator.
   [*Hint.* To verify that $U_\tau D U_\tau^{-1}$ is trace-class see Appendix A.]

---

*Notice that the weak continuity of U_τ corresponds to the fact that for every $\varphi, \psi \in \mathcal{K}$, the function $(\varphi, U_\tau \psi) = \Sigma_k (\varphi, e^{-iH_k\tau} P_k \psi)$ is continuous in τ because each term of the sum is continuous and the sum converges uniformly in τ.

2. Show that if $D=\text{u-lim}_{n\to\infty}\sum_{i=1}^{n}w_i D_i$, then

$$U_\tau DU_\tau^{-1}=\text{u-lim}_{n\to\infty}\sum_{i=1}^{n}w_i U_\tau D_i U_\tau^{-1}.$$

[*Hint.* The multiplication of bounded operators is continuous in the uniform topology of operators.]

3. Prove (6.1.7).
[*Hint.* Show that $U_\tau P^{\hat{\psi}}U_\tau^{-1}$ and $P^{\hat{\phi}}$ with $\phi=U_\tau\psi$ have identical matrix elements between any two vectors of $\mathcal{H}$.]

4. Show that if, for every pair of unit vectors $\phi,\psi\in\mathcal{H}$, the function $\tau\mapsto(\phi,U_\tau\psi)$ is continuous, then also the function $\tau\mapsto\text{tr}(U_\tau DU_\tau^{-1}P)$ is continuous for every. density operator D and every projector P.
[*Hint.* Decompose D and P on orthogonal bases, use (2.3.2), and remark that one has to do with uniformly convergent series.]

5. Suppose that the Hamiltonian H has a simple point spectrum with eigenvalues $\lambda_1,\lambda_2,\ldots$ and eigenvectors $\phi_1,\phi_2,\ldots$. Show that if the state at time 0 is represented by $\psi(0)=c_1\phi_1+c_2\phi_2+\cdots$, then it evolves, at time t, into $\psi(t)=c_1\exp(-i\lambda_1 t)\phi_1+c_2\exp(-i\lambda_2 t)\phi_2+\cdots$, and this vector satisfies (6.2.1).

6. Let H_0 be the free-particle Hamiltonian of Section 4.4. Show that the pure state $\psi(x)$ evolves into

$$e^{-iH_0 t}\psi(x)=(4\pi it)^{-1/2}\int_{\mathbb{R}}\exp\frac{i(x-y)^2}{4t}\psi(y)\,dy$$

[if $\psi(x)$ is not an element of $L^1(\mathbb{R})$, then the integral has to be understood as $\text{l.i.m.}_{\lambda\to\infty}\int_{-\lambda}^{+\lambda}$].
[*Hint.* See Chapter 9, Section 7 of Reference 7.]

7. Suppose that in the Hilbert space $\mathbb{C}^2$ (see Section 4.2) a superselection rule determines the coherent one-dimensional subspaces, and denote by P^+ and P^- the corresponding projectors. Show that the Hamiltonian

$$H=\begin{pmatrix}0 & 1\\ 1 & 0\end{pmatrix}\qquad(\text{in the }\langle P^+,P^-\rangle\text{ basis})$$

makes the pure state P^+ evolve into $\cos^2 t\,P^+ +\sin^2 t\,P^-$, and P^- into $\sin^2 t\,P^+ +\cos^2 t\,P^-$.
[*Hint.* Use (6.2.1) or (6.1.7).]

References

1. B. Davies, *Quantum Theory of Open Systems*, Academic Press, London, 1976.
2. R. Kadison, *Ann. Math.* 54 (1951) 325.
3. G. W. Mackey, *Mathematical Foundations of Quantum Mechanics*, Benjamin, Advanced Book Program, Reading, Mass., 1963.
4. B. Simon, in *Studies in Mathematical Physics—Essays in Honour of Valentine Bargmann*, E. H. Lieb, B. Simon, and A. S. Wightman, eds., Princeton University Press, Princeton, N.J., 1976.
5. M. Reed and B. Simon, *Methods of Modern Mathematical Physics*, Vol. I, Academic Press, New York, 1972.
6. T. F. Jordan, *Linear Operators for Quantum Mechanics*, Wiley, New York, 1969.
7. M. Reed and B. Simon, *Methods of Modern Mathematical Physics*, Vol. II, Academic Press, New York, 1975.

CHAPTER 7

Compound Systems

7.1 Composition of Different and of Identical Systems

It is a general tendency of physics to interpret physical systems as composed of simpler, more elementary subsystems. The question "What is it made of?" is a very tempting and popular one. Though this attitude does not reflect a logical necessity, it has proved to be successful in significant circumstances: we think, e.g., of macroscopic matter as composed of atoms and molecules, of molecules as composed of atoms, of atoms as composed of nuclei and electrons, and of nuclei as composed of protons and neutrons. Loosely speaking, one might expect that this attitude will be successful whenever the interaction that packs the constituent subsystems together is not strong enough to make them lose their identity.

Here, we are concerned with the problem of constructing the quantum-mechanical formalism of a compound system knowing the formalisms of the constituent subsystems. Thus our first problem is to specify the Hilbert space of the compound system. For the sake of simplicity, let us restrict ourselves to the case of a system composed of only two subsystems.

The prescriptions of quantum theory vary according to whether one is faced with the composition of identical or nonidentical subsystems.

7.1.1 *Composition of Nonidentical Subsystems*

Let $\mathcal{H}_1$ and $\mathcal{H}_2$ be the Hilbert spaces of the component subsystems. Then the formalism of quantum theory assigns to the compound system the Hilbert space $\mathcal{H}_1 \otimes \mathcal{H}_2$, i.e., the tensor product of $\mathcal{H}_1$ and $\mathcal{H}_2$.

We refer to classical works[1] for an exhaustive treatment of tensor products of Hilbert spaces; here we just recall a few facts. A Hilbert space $\mathcal{H}$ is said to be the tensor product of $\mathcal{H}_1$ and $\mathcal{H}_2$ if there exists a bilinear

ENCYCLOPEDIA OF MATHEMATICS and Its Applications, Gian-Carlo Rota (ed.).
Vol. 15: E. G. Beltrametti and G. Cassinelli, The Logic of Quantum Mechanics

ISBN 0-201-13514-0

mapping $(\cdot \otimes \cdot)$ from the cartesian product $\mathcal{H}_1 \times \mathcal{H}_2$ (i.e. the set of all ordered pairs $\langle \varphi, \psi \rangle$ with $\varphi \in \mathcal{H}_1$ and $\psi \in \mathcal{H}_2$) into $\mathcal{H}$ such that:

(i) the set $\{\varphi \otimes \psi : \varphi \in \mathcal{H}_1, \psi \in \mathcal{H}_2\}$ spans $\mathcal{H}$,
(ii) $((\varphi_1 \otimes \psi_1), (\varphi_2 \otimes \psi_2))_{\mathcal{H}} = (\varphi_1, \varphi_2)_{\mathcal{H}_1} (\psi_1, \psi_2)_{\mathcal{H}_2}$ for all $\varphi_1, \varphi_2 \in \mathcal{H}_1$ and all $\psi_1, \psi_2 \in \mathcal{H}_2$.*

The tensor product is unique up to unitary equivalence. If $\{\varphi_i\}$ and $\{\psi_i\}$ are orthonormal bases for $\mathcal{H}_1$ and $\mathcal{H}_2$ respectively, then $\{\varphi_i \otimes \psi_j\}$ is an orthonormal basis for $\mathcal{H}_1 \otimes \mathcal{H}_2$. Of course, a vector of $\mathcal{H}_1 \otimes \mathcal{H}_2$ need not be of the form $\varphi \otimes \psi$ ($\varphi \in \mathcal{H}_1$, $\psi \in \mathcal{H}_2$): what we are saying is that it can always be expanded as a linear combination of vectors of that form.

Having specified what the Hilbert space of the compound system is, the prescriptions about states and physical quantities follow naturally. In the absence of superselection rules one has to follow the pattern discussed in Section 1.1; in presence of superselection rules, the pattern discussed in Section 5.4 applies unchanged, but a highly nontrivial problem is constituted by the relations between the superselection rules of the compound system and those of the component subsystems (we shall return to this point in Section 7.6).

Thus, in absence of superselection rules, the states of the compound system correspond one-to-one to the density operators on $\mathcal{H}_1 \otimes \mathcal{H}_2$; in particular the pure states correspond to the one-dimensional subspaces of $\mathcal{H}_1 \otimes \mathcal{H}_2$. Similarly, the physical quantities of the compound system are in one-to-one correspondence with the self-adjoint operators on $\mathcal{H}_1 \otimes \mathcal{H}_2$.

We introduce now a relevant class of operators on $\mathcal{H}_1 \otimes \mathcal{H}_2$. If A_1 and A_2 are bounded self-adjoint operators on $\mathcal{H}_1$ and $\mathcal{H}_2$, respectively, then define the self-adjoint operator $A_1 \otimes A_2$ on $\mathcal{H}_1 \otimes \mathcal{H}_2$ as the operator that transforms the vectors of the form $\varphi \otimes \psi$ according to

$$(A_1 \otimes A_2)(\varphi \otimes \psi) = A_1 \varphi \otimes A_2 \psi, \qquad \varphi \in \mathcal{H}_1, \quad \psi \in \mathcal{H}_2,$$

and is extended by linearity to the whole of $\mathcal{H}_1 \otimes \mathcal{H}_2$ (see Exercises 1–4). This definition does not work when A_1 (A_2) is unbounded, for then $A_1 \varphi$ ($A_2 \psi$) is no longer defined for all $\varphi \in \mathcal{H}_1$ ($\psi \in \mathcal{H}_2$); nevertheless, it admits a generalization to the unbounded case, at the cost of some technical care[2] which we do not need to go into. Physical quantities of the compound system that are represented by operators of the form $A_1 \otimes A_2$ have a typical feature: their expectation values are "multiplicative" with respect to the subsystems when evaluated in vector states of the form $\varphi \otimes \psi$. Explicitly,

---

*For clarity we have made explicit, by subscripts, the Hilbert spaces the scalar products refer to. In the sequel, however, this specification should be deduced from the context.

using the notation of Section 3.1,

$$\mathscr{E}(A_1 \otimes A_2, \varphi \otimes \psi) = \mathscr{E}(A_1, \varphi) \cdot \mathscr{E}(A_2, \psi).$$

Of course, there are interesting physical quantities of the compound system that are not represented by operators of this kind: e.g., "additive" quantities like the total energy, or the total angular momentum, whose expectation values are obviously nonmultiplicative. Nevertheless, the von Neumann algebra $\mathbb{B}(\mathcal{H}_1 \otimes \mathcal{H}_2)$ of all bounded operators on $\mathcal{H}_1 \otimes \mathcal{H}_2$ is generated by operators of the form $A_1 \otimes A_2$, with A_1 and A_2 running respectively over the sets of all self-adjoint operators on $\mathcal{H}_1$ and on $\mathcal{H}_2$. In particular, if $\mathcal{C}_1$ and $\mathcal{C}_2$ are complete sets of commuting self-adjoint operators on $\mathcal{H}_1$ and $\mathcal{H}_2$ respectively, then the set*

$$\{A_1 \otimes I_2, I_1 \otimes A_2 : A_1 \in \mathcal{C}_1, A_2 \in \mathcal{C}_2\}, \tag{7.1.1}$$

where I_1 and I_2 are the identities in $\mathcal{H}_1$ and $\mathcal{H}_2$, is complete in $\mathcal{H}_1 \otimes \mathcal{H}_2$ (see Exercise 5), thus representing a complete set of compatible physical quantities of the compound system. Notice that A_1 and $A_1 \otimes I_2$ actually represent the same physical quantity (and the same for A_2 and $I_1 \otimes A_2$); the only difference is that when we write A_1 we refer to system 1 as a physical system by itself, and when we write $A_1 \otimes I_2$ we refer to system 1 as a part of a compound system.

7.1.2 Composition of Identical Subsystems

As emphasized in any textbook of quantum mechanics, the notion of identity plays, for quantum systems, a crucial role which has no counterpart in classical ones. Classical systems, though identical, can always be distinguished by attaching to them, so to speak, different labels, or proper names, which reflect different extrinsic properties. (The old philosophical debate on the problem of identity should perhaps be recalled; we refer to Max Jammer[3] for an overview of this issue). These labels may reflect different positions in physical space at a given initial time, and, since the classical dynamical evolution does not exclude the possibility of continuously following the trajectories of the systems, the significance of the labels is preserved in time. This labeling procedure does not work, in general, for identical quantum systems, due to the nonvanishing variance of the space

---

*In case A_1 is unbounded, the operator $A_1 \otimes I_2$ can be explicitly defined as the operator on $\mathcal{H}_1 \otimes \mathcal{H}_2$ whose spectral decomposition is

$$E \mapsto P_{A_1}(E) \otimes I_2, \qquad E \in \mathscr{B}(\mathbb{R}),$$

where $P_{A_1}(E) \otimes I_2$ falls under our previous definition because we have now to do with bounded operators. A similar remark holds for $I_1 \otimes A_2$.

positions, so that overlappings of the wave functions may occur, a fact that prevents the assignment of proper names.

Let $\mathcal{H}$ be the Hilbert space associated to each of the identical subsystems to be composed. Then, the formalism of quantum theory assigns to the resulting compound system one of two proper subspaces of $\mathcal{H} \otimes \mathcal{H}$: either the "symmetrical tensor product" $\mathcal{H} \textcircled{s} \mathcal{H}$, or the "antisymmetrical tensor product" $\mathcal{H} \textcircled{a} \mathcal{H}$. Let us recall the definitions.

DEFINITION 7.1.1 (The symmetrical tensor product). Consider, in $\mathcal{H} \otimes \mathcal{H}$, the set of all vectors of the form $\varphi \otimes \psi + \psi \otimes \varphi$. The closed subspace they span is called the *symmetrical tensor product*, and is denoted $\mathcal{H} \textcircled{s} \mathcal{H}$. If π denotes the switching operator in $\mathcal{H} \otimes \mathcal{H}$, which maps an element of the form $\varphi \otimes \psi$ into $\psi \otimes \varphi$ and is extended by linearity to the whole of $\mathcal{H} \otimes \mathcal{H}$, we can say that $\mathcal{H} \textcircled{s} \mathcal{H}$ consists precisely of those vectors of $\mathcal{H} \otimes \mathcal{H}$ that are invariant under π.

DEFINITION 7.1.2 (The antisymmetrical tensor product). Consider, in $\mathcal{H} \otimes \mathcal{H}$, the set of all vectors of the form $\varphi \otimes \psi - \psi \otimes \varphi$. The closed subspace they span is called the *antisymmetrical tensor product*, and is denoted $\mathcal{H} \textcircled{a} \mathcal{H}$. It is precisely the subspace of $\mathcal{H} \otimes \mathcal{H}$ formed by all vectors that undergo a change in sign under the action of π, that is, under the switching of the arguments.

Notice that $\mathcal{H} \textcircled{s} \mathcal{H}$ and $\mathcal{H} \textcircled{a} \mathcal{H}$ are orthogonal complements in $\mathcal{H} \otimes \mathcal{H}$. In fact, the switch operator π is both unitary and self-adjoint; hence it has just two eigenvalues, $+1$ and -1, and the corresponding eigenspaces are precisely $\mathcal{H} \textcircled{s} \mathcal{H}$ and $\mathcal{H} \textcircled{a} \mathcal{H}$.

Once the Hilbert space of the compound system is specified, the prescriptions about states and physical quantities follow the general rule: states correspond to density operators, and physical quantities correspond to self-adjoint operators. Clearly, the operators in $\mathcal{H} \textcircled{s} \mathcal{H}$ or in $\mathcal{H} \textcircled{a} \mathcal{H}$ are symmetrical with respect to the two subsystems, namely, they must commute with the switch operator π.

In the fact that the composition of identical subsystems calls into play (both in the symmetrical and the antisymmetrical cases) state vectors that have the form of quantum superpositions, we see clearly the departure from classical physics of the quantum description of identical subsystems.

Notice that the prescription of restricting $\mathcal{H} \otimes \mathcal{H}$ either to the symmetrical or to the antisymmetrical tensor product, when identical subsystems are dealt with, is not to be regarded as a new, independent assumption of quantum theory: it follows from proper use of the notion of identity.[4]

The rule for deciding, in concrete situations, which one of the two possibilities $\mathcal{H} \textcircled{s} \mathcal{H}$ or $\mathcal{H} \textcircled{a} \mathcal{H}$ is to be chosen is the famous rule discovered by W. Pauli. It is referred to as the *Pauli principle* or, more generally, the *spin-statistics connection*, and asserts that whenever the intrinsic spin of the identical subsystems is an integer, then $\mathcal{H} \textcircled{S} \mathcal{H}$ is the correct choice, whereas whenever the spin of the identical subsystems is a half integer, then $\mathcal{H} \textcircled{A} \mathcal{H}$ is correct. This rule, known since the early days of quantum mechanics (at least in the weaker form of "exclusion principle" of atomic physics*), is a basic fact about the structure of matter. Despite its spectacular effects in essentially nonrelativistic physical systems, the only available theoretical deduction of the spin-statistics connection belongs to a relativistic framework.[5] This is a baffling gap in nonrelativistic quantum theory.

Before ending this section let us add a remark that refers equally to the composition of nonidentical or of identical subsystems. In the common terminology the expression "compound system" sometimes does not match exactly with the formalism above. Take as an example the deuteron as composed of a proton and a neutron: it is implicit in the use of the word "deuteron" that it is a bound system, with given binding energy, given spin, given magnetic moment, and so on. Thus the Hilbert space associated with the deuteron is not the whole of $\mathcal{H}_p \otimes \mathcal{H}_n$ (p and n referring to proton and neutron), but only a subspace of it. We are allowed to say that $\mathcal{H}_p \otimes \mathcal{H}_n$ corresponds to the combination of a proton and a neutron provided we understand the compound system as encompassing all situations that can be thought of as made of a proton and a neutron.

7.2 States of Compound and Component Systems

The problem here considered is that of relating the state D of the compound system to the states D_1 and D_2 of the component subsystems. Let us suppose for simplicity that the two subsystems are different and that no superselection rule is present. We have to express the requirement that each subsystem preserve its distinctive marks when it constitutes part of a compound system, so that every physical quantity pertaining to one subsystem can be equivalently viewed as pertaining either to an autonomous system or to a part of the compound system. Formally, we are led to require that, for every self-adjoint operator A_1 on $\mathcal{H}_1$ and A_2 on $\mathcal{H}_2$,

$$
\begin{aligned}
\operatorname{tr}\big(D_1 P_{A_1}(E)\big) &= \operatorname{tr}\big(D\big(P_{A_1}(E) \otimes I_2\big)\big) && \text{for all} \quad E \in \mathcal{B}(\mathbb{R}), \\
\operatorname{tr}\big(D_2 P_{A_2}(F)\big) &= \operatorname{tr}\big(D\big(I_1 \otimes P_{A_2}(F)\big)\big) && \text{for all} \quad F \in \mathcal{B}(\mathbb{R}).
\end{aligned}
\tag{7.2.1}
$$

In fact this means that the probability distribution of the physical quantity

---

*In atomic physics the typical situation is the one in which the subsystems are electrons, that is, spin-$\frac{1}{2}$ particles, so that $\mathcal{H} \textcircled{A} \mathcal{H}$ applies; the fact that $\varphi \otimes \psi - \psi \otimes \varphi$ reduces to the null vector if $\varphi = \psi$ is often epitomized by saying that there is an "exclusion principle" preventing two electrons from being in the same state.

represented by A_1 (A_2) in the state D_1 (D_2) equals the probability distribution of the physical quantity represented by $A_1 \otimes I_2$ ($I_1 \otimes A_2$) in the state D. Since the conditions (7.2.1) must hold for all self-adjoint operators A_1 and A_2 (we are assuming that no superselection rule is present), $P_{A_1}(E)$ and $P_{A_2}(F)$ run over all projectors on $\mathfrak{K}_1$ and $\mathfrak{K}_2$, respectively, so that we can rephrase these conditions in the form

$$\begin{aligned}
\operatorname{tr}(D_1 P_1) &= \operatorname{tr}(D(P_1 \otimes I_2)) && \text{for every projector } P_1 \text{ on } \mathfrak{K}_1, \\
\operatorname{tr}(D_2 P_2) &= \operatorname{tr}(D(I_1 \otimes P_2)) && \text{for every projector } P_2 \text{ on } \mathfrak{K}_2.
\end{aligned} \tag{7.2.2}$$

Suppose first that the states D_1 and D_2 of the subsystems are given. Then we have the following facts:

(1) The choice $D = D_1 \otimes D_2$ is always a possible solution of (7.2.2) (Exercise 6).

(2) The choice $D = D_1 \otimes D_2$ is not, in general, the only solution of (7.2.2): this means that, in general, knowledge of the states of the subsystems is not sufficient to determine the state of the compound system (Exercise 7).

(3) If the subsystems are in pure states, then $D = D_1 \otimes D_2$ is the only solution; therefore, in this particular case, the state of the compound system is uniquely determined by the states of the subsystems (Exercise 8).

Consider now the reverse problem, and suppose that the state D of the compound system is given. Then we have the following facts:

(4) D_1 and D_2 are uniquely determined by D: in other words, knowledge of the state of the compound system is sufficient to determine the actual states of the subsystems (Exercise 9).

(5) If D is a mixture, say $D = wD' + (1-w)D''$ ($0 < w < 1$), then also D_1 and D_2 are mixtures and have the form $D_1 = wD_1' + (1-w)D_1''$ and $D_2 = wD_2' + (1-w)D_2''$, where D_1', D_2' (D_1'', D_2'') are the states of the subsystems corresponding to the state D' (D'') of the compound system (Exercise 10). Therefore, we can always reduce the problem to the case in which the state D of the compound system is pure.

(6) If D is a pure state represented by a state vector of the general form

$$\sum_{i,j} \lambda_{ij} \varphi_i \otimes \psi_j, \qquad \varphi_i \in \mathfrak{K}_1, \quad \psi_j \in \mathfrak{K}_2,$$

then (see Exercise 11) D_1 and D_2 are the density operators defined by

$$\begin{aligned}
D_1 \varphi &= \sum_{i,j} c_{ij}(\varphi_i, \varphi)\varphi_j, & c_{ij} &= \sum_k \bar{\lambda}_{ik}\lambda_{jk}, & \varphi \in \mathfrak{K}_1, \\
D_2 \psi &= \sum_{i,j} d_{ij}(\psi_i, \psi)\psi_j, & d_{ij} &= \sum_k \bar{\lambda}_{ki}\lambda_{kj}, & \psi \in \mathfrak{K}_2.
\end{aligned} \tag{7.2.3}$$

Notice that, in general, these expressions do not give D_1 and D_2 directly in the standard form of convex combinations; of course, that form can be achieved by use of the spectral theorem.

(7) If D is a pure state represented by a state vector of the form

$$\sum_i \lambda_i \varphi_i \otimes \psi_i, \qquad \varphi_i \in \mathcal{K}_1, \quad \psi_i \in \mathcal{K}_2,$$

then D_1 and D_2 are given by

$$D_1 = \sum_i |\lambda_i|^2 P^{\hat{\varphi}_i},$$

$$D_2 = \sum_i |\lambda_i|^2 P^{\hat{\psi}_i}.$$

This follows immediately by the foregoing item (6).

(8) If D is a pure state represented by a state vector of the form $\varphi \otimes \psi$, with $\varphi \in \mathcal{K}_1$ and $\psi \in \mathcal{K}_2$, then the states of the subsystems also are pure and are represented by the state vectors φ and ψ respectively. This is an immediate consequence of previous item (6) or (7). Comparing with item (3) above, we conclude that the states of the subsystems are pure if and only if the state of the compound system is a pure state of the form $\varphi \otimes \psi$.

Among the facts mentioned above there is one that is typical of quantum systems; it is contained in items (6) and (7) and consists of the following: we can have the compound system in a pure state with the subsystems in nonpure states. This occurs when the state of the compound system has the form of a superposition of pure states of tensor-product form, thus revealing the quantum nature of the phenomenon. In classical mechanics a state of this sort is off limits, and nonpure states of the subsystems can come only from nonpure states of the compound system.

7.3 On Correlations between Subsystems

The formalism of compound systems accommodates in a natural way the notion of correlation between physical quantities pertaining to distinct physical systems. For the sake of simplicity, suppose still that no superselection rule is present, so that physical quantities correspond one-to-one to self-adjoint operators in the appropriate Hilbert spaces. Also suppose that the two systems S_1 and S_2 we are dealing with are nonidentical. If A_1 and A_2 are self-adjoint operators acting, respectively, on the Hilbert spaces $\mathcal{K}_1$ (of S_1) and $\mathcal{K}_2$ (of S_2), then, thinking of S_1 and S_2 as constituents of a compound system with Hilbert space $\mathcal{K}_1 \otimes \mathcal{K}_2$, we define the coefficient of correlation between A_1 and A_2, given the state D of the compound system, as

$$\rho(A_1, A_2; D) = \frac{\mathcal{E}(A_1 \otimes A_2, D) - \mathcal{E}(A_1, D_1)\mathcal{E}(A_2 D_2)}{\left[\mathcal{V}(A_1, D_1)\mathcal{V}(A_2, D_2)\right]^{1/2}}, \qquad (7.3.1)$$

according to classical probability theory. As seen in Section 7.2, the states D_1 and D_2 of the subsystems are uniquely determined by D, so that the arguments of expectations and variances here occurring are all well defined. For simplicity suppose that A_1 and A_2 are bounded, so that expectations and variances can be evaluated by the formulae (3.1.4), (3.1.6), (3.1.8), (3.1.9) of Chapter 3. Of course, once the correlation coefficient is known, the probability distributions of A_1 and A_2 become connected by the rules of classical probability theory.[6]

The fact that the foregoing notion of correlation involves knowledge of the state D of the compound system emphasizes the assertion, already outlined in Section 7.2, that D contains richer information than that contained in the reduced states D_1 and D_2; recall that D determines D_1 and D_2, but usually not the converse.

Let us add a few comments.

(i) Suppose D to be of the tensor-product form $D = D_1 \otimes D_2$. Then, noticing that $\mathrm{tr}((D_1 \otimes D_2)(A_1 \otimes A_2)) = \mathrm{tr}(D_1 A_1)\mathrm{tr}(D_2 A_2)$ (see Exercises 2, 4), we have that the coefficient of correlation between any two physical quantities vanishes. The two subsystems $\mathbb{S}_1$ and $\mathbb{S}_2$ behave independently. This fact solves the following consistency problem. Suppose the two systems $\mathbb{S}_1$ and $\mathbb{S}_2$ independent (two particles far apart and having never interacted), and suppose their states D_1, D_2 are given; then whether to treat them independently or as parts of a compound system must reduce to a matter of convention. This is indeed the case if we agree that, in case we want to view them as forming a compound system, we have it in the state $D = D_1 \otimes D_2$. Notice that, in order to preserve for all times the tensor-product form of D, the corresponding evolution operator U_t (see Section 6.1) must be of the form $U_t = U_t^{(1)} \otimes U_t^{(2)}$, and this is precisely the condition of no interaction between $\mathbb{S}_1$ and $\mathbb{S}_2$.

(ii) Suppose D to be a pure state. Should it be (the projector associated with a vector) of the tensor-product form $\varphi \otimes \psi$ ($\varphi \in \mathcal{H}_1, \psi \in \mathcal{H}_2$), we would have the case discussed above, with all correlations disappearing. But a quantum system can have pure states of another kind, namely states that are linear superpositions of the general form

$$\sum_{i,j} \lambda_{ij} \varphi_i \otimes \psi_j, \qquad \varphi_i \in \mathcal{H}_1, \ \psi_j \in \mathcal{H}_2, \qquad (7.3.2)$$

and in this case it is definitely possible to have nontrivial correlations, as we shall see in the next section on a concrete example. This possibility is open to quantum systems, but not to classical ones, since the required state is a quantum superposition of pure states. In classical mechanics we can get correlations only if the state of the compound system is nonpure (the pure states being necessarily of the tensor-product form). Thus, the occurrence of correlations between the subsystems, when the state of the compound system is pure, has to be considered as a distinguishing feature of quantum mechanics.

(iii) As seen in Section 7.2, a pure state D of the form (7.3.2) determines nonpure states of the subsystems. Thus we see that the quantum phenomenon of the presence of correlations between subsystems with the compound system in a pure state occurs jointly with the quantum phenomenon of the compound system in a pure state with subsystems in nonpure states.

(iv) Notice that, though quantum mixtures have a nonunique decomposability into pure states, the coefficient of correlation is independent of the particular decomposition of D_1, D_2, D one choses. Notice also that in case there are superselection rules, so that the correspondence between density operators and states is many-to-one (see Section 5.4), the choice of the density operator representing a given state has no influence on (7.3.1).

7.4 An Example, the E.P.R. Problem

In this section we consider a simple compound system that illustrates some of the facts discussed above and amounts to a simplified version of an old problem raised by Einstein, Podolski, and Rosen[7] to question the completeness of quantum theory (it is often called the E.P.R. paradox).

The physical system is composed of two nonidentical spin-$\frac{1}{2}$ subsystems, for definiteness an electron and a positron. We write φ_+ and φ_- respectively for the spin-up and the spin-down state of the electron, and ψ_+, ψ_- for the analogous states of the positron. We refer to Section 4.2 for more details on the states of spin-$\frac{1}{2}$ systems: let us only recall that, writing $J_3^{(e)}$ and $J_3^{(p)}$ for the operators representing the spin components (along the x_3-axis) of the electron and the positron, we have

$$J_3^{(e)}\varphi_+ = \tfrac{1}{2}\varphi_+, \qquad J_3^{(e)}\varphi_- = -\tfrac{1}{2}\varphi_-,$$
$$J_3^{(p)}\psi_+ = \tfrac{1}{2}\psi_+, \qquad J_3^{(p)}\psi_- = -\tfrac{1}{2}\psi_-.$$

We assume that the electron and the positron, after a period of mutual interaction, fly away in opposite directions, thus becoming completely separated and no longer interacting. Since we have in mind to consider only physical quantities relative to the spin, we only need to specify the spin part of the state of the compound system: let us suppose that it is the pure state given by

$$\Phi = \frac{1}{\sqrt{2}}(\varphi_+ \otimes \psi_- - \varphi_- \otimes \psi_+), \qquad (7.4.1)$$

the so-called singlet state of the electron-positron pair, representing the spin-0 configuration of the pair.* More specifically, we suppose that the state (7.4.1) is left unchanged by the dynamical evolution of the pair, and hence that it applies also to the situation in which the electron and the positron become spatially separated.

As we know, the state of the compound system determines the state of the subsystems, via the prescriptions (7.2.3). Now, writing $D^{(e)}$ and $D^{(p)}$ for the state of the electron and of the positron, we get (see Exercise 7)

$$D^{(e)} = \tfrac{1}{2}P^{\hat{\varphi}_+} + \tfrac{1}{2}P^{\hat{\varphi}_-}, \qquad D^{(p)} = \tfrac{1}{2}P^{\hat{\psi}_+} + \tfrac{1}{2}P^{\hat{\psi}_-}. \qquad (7.4.2)$$

We can easily surmise that the information coded in the state (7.4.1) of the compound system establishes nontrivial correlations between the subsystems. We consider the physical quantities represented by $J_3^{(e)}$ and $J_3^{(p)}$, that is, the components of electron's and the positron's spin along the x_3-axis; their correlation, as defined by (7.3.1), is (Exercise 12)

$$\rho\left(J_3^{(e)}, J_3^{(p)}; P^{\hat{\Phi}} \right) = -1, \qquad (7.4.3)$$

which means that whenever the component of the electron's spin is found to be $+\tfrac{1}{2}$ (respectively, $-\tfrac{1}{2}$), then the component of positron's spin must necessarily be found to be $-\tfrac{1}{2}$ (respectively, $+\tfrac{1}{2}$). Roughly speaking, the electron and the positron have opposite spin orientations.

Let us outline some features which are contained, implicitly or explicitly, in what is said above.

(1) The electron and the positron are in nonpure states [see (7.4.2)], while the compound system they form is in a pure state [see (7.4.1)].

(2) If we choose, in the Hilbert space of the electron, another orthonormal basis $\{\chi_+, \chi_-\}$, and we write $\{\eta_+, \eta_-\}$ for its counterpart in the Hilbert space of the positron, then, expanding φ_+, φ_- and ψ_+, ψ_- in the new bases, the vector state (7.4.1) takes the form

$$\Phi = \frac{1}{\sqrt{2}}(\chi_+ \otimes \eta_- - \chi_- \otimes \eta_+).$$

We can, for instance, think of χ_+ and χ_- (η_+ and η_-) as the spin-right and the spin-left state of the electron (positron), that is, the eigenstates of $J_1^{(e)}$ ($J_1^{(p)}$), the spin component along the x_1-axis.

---

*The reader will realize that what follows could be rephrased if, in place of (7.4.1), one assumed the triplet state $\Phi = (1/\sqrt{2})(\varphi_+ \otimes \psi_- + \varphi_- \otimes \psi_+)$, that is, the spin-1 configuration with third component equal to zero.

(3) The mixtures $D^{(e)}$ and $D^{(p)}$ have nonunique decomposability into pure states. The expression (7.4.2) provides a particular decomposition, but we can replace $\{\varphi_+, \varphi_-\}$ and $\{\psi_+, \psi_-\}$ by any other basis: for instance, we can write

$$D^{(e)} = \tfrac{1}{2} P^{\hat{x}_+} + \tfrac{1}{2} P^{\hat{x}_-}, \qquad D^{(p)} = \tfrac{1}{2} P^{\hat{\eta}_+} + \tfrac{1}{2} P^{\hat{\eta}_-}.$$

Notice that $D^{(e)}$, as well as $D^{(p)}$, is a completely unpolarized state (see Section 4.2), and it does not admit the ignorance interpretation of classical mixtures (see Section 2.4).

(4) In view of (7.4.3), we see that our electron-positron pair exhibits the quantum phenomenon of the presence of correlations between the subsystems with the compound system in a pure state.

(5) The electron and the positron remain correlated, after they have interacted to form a nontrivial compound system, even if they are quite far apart and no longer interacting. There is a sort of nonseparability for two subsystems that in their past have formed a compound system: they remain correlated in a way that is written in the state of the compound system they formed.

(6) The -1 value of the correlation coefficient $\rho(J_3^{(e)}, J_3^{(p)}; P^{\hat{\Phi}})$ entails that once the spin component along the x_3-axis is measured on one subsystem, say the electron, and a certain value is found, say $+\tfrac{1}{2}$, then the result of a future analogous measurement on the positron is predicted with certainty to be $-\tfrac{1}{2}$, no matter how far it is and without its interacting in any way with the electron.

(7) Taking into account items (2) and (3) above, it is evident that not only are the spin components along the x_3-axis correlated, but so are the spin components along any other axis. We can take, for instance, the spin components along the x_1-axis and find

$$\rho\left(J_1^{(e)}, J_1^{(p)}; P^{\hat{\Phi}}\right) = -1.$$

This means that, once the electron is found to be spin-right, then we can predict with certainty that the positron will be found to be spin-left.

(8) Comparing items (6) and (7), we see that by taking different decisions on what to measure on the electron, we can predict with certainty, and without in any way interacting with the positron, either the value of the x_1-component or the value of the x_3-component of the positron spin. What makes this conclusion puzzling is that the components of the spin along different axes are noncompatible physical quantities (and indeed this is the main point raised by Einstein, Podolski, and Rosen). But, we believe, the conclusion is puzzling only as long as we interpret classically the nonpure state of the positron, that is, according to the ignorance interpretation of mixtures (see Section 2.4); this interpretation is however ruled out, as stated in item (3) above.

The foregoing list of features that characterize the behavior of our electron-positron pair contains a number of characteristically quantum facts. Each of these facts could be taken as a facet of the E.P.R. problem (and we do not claim to have exhausted all the facets). We shall not pursue these points further; of historicial interest are Bohr's reply[8] to Einstein, Podolski, and Rosen, and Einstein's rejoinder[9].

Let us finally remark that, at first sight, one might feel disturbed by the fact, noticed in items (5) and (6), that an observation made on one subsystem makes instantaneously certain the result of an analogous future measurement on the other subsystem, no matter how far it is and without in any way interacting with it. However, the innocuousness of this fact becomes evident if one thinks of familiar classical situations of correlated systems: we are dealing with the usual "reduction of probability distributions" of classical probability theory.

7.5 On the Time Evolution of Subsystems

In view of the rules that determine the states of the subsystems given the state of the compound system [see in particular items (6), (7), and (8) of the last section], the following remarkable possibility arises: while the compound system undergoes a unitary dynamical evolution, in the sense discussed in Section 6.1, with pure states evolving into pure states, the constituent subsystems can inherit a dynamical evolution of quite a different nature, with pure states evolving into nonpure ones, and vice versa.

As an explicit example, consider again, as in the last section, a physical system composed of two nonidentical spin-$\frac{1}{2}$ subsystems, but suppose now that it is in a spin-1 configuration. We shall pay attention only to the spin coordinates. Suppose that, at time $t=0$, the state of the compound system is the state vector

$$\Phi(0)=\varphi_+\otimes\psi_+ , \qquad (7.5.1)$$

and that it evolves in time according to

$$\Phi(t)=\cos\omega t\,\varphi_+\otimes\psi_+ +\sin\omega t\,\frac{1}{\sqrt{2}}(\varphi_+\otimes\psi_- +\varphi_-\otimes\psi_+); \qquad (7.5.2)$$

in other words, the state of the compound system oscillates between the state with x_3-component equal to 1 and the state with z-component equal to 0, the frequency of the oscillation being $\omega/2\pi$. This evolution can be derived from a unitary dynamical group of the kind discussed in Section 6.1 (see Exercise 13). Now consider the state of one subsystem, say D_1; it comes from $D=P^{\Phi(t)}$ by use of (7.2.3). Here we omit the expression of D_1 as a function of time (Exercise 14), but only notice that whenever $\sin\omega t=0$ we have

$$D_1 = P^{\hat{\varphi}_+},$$

while whenever $\cos \omega t = 0$ we have (see Exercise 7)

$$D_1 = \tfrac{1}{2} P^{\hat{\varphi}_+} + \tfrac{1}{2} P^{\hat{\varphi}_-}.$$

Thus we see that the dynamics inherited by the subsystem makes it evolve back and forth from pure to nonpure states. The dynamical evolution of the subsystem is still continuous and also reversible, but it is quite apart from the dynamics described by a unitary evolution operator.

7.6 Superselection Rules and Compound Systems

The existence of superselection rules in the subsystems certainly induces superselection rules in the compound system. Consider for instance the composition of two nonidentical subsystems with Hilbert spaces $\mathcal{H}_1$ and $\mathcal{H}_2$, and suppose that there are in $\mathcal{H}_1$ superselection rules, so that there are density operators on $\mathcal{H}_1$ that do not correspond to physical states, as discussed in Chapter 5. If D_1 is one of them, a density operator on $\mathcal{H}_1 \otimes \mathcal{H}_2$ of the form $D_1 \otimes D_2$ cannot represent a physical state of the compound system, for it would then determine for the first subsystem a state precisely given by D_1. The coherent subspaces of $\mathcal{H}_1 \otimes \mathcal{H}_2$ must be patterned so as to make the states of the subsystems fit in with the superselection rules operating for the subsystems.

However, the knowledge of the coherent subspaces of $\mathcal{H}_1$ and $\mathcal{H}_2$ is generally not sufficient to determine uniquely the coherent subspaces of $\mathcal{H}_1 \otimes \mathcal{H}_2$. This fact is suggested by some simplified examples: consider for instance the combination of a pion with a nucleon, and compare it with the combination of a pion with a muon. Let us take into account only the superselection rules due to electric charge conservation, and disregard other superselection rules, here present, like the ones related to baryonic and leptonic charge. We have the same situation for the coherent subspaces of the subsystems: $\mathcal{H}_1$ has the three one-dimensional subspaces π^+, π^0, π^-, while $\mathcal{H}_2$ has two one-dimensional subspaces, which are denoted p, n if it refers to the nucleon and μ^+, μ^- if it refers to the muon. But the coherent subspaces of the compound system are different in the two cases (recall that we refer only to electric charge): there are four coherent subspaces for the pion-nucleon system ($\pi^+ \otimes p$, $\pi^- \otimes n$, and the two-dimensional subspaces spanned by $\{\pi^0 \otimes p, \pi^+ \otimes n\}$ and by $\{\pi^- \otimes p, \pi^0 \otimes n\}$) and five coherent subspaces for the pion-muon system ($\pi^+ \otimes \mu^+$, $\pi^0 \otimes \mu^+$, $\pi^0 \otimes \mu^-$, $\pi^- \otimes \mu^-$, and the two-dimensional subspace spanned by $\{\pi^+ \otimes \mu^-, \pi^- \otimes \mu^+\}$).

When the problem of the superselection rules of a compound system is considered, it is tempting to ask whether the very mechanism of combining subsystems can cause, at least in the limit of very large numbers of subsystems, the birth of superselection rules of the compound system. As a matter of fact, a macroscopic body, formed by exceedingly many microscopic subsystems that are strictly quantum and hence free of superselection rules, usually behaves as a classical system,* thus displaying a huge number of superselection rules. Despite the great conceptual interest of this question, no general answer is known. Of course, should one have a theory predicting the appearance of superselection rules as an effect of the combination of subsystems with purely quantum behavior, one would have at hand the possibility of a deep understanding of the "classical limit" of quantum mechanics. In particular, this would open the way to a complete theory of the measurement process, that is, a unified theory describing the quantum physical system to be measured, the macroscopic classical measuring instrument, and their mutual interaction. We shall return to this point in next chapter.

Exercises

1. If $A_1 \in \mathbb{B}(\mathcal{H}_1)$, $A_2 \in \mathbb{B}(\mathcal{H}_2)$, then $A_1 \otimes A_2$ is bounded.

[*Hint.* Notice that $\|(A_1 \otimes A_2)(\varphi \otimes \psi)\| \leqslant \|A_1\| \cdot \|A_2\| \cdot \|\varphi \otimes \psi\|$, and this inequality is extended to every vector in $\mathcal{H}_1 \otimes \mathcal{H}_2$ by using the fact that a basis in $\mathcal{H}_1 \otimes \mathcal{H}_2$ is formed by vectors having the tensor-product form.]

2. If $A_1, B_1 \in \mathbb{B}(\mathcal{H}_1)$, $A_2, B_2 \in \mathbb{B}(\mathcal{H}_2)$, and $\lambda \in \mathbb{C}$, then

$$\lambda(A_1 \otimes A_2) = (\lambda A_1) \otimes A_2 = A_1 \otimes (\lambda A_2),$$
$$(A_1 + B_1) \otimes A_2 = A_1 \otimes A_2 + B_1 \otimes A_2,$$
$$(A_1 \otimes A_2)(B_1 \otimes B_2) = A_1 B_1 \otimes A_2 B_2.$$

[*Hint.* Check these properties on vectors of the form $\varphi \otimes \psi$, refer to the structure of a basis in $\mathcal{H}_1 \otimes \mathcal{H}_2$, and use linearity and continuity.]

3. If $A_1 \in \mathbb{B}(\mathcal{H}_1)$ and $A_2 \in \mathbb{B}(\mathcal{H}_2)$ are self-adjoint, then $A_1 \otimes A_2$ is also self-adjoint.

[*Hint.* Notice that $(\varphi \otimes \psi, (A_1 \otimes A_2)(\varphi \otimes \psi))$ is real for every $\varphi \in \mathcal{H}_1$ and $\psi \in \mathcal{H}_2$; then refer to the structure of a basis in $\mathcal{H}_1 \otimes \mathcal{H}_2$ and use linearity and continuity.]

4. If $A_1 \in \mathbb{B}(\mathcal{H}_1)$ and $A_2 \in \mathbb{B}(\mathcal{H}_2)$ are trace-class, then $A_1 \otimes A_2$ is also trace-class and $\mathrm{tr}(A_1 \otimes A_2) = \mathrm{tr}(A_1) \cdot \mathrm{tr}(A_2)$.

---

*Let us emphasize that macroscopic bodies are usually classical, but not necessarily. A pot of liquid helium is macroscopic but displays quantum behavior.

[*Hint*. Refer to the fact that a bounded operator A is trace-class if and only if there exists in its Hilbert space a basis $\{\varphi_i\}$ such that the series $\Sigma_i \|A\varphi_i\|$ converges.]

5. Prove that (7.1.1) is a complete set in $\mathcal{H}_1 \otimes \mathcal{H}_2$.
[*Hint*. Refer to Section 3.2 and show that if φ is a cyclic vector of $\mathcal{C}_1''$ and ψ is a cyclic vector of $\mathcal{C}_2''$, then $\varphi \otimes \psi$ is a cyclic vector of the von Neumann algebra generated by (7.1.1).]

6. Show that $D = D_1 \otimes D_2$ is a density operator (if D_1 and D_2 are density operators) and that it is always a solution of (7.2.2).
[*Hint*. Refer to Exercises 4 and 2.)

7. Take $\mathcal{H}_1 = \mathcal{H}_2 = \mathbf{C}^2$; let $\{\varphi_+, \varphi_-\}$ be an orthonormal basis in $\mathcal{H}_1$, and $\{\psi_+, \psi_-\}$ an orthonormal basis in $\mathcal{H}_2$. Suppose

$$D_1 = \tfrac{1}{2}P^{\hat{\varphi}_+} + \tfrac{1}{2}P^{\hat{\varphi}_-} \quad \text{and} \quad D_2 = \tfrac{1}{2}P^{\hat{\psi}_+} + \tfrac{1}{2}P^{\hat{\psi}_-},$$

and verify that, writing $\Phi_1 = \varphi_+ \otimes \psi_-$, $\Phi_2 = \varphi_- \otimes \psi_+$, $\Phi_3 = (1/\sqrt{2})(\varphi_+ \otimes \psi_- + \varphi_- \otimes \psi_+)$, $\Phi_4 = (1/\sqrt{2})(\varphi_+ \otimes \psi_- - \varphi_- \otimes \psi_+)$, the following are solutions of (7.2.2):

$$D = \tfrac{1}{2}P^{\Phi_1} + \tfrac{1}{2}P^{\Phi_2}, \qquad D = P^{\Phi_3}, \qquad D = P^{\Phi_4}.$$

[*Hint*. Refer to (4.2.2) of Chapter 4, and choose for φ_+, φ_- (as well as ψ_+, ψ_-) the eigenvectors of σ_z.]

8. If $D_1 = P^{\hat{\varphi}}$ and $D_2 = P^{\hat{\psi}}$, then $D = P^{\hat{\varphi}} \otimes P^{\hat{\psi}} = \widehat{P^{\varphi \otimes \psi}}$ is the only solution of (7.2.2).
[*Hint*. Choose a basis $\{\varphi_i \otimes \psi_j\}$ in $\mathcal{H}_1 \otimes \mathcal{H}_2$ with $\varphi_1 = \varphi$ and $\psi_1 = \psi$; take $P_1 = P^{\hat{\varphi}}$ and $P_2 = P^{\hat{\psi}}$ in (7.2.2), and deduce $(\varphi_i \otimes \psi_j, D(\varphi_i \otimes \psi_j)) = \delta_{i,1}\delta_{j,1}$. Inasmuch as $0 \leqslant D^2 \leqslant D$, notice that $(\varphi_i \otimes \psi_j, D^2(\varphi_i \otimes \psi_j)) = \|D(\varphi_i \otimes \psi_j)\|^2 = 0$ if $i, j \neq 1$.]

9. Show that D_1 and D_2 are uniquely determined by D.
[*Hint*. If D_1 and D_1' satisfy (7.2.2), we get $\mathrm{tr}(D_1 P_1) = \mathrm{tr}(D_1' P_1)$ for every projector P_1 on $\mathcal{H}_1$.]

10. If $D = wD' + (1-w)D''$, then $D_1 = wD_1' + (1-w)D_1''$ (and similarly for D_2), where D_1' and D_1'' are states of subsystem 1 determined by the states D' and D'' of the compound system.
[*Hint*. Use the linearity of the trace.]

11. Deduce the formulae (7.2.3).
[*Hint*. Choose a basis in $\mathcal{H}_1 \otimes \mathcal{H}_2$ containing $\Sigma_{ij}\lambda_{ij}\varphi_i \otimes \psi_j$ as an element.]

12. Verify (7.4.3).
[*Hint*. Use (3.1.4), (3.1.6), and (3.1.9) of Chapter 3.]

13. Show that there is a unitary dynamical group $t \mapsto U_t$ that maps (7.5.1) into (7.5.2).

[*Hint*. In the basis

$$\left\{ \varphi_+ \otimes \psi_+, \ \frac{1}{\sqrt{2}}(\varphi_+ \otimes \psi_- + \varphi_- \otimes \psi_+), \ \varphi_- \otimes \psi_-, \ \frac{1}{\sqrt{2}}(\varphi_+ \otimes \psi_- - \varphi_- \otimes \psi_+) \right\}$$

of $\mathcal{H}_1 \otimes \mathcal{H}_2$, take the matrix form

$$U(t) = \begin{pmatrix} \cos \omega t & -\sin \omega t & 0 & 0 \\ \sin \omega t & \cos \omega t & 0 & 0 \\ 0 & 0 & 1 & 0 \\ 0 & 0 & 0 & 1 \end{pmatrix}$$

for $U(t)$.]

14. Deduce from (7.5.2) the expression for D_1.

[*Hint*. Use (7.2.3) to deduce, in the $\{\varphi_+, \varphi_-\}$ basis, the matrix form

$$D_1 = \begin{pmatrix} \frac{1}{2}(1 + \cos^2 \omega t) & \frac{1}{\sqrt{2}} \sin \omega t \cos \omega t \\ \frac{1}{\sqrt{2}} \sin \omega t \cos \omega t & \frac{1}{2} \sin^2 \omega t \end{pmatrix};$$

then diagonalize this matrix.]

References

1. J. Dixmier, *Les Algèbres d'Opérateurs dans l'Espace Hilbertien*, Gauthiers-Villars, Paris, 1957.
2. M. Reed and B. Simon, *Methods of Modern Mathematical Physics*, Vol. I, Academic Press, New York, 1972.
3. M. Jammer, *The Philosophy of Quantum Mechanics*, Wiley, New York, 1974.
4. J. M. Jauch, *Foundations of Quantum Mechanics*, Addison-Wesley, Advanced Book Program, Reading, Mass., 1968.
5. See, e.g., R. F. Streater and A. S. Wightman, *PCT, Spin and Statistics, and All That*, Benjamin Cummings, Advanced Book Program, Reading, Mass., 1964.
6. See. e.g., H. Cramer, *Mathematical Methods of Statistics*, Princeton University Press, Princeton, N.J., 1946.
7. A. Einstein, B. Podolsky, and N. Rosen, *Phys. Rev.* 47 (1935) 777.
8. N. Bohr, *Phys. Rev.* 48 (1935) 696.
9. A. Einstein, *J. Franklin Inst.* 221 (1936) 349.

CHAPTER 8

Elementary Analysis of the Measurement Process

8.1 Measuring Instruments; Idealizations; Projection Postulate

In order to give to the formalism of quantum mechanics the status of a physical theory, one must assume the *measurability* of what we have called "physical quantities"; this means that, given a physical quantity represented by the operator A, there must exist an experimental macroscopic device M, equipped with a reading scale and with instructions on how to couple it with the physical system S, such that, for any state D of S, the probability of having the pointer of M, after interaction with S, within the interval E of the reading scale is precisely given by $\mathrm{tr}(DP_A(E))$. Thus, in particular, if A has a point spectrum and λ_i is an eigenvalue, the device M must be such that the interaction with S in a state D belonging to the λ_i-subspace causes with certainty the appearance of the pointer of M at the position λ_i on the reading scale.

It is natural to call an experimental device M of this sort a measuring instrument for A, though we do not claim that this characterization exhausts all properties of an actual measuring instrument (in Section 8.3 we shall indeed encounter another crucial property).

Now, obvious requirements of completeness and homogeneity of the theory suggest considering the measuring instruments themselves as physical systems, and it is natural to agree that different positions of the pointer of the instrument M label different states of the system M. But here it becomes compulsory to recognize that a measuring instrument is not a purely quantum system, as evident from the fact that we cannot give physical meaning to the superposition of two different positions of an actual, macroscopic pointer. Thus, the measuring instruments belong to that class

ENCYCLOPEDIA OF MATHEMATICS and Its Applications, Gian-Carlo Rota (ed.). Vol. 15: E. G. Beltrametti and G. Cassinelli, The Logic of Quantum Mechanics

ISBN 0-201-13514-0

of physical systems to which the formalism of quantum mechanics without superselection rules does not apply. But, on the other hand, the inclusion of superselection rules makes the formalism of quantum mechanics so flexible that no physical system is known that irrecoverably escapes from it. Hence, once a measuring instrument is considered as a physical system, we can have recourse to the formalism of quantum mechanics after recognizing that states labeled by different positions of the pointer are separated by superselection rules. Of course, the acknowledgment of these superselection rules does not overcome the problem of their origin: in particular, there remains the crucial problem (see Section 7.6) of whether the description of a macroscopic measuring instrument as made of purely quantum subsystems (its atoms and molecules) could entail its superselection rules as a very consequence of aggregating all these subsystems.[1-3] Notice, however, that we do not claim that all macroscopic systems must have, like measuring instruments, a family of states labeled by real numbers and separated by superselection rules; what we claim is that when this is not the case the system has nothing to do with a measuring instrument.

A familiar idealization of the measurement process amounts to supposing that the measuring instruments are "nondestructive". This means that the physical system S and the measuring instrument M, initially separated, will then form a compound system, both preserving their own identity during the mutual interaction, and become separated again after a certain time, so that it makes sense to speak of final state of S and of M after the measurement.

It must be stressed that this idealization is not a logical necessity of quantum theory, for the validity of quantum mechanics does not depend upon the particular choice of the measuring instruments used to verify the statistical predictions of the theory. And we know that many actual measuring instruments do violate this idealization: there are "destructive" instruments, such as the photon spectrometer which measures the energy of a photon by absorbing it so that the photon is destroyed by the very act of measurement. In other words, there is no logical necessity in quantum mechanics of speaking of the final state of S after the measurement.

Clearly, the idealization above is not strong enough to make the state of the system S after the measurement unambiguously determined by the initial state of S and by the physical quantity one is measuring. In fact a physical quantity can admit many measuring instruments, of the nondestructive kind, which affect the state of S in different ways. To reduce, or even to eliminate, this ambiguity, it is usual to have recourse to a further idealization of the measurement process, generally referred to as von Neumann's or Lüders's projection postulate.

In von Neumann's formulation,[4] the projection postulate says that if a physical quantity is measured twice in succession, then the same value is obtained each time. In case the operator A representing the physical quantity has a point spectrum and λ_i is an eigenvalue, this amounts to assuming that after a measurement with result λ_i the state of $\mathbb{S}$ belongs to the λ_i-eigenspace of A. Notice that, by this postulate, the final state of $\mathbb{S}$ is uniquely determined only if λ_i is a nondegenerate eigenvalue of A, for in this case the corresponding eigenspace is one-dimensional; but when λ_i is degenerate, so that the corresponding eigenspace has dimension $\geqslant 2$, the final state of $\mathbb{S}$ is left undetermined and will still depend upon the actual measuring equipment.

A more stringent version of von Neumann's projection postulate has been proposed by Lüders:[5] it adds the requirement that if the initial state of $\mathbb{S}$ is the vector state φ and the observed value λ_i is degenerate, then the final state of $\mathbb{S}$ is the normalized projection of φ onto the eigenspace belonging to λ_i. Lüders's projection postulate is generally completed by assuming that the nonpure initial states of $\mathbb{S}$ are affected by the measuring instrument in such a way that their convex structure is preserved: thus it can be summarized by saying that the observation of the result λ_i induces the following transformation of the initial state D:

$$D \mapsto \frac{P_A(\{\lambda_i\}) D P_A(\{\lambda_i\})}{\mathrm{tr}\left(D P_A(\{\lambda_i\})\right)}$$

If A has not a purely discrete spectrum and if the observed value is in a subset E of the spectrum of A, then this formula admits the obvious generalization

$$D \mapsto \frac{P_A(E) D P_A(E)}{\mathrm{tr}\left(D P_A(E)\right)}. \tag{8.1.1}$$

It seems worth stressing that von Neumann's as well as Lüders's projection postulates (despite the name, which reflects a historical tradition) have not the status of postulates of quantum theory, necessary to its internal coherence: they are nothing more than the definitions of special classes of measuring instruments.

Though the majority of the measuring instruments used in practice do not satisfy either Lüders's or von Neumann's projection postulate, there is no definite example of a physical quantity not admitting at least one measuring instrument satisfying, even if only approximately, these postulates. Following a terminology proposed by Pauli,[6] the measuring instruments that obey von Neumann's projection postulate are referred to as "first-kind instruments". Those that match the stronger postulate of Lüders are often called "ideal and of first kind".

The prescription of using only instruments that satisfy Lüders's postulate allows one to specify uniquely the state of the system after a measurement of a physical quantity with a given result: thus it becomes meaningful to speak of the probability distribution of a physical quantity given the result of a previous measurement of another physical quantity. In other words, the prescription of using only ideal first-kind instruments allows the introduction and the interpretation, within quantum theory, of the concept of conditional probabilities (this issue will be deepened in Chapters 16 and 26).

8.2 States of the Measured System and of the Measuring Instrument

In this section we formalize the ideas sketched in the preceding section. Let $\mathcal{H}_S$ and $\mathcal{H}_M$ denote, respectively, the Hilbert space of the measured physical system S and that of the measuring instrument M. For simplicity we shall hereafter assume that the operator A on $\mathcal{H}_S$ representing the physical quantity under discussion has a simple point spectrum. Let $\lambda_1, \lambda_2, \ldots$ be its eigenvalues and $\varphi_1, \varphi_2, \ldots \in \mathcal{H}_S$ the corresponding normalized eigenvectors. The reading scale of M will consist precisely of the values $\lambda_1, \lambda_2, \ldots$: these numbers label the possible positions of the pointer of M and also a distinguished family of states of M that we call "indicator states" of M. We write $\Phi_1, \Phi_2, \ldots$ for these states, and interpret Φ_i as the state of M when the position of the pointer is λ_i. It will be useful to introduce also the rest state of the instrument, that is, the state of M before interaction with S: we denote it by Φ_0. As stated in the preceding section, the very notion of measuring instrument requires that any two among the states $\Phi_0, \Phi_1, \Phi_2, \ldots$ be separated by superselection rules. Thus we can view these states as the pure states of M.

In the discussion to follow we shall assume that M is a nondestructive measuring instrument; however, we do not assume that it meets von Neumann's or Lüders's projection postulate, for this would only introduce a modest simplification of notation. S and M will form a compound system $S + M$ whose Hilbert space will be (we refer to Section 7.1 and remark that S and M are certainly not identical)

$$\mathcal{H}_{S+M} = \mathcal{H}_S \otimes \mathcal{H}_M.$$

Since $\mathcal{H}_M$ is endowed with superselection rules, we know that $\mathcal{H}_{S+M}$ certainly inherits superselection rules, whose pattern, however, is not determined by the general formalism of quantum mechanics, but depends on the actual physical situation, according to the discussion of Section 7.6.

Consider now the evolution of the states of S, M, and $S + M$, and assume the simplifying hypothesis that S has no superselection rules and that its initial state is pure. Suppose first that, before interacting with M, S is precisely in one of the eigenstates of A, say φ_i, while M is in its rest state Φ_0; therefore the initial state of $S + M$ is uniquely determined (see Section 7.2) as $\varphi_i \otimes \Phi_0$. After the interaction, by the definition of the measuring instrument, the state of M has to be Φ_i; hence the state of S + M will take the tensor-product form $\psi_i^{(M)} \otimes \Phi_i$, with $\psi_i^{(M)}$ representing the final state of S. The notation alludes to a pure final state of S, with $\psi_i^{(M)}$ some normalized vector of $\mathcal{H}_S$: if not, $\psi_i^{(M)}$ should be replaced by a density operator and the notation changed accordingly. The index M emphasizes the fact that the final state of S will depend, in general, not only on the initial state of S but also on the actual instrument M. Recall that we are not assuming the projection postulate; should we assume it, we would just have to replace $\psi_i^{(M)}$ by φ_i.

Suppose now that the initial state of S, though still pure, is not an eigenstate of the physical quantity to be measured, thus being represented by a vector $\Sigma_i c_i \varphi_i$; then the initial state of S + M will be of the form $(\Sigma_i c_i \varphi_i) \otimes \Phi_0$ and will evolve, by linearity, into the vector $\Sigma_i c_i (\psi_i^{(M)} \otimes \Phi_i)$. However, it must be stressed that, since the compound system S + M is certainly endowed with superselection rules, we are not, *a priori*, guaranteed that the vector $\Sigma_i c_i (\psi_i^{(M)} \otimes \Phi_i)$ represents a pure state of S + M: it might not belong to a coherent subspace of $\mathcal{H}_{S+M}$, in which case it should be replaced by the appropriate mixture, as discussed in Sections 5.4 and 6.4. We must indeed expect that the superselection rules of S + M will be such as to cause the replacement of $\Sigma_i c_i (\psi_i^{(M)} \otimes \Phi_i)$ by a mixture, for there would be little physical meaning in a state of S + M that superposed different states of M, i.e., different positions of the pointer (we return to this point in the next section).

No matter whether $\Sigma_i c_i (\psi_i^{(M)} \otimes \Phi_i)$ represents a pure or a nonpure state of S + M, it follows by the standard methods described in Section 7.2 that the reduced states of S and M are given, respectively, by (Exercise 1)

$$D_S = \sum_i |c_i|^2 P^{\hat{\psi}_i^{(M)}} \quad \text{and} \quad D_M = \sum_i |c_i|^2 P^{\hat{\Phi}_i}. \tag{8.2.1}$$

Thus we see that S and M evolve into nonpure states, and D_M fits the assumed superselection rules of the instrument M.

Of course, there is a relevant correlation between S and M, coded in the state $\Sigma_i c_i (\psi_i^{(M)} \otimes \Phi_i)$ of S + M. Specifically, consider, for a given index k, the projectors $P^{\hat{\psi}_k^{(M)}}$ and $P^{\hat{\Phi}_k}$, and look at them as representing, respectively, the two-valued physical quantities "S is in the state $\psi_k^{(M)}$" and "M is the state Φ_k"; then the coefficient of correlation between $P^{\hat{\psi}_k^{(M)}}$ and $P^{\hat{\Phi}_k}$ reads, according to Section 7.3,

$$\rho = \frac{\mathcal{E}\left(P^{\hat{\psi}_k^{(M)}} \otimes P^{\hat{\Phi}_k}, \sum_i c_i \left(\psi_i^{(M)} \otimes \Phi_i \right) \right) - \mathcal{E}\left(P^{\hat{\psi}_k^{(M)}}, D_S \right) \mathcal{E}\left(P^{\hat{\Phi}_k}, D_M \right)}{\left[\mathcal{V}\left(P^{\hat{\psi}_k^{(M)}}, D_S \right) \mathcal{V}\left(P^{\hat{\Phi}_k}, D_M \right) \right]^{1/2}},$$

$$\tag{8.2.2}$$

and a straightforward calculation (see Exercise 2) yields

$$\rho = 1.$$

In view of the interpretation of $P^{\hat{\psi}_k^{(M)}}$ and $P^{\hat{\Phi}_k}$, this result confirms what we had to expect: whenever the measuring instrument $\mathbb{M}$ jumps into the state Φ_k, then necessarily the system $\mathbb{S}$ is left in the state $\psi_k^{(M)}$.

Notice that, due to the superselection rules of $\mathbb{M}$, the mixture $D_{\mathbb{M}} = \Sigma_i |c_i|^2 P^{\hat{\Phi}_i}$ has a unique decomposition into pure states and thus admits the familiar "ignorance interpretation": the actual state of $\mathbb{M}$ is in $\{\Phi_i\}$ and the probability of being Φ_i is $|c_i|^2$. In view of the correlation between $\mathbb{S}$ and $\mathbb{M}$, we are then also led to interpret the mixture $D_{\mathbb{S}} = \Sigma_i |c_i|^2 P^{\hat{\psi}_i^{(M)}}$ unambiguously by saying that $\mathbb{S}$ is actually in one of the states $\psi_i^{(M)}$, with probability $|c_i|^2$. Thus we see that the typical pathology of quantum mixtures, consisting in the nonunique decomposition into pure states and in the attached difficulty of the "ignorance interpretation", is here overcome, thanks to the superselection rules that are implicit in the very notion of a measuring instrument.

Of course, (8.2.1) gives the states of $\mathbb{S}$ and $\mathbb{M}$ after their mutual interaction but before one records the actual position of the pointer of $\mathbb{M}$. Once this position is recorded, $\mathbb{M}$ collapses into one of the indicator states and $\mathbb{S}$ into the states determined by the correlation with $\mathbb{M}$. But, in view of what was said before, this collapse is a purely classical one: it is nothing else than the reduction of ignorance caused by the information about the position of the pointer.

8.3 A Consistency Problem

The measuring process calls into play two entities: the observed physical system and the measuring instrument; but the boundary between the two is arbitrary to a large extent. Indeed, physical evidence says that the same information achieved by the measuring instrument $\mathbb{M}$ acting on $\mathbb{S}$ can be achieved by a suitable measuring instrument $\mathbb{M}'$ acting on $\mathbb{S}' \equiv \mathbb{S} + \mathbb{M}$, as well as by a suitable measuring instrument $\mathbb{M}''$ acting on $\mathbb{S}'' \equiv \mathbb{S} + \mathbb{M} + \mathbb{M}'$, and so on. Thus we have a chain, each link of which corresponds to a different positioning of the boundary between what is observed and the measuring instrument. Since the choice of a link in the chain must reduce to a matter of convention, we need the descriptions relative to the various links to be fully equivalent, and this equivalence must be incorporated in the very notion of a measuring instrument. To point at the pertinent property of measuring instruments that resolves this consistency problem is the purpose of this section.

We still adopt the simplifying assumptions of the last section: the physical quantity is represented by an operator A with simple spectrum, $\mathbb{M}$

is nondestructive, S is strictly quantum, and its initial state is pure. But, to simplify the discussion further, we add the assumption that M meets the projection postulate. The reader will notice that, having assumed an A with simple spectrum, von Neumann's version of the projection postulate coincides with Lüders's.

For the sake of definiteness let us compare the first two links of the chain: M as measuring instrument acting on S, and M' as measuring instrument acting on $\mathsf{S}' \equiv \mathsf{S} + \mathsf{M}$.

(i) M *acting on* S. We have only to refer to the last section, with $\psi_i^{(\mathsf{M})}$ replaced by φ_i, thus obtaining the following scheme: if the initial state of S is $\Sigma_i c_i \varphi_i$, and that of M is Φ_0, then $\mathsf{S} + \mathsf{M}$ evolves from the initial state $(\Sigma_i c_i \varphi_i) \otimes \Phi_0$ into

$$\sum_i c_i (\varphi_i \otimes \Phi_i), \tag{8.3.1}$$

or into an equivalent mixture if the superselection rules of $\mathsf{S} + \mathsf{M}$ so demand; in any case the final states of S and M are the mixtures

$$D_{\mathsf{S}} = \sum_i |c_i|^2 P^{\hat{\varphi}_i}, \qquad D_{\mathsf{M}} = \sum_i |c_i|^2 P^{\hat{\Phi}_i}. \tag{8.3.2}$$

(ii) M' *acting on* $\mathsf{S}' = \mathsf{S} + \mathsf{M}$. Since M' is assumed to provide an alternative, but equivalent, way of measuring what M measures, to each position of the pointer of M there must correspond a position of the pointer of M'. These positions of the pointer of M' will label the indicator states of M', which we denote $\Phi_0', \Phi_1', \Phi_2', \ldots$, with Φ_0' standing for the rest state. Again we must assume, as we did for M, that in the Hilbert space $\mathcal{H}_{\mathsf{M}'}$ of M', these indicator states are separated by superselection rules. Now, translating to S', M', and $\mathsf{S}' + \mathsf{M}'$ what was said above, we get the following scheme. If the initial state of S' is $(\Sigma_i c_i \varphi_i) \otimes \Phi_0$ (as a result of the fact that the initial state of S is $\Sigma_i c_i \varphi_i$ and that of M is Φ_0) and the initial state of M' is Φ_0', then $\mathsf{S}' + \mathsf{M}'$ evolves from the state $(\Sigma_i c_i \varphi_i) \otimes \Phi_0 \otimes \Phi_0'$ into

$$\sum_i c_i (\varphi_i \otimes \Phi_i \otimes \Phi_i') \tag{8.3.3}$$

or into an equivalent mixture if the superselection rules of $\mathsf{S}' + \mathsf{M}'$ so demand; in any case M' evolves into the mixture $\Sigma_i |c_i|^2 P^{\hat{\Phi}_i'}$ and $\mathsf{S}' = \mathsf{S} + \mathsf{M}$ into the mixture

$$\sum_i |c_i|^2 P^{\widehat{\varphi_i \otimes \Phi_i}}, \tag{8.3.4}$$

which determines for S and M the same mixtures given in (8.3.2).

Now, comparing the two schemes (i) and (ii), we see that our consistency problem reduces uniquely to the requirement of full equivalence between the final states of $S + M$ expressed by (8.3.1) and (8.3.4). This resolves the difficulty left open before: (8.3.1) cannot represent a pure state of $S + M$, a fact that leaves us free from the embarrassing idea of having a state that superposes different positions of the instrument pointer. At this point, we can easily see which intrinsic property a measuring instrument must possess in order to solve the consistency problem. According to the rules that specify the correspondence between states and density operators in the presence of superselection rules, we conclude that in order to make (8.3.1) equivalent to (8.3.4) we must assume for the compound system $S + M$ the existence of superselection rules such that the coherent subspaces $\mathfrak{M}_k^{(S+M)}$ of the Hilbert space $\mathcal{H}_{S+M} = \mathcal{H}_S \otimes \mathcal{H}_M$ take the form (Exercise 3)

$$\mathfrak{M}_k^{(S+M)} = \mathcal{H}_S \otimes \hat{\Phi}_k, \qquad k = 1, 2, \ldots, \qquad (8.3.5)$$

Recall that we are assuming, for simplicity, that S is purely quantum (that is, free of superselection rules), so that we can read this formula as saying that the coherent subspaces of the compound system $S + M$ must be products of the coherent subspaces of the subsystems S and M. Thus we realize that in order to qualify M as a measuring instrument, we have to assume not only its own superselection rules, but also the fact that, coupling it with S, we get a compound system $S + M$ that inherits superselection rules according to the product rule (8.3.5).*

There are some theoretical models[1-3] of the measuring process that try to approach explicitly the description of measuring instruments as formed by large numbers of component subsystems; they seem to agree, in the limit of long interaction time between S and M, on the superselection pattern said above. Let us also notice, as a bibliographical remark, that the propagation of superselection rules from M to $S + M$ according to (8.3.5) is in some way acknowledged by Landau and Lifshitz's textbook on quantum mechanics.[7]

8.4 Schrödinger's Cat

There are examples of physical situations in which too crude a use of the formalism of quantum mechanics would imply grotesque facts. These examples, generally related to the measurement process, have been proposed as blemishes of quantum mechanics and are often called, tendentiously, "paradoxes" of quantum mechanics. Among them we might mention the celebrated "Schrödinger's cat"[8] and "Wigner's friend";[9] here we shall be

---

*Of course, going to $S' + M'$, we would find that (8.3.3) has to be equivalent to a mixture and that the coherent subspaces of $S' + M' = S + M + M'$ must take the form $\mathfrak{M}^{(S'+M')} = \mathcal{H}_S \otimes \hat{\Phi}_k \otimes \hat{\Phi}'_k$, and so on, going up along the chain $M, M', M'', \ldots$.

concerned only with the first, which looks simpler, for Wigner's friend adds to the issue of Schrödinger's cat the element of consciousness of the observer, an element whose analysis in terms of quantum mechanics seems impossible at present.

Schrödinger's cat is an unlucky cat placed in a closed box together with a small amount of radioactive substance whose first decay causes, via a suitable amplification device, the diffusion in the box of a poison gas, killing the cat. The question is: which will be the state of the whole system contained in the box after a time interval in which the probability of getting the first radioactive decay is $\frac{1}{2}$ (so that the probability of getting no decay is also $\frac{1}{2}$)?

For the sake of simplicity let us focus attention on the essential elements, the decaying nucleus and the cat, leaving aside the description of the intermediary elements such as the amplification device and the poison gas. Let φ_u and φ_d represent, respectively, the undecayed and decayed state of the radioactive nucleus, and let Φ_a and Φ_d designate, respectively, the alive and dead state of the cat. Then, should one treat cat and radioactive nucleus as purely quantum systems, without superselection rules, the answer to the foregoing question would be

$$c_1\varphi_u \otimes \Phi_a + c_2\varphi_d \otimes \Phi_d, \qquad |c_1|^2 = |c_2|^2 = \frac{1}{2}, \qquad (8.4.1)$$

an unlikely answer, for this state is a superposition of states corresponding to a live and a dead cat, a situation never experienced. Hence the claimed paradox. Of course, all that is consequence of an oversimplified hypothesis. Indeed, a cat, as such, is a physical system for which the formalism of quantum mechanics without superselection rules is simply wrong: the impossibility of superposing the live and dead states has to be taken as part of the very "definition" of a cat. Thus, the analysis, within quantum theory, of Schrödinger's cat needs to assume from the outset that the cat has superselection rules that separate the states Φ_a and Φ_d. True, what remains an open problem is whether these superselection rules can find their origin in viewing the cat as composed of atoms and molecules, but this is another question.

Once we acknowledge the superselection rules of the cat, we know that also the whole content of the box, the system composed of the cat and the radioactive nucleus, will inherit superselection rules, though their precise pattern is not determined by the general formalism of quantum mechanics (as seen in Section 7.6). However, we can remark that the cat acts as a measuring instrument for the decay of the radioactive nucleus, so that we can make reference to the discussion of the preceding section, and conclude that the propagation of the superselection rules to the compound system must follow the product rule (8.3.5) with S standing for the radioactive nucleus and M for the cat. Hence the vector (8.4.1) does not represent a pure state but must properly be replaced by the mixture

$$\tfrac{1}{2}P^{\widehat{\varphi_u \otimes \Phi_a}} + \tfrac{1}{2}P^{\widehat{\varphi_d \otimes \Phi_d}},$$

which, in view of cat's superselection rules, is uniquely decomposable into pure states and admits the usual ignorance interpretation: there is probability $\tfrac{1}{2}$ of having the nucleus undecayed and the cat alive, and probability $\tfrac{1}{2}$ of having the nucleus decayed and the cat dead. This is precisely what conforms to our direct experience, and the "paradoxical" aspects of the issue are avoided.

Rather than a paradox of quantum mechanics, Schrödinger's cat is a convincing didactic tool to outline the fact that the superposition of pure states is a true novelty and a distinctive mark of quantum theory, so that we come to grotesque conclusions whenever we improperly apply the notion of superposition to physical systems that are not purely quantum.

Exercises

1. Show that if the compound system $\mathbb{S} + \mathbb{M}$ is in the state $\sum_i c_i \psi_i^{(\mathsf{M})} \otimes \Phi_i$, then $\mathbb{S}$ and $\mathbb{M}$ are in the nonpure states (8.2.1).
[*Hint.* Refer to Section 7.2, item (7).]

2. Evaluate (8.2.2).
[*Hint.* Refer to Section 3.1 and observe that the superselection rules acting in $\mathcal{H}_{\mathsf{M}}$ imply, in particular, that the set $\{\Phi_i\}$ is orthogonal. Notice that the result $\rho = 1$ follows without the need of assuming orthogonality for the set $\{\psi_i^{(\mathsf{M})}\}$).

3. Show that (8.3.5) implies and is implied by the equivalence of (8.3.1) and (8.3.4).
[*Hint.* Refer to Sections 5.4 and 6.4.]

References

1. K. Hepp, *Helv. Phys. Acta* 45 (1972) 237.
2. J. S. Bell, *Helv. Phys. Acta* 48 (1975) 93.
3. J. M. Levy-Leblond, in *Quantum Mechanics, a Half Century Later*, J. Leite Lopes and M. Paty, eds., Reidel, Dordrecht, 1977.
4. J. von Neumann, *Mathematical Foundations of Quantum Mechanics*, Princeton University Press, Princeton, N.J., 1955.
5. G. Lüders, *Ann. Physik* 8 (1951) 322.
6. W. Pauli, *Die Allgemeinen Prinzipien der Wellenmechanik*, in *Handbuch der Physik*, Vol. V, Springer, Berlin, Göttingen, Heidelberg, 1958 (Part I, pp. 1–168).
7. L. D. Landau and E. M. Lifshitz, *Quantum Mechanics—Non-relativistic Theory*, Pergamon Press, Oxford, 1965.
8. E. Schrödinger, *Naturwiss.* 48 (1935) 52.
9. E. P. Wigner, in *The Scientist Speculates*, I. J. Good, ed., Heinenmann, London, 1962.

CHAPTER 9

Mathematical Structures Emerging from the Hilbert-Space Formulation of Quantum Mechanics

9.1 The Ordered Structure of Projectors

Within the formalism of Hilbert-space quantum mechanics one can isolate several mathematical substructures. To see how they contribute to shape the whole edifice of quantum theory and how they are mutually related constitutes the main concern of what is commonly called the study of the mathematical foundations of quantum mechanics. In the present chapter we shall mention a number of these mathematical substructures, postponing to Part II an analysis of them and of their mutual relations.

In this section we consider the order structure pertaining to the set $\mathcal{P}(\mathcal{H})$ of all projection operators on the Hilbert space $\mathcal{H}$ of the physical system.* For simplicity we shall exclude superselection rules, and we shall also do so tacitly in the rest of the chapter. The elements of $\mathcal{P}(\mathcal{H})$ can be viewed as physical quantities of a special kind, namely, those that can take only two values, 0 or 1: they play the role of "propositions" of the physical system according to the intuitive notion of "proposition" outlined in the Preface. Conversely, the spectral theorem suggests viewing physical quantities as collections of elements of $\mathcal{P}(\mathcal{H})$.

We say that a projector P_1 is *smaller than* a projector P_2, and write $P_1 \leqslant P_2$, whenever

$$P_1 P_2 = P_2 P_1 = P_1. \qquad (9.1.1)$$

*The structure of $\mathcal{P}(\mathcal{H})$ will be examined in some detail in Section 10.3.

ENCYCLOPEDIA OF MATHEMATICS and Its Applications, Gian-Carlo Rota (ed.). Vol. 15: E. G. Beltrametti and G. Cassinelli, The Logic of Quantum Mechanics

ISBN 0-201-13514-0

That $\leqslant$ is indeed an order relation follows from the obvious check of its transitivity, reflexivity, and antisymmetry. This ordering admits equivalent characterizations; for instance, (9.1.1) is fully equivalent to requiring that the range of P_1 be set-theoretically included in the range of P_2.

Thus, $\mathcal{P}(\mathcal{H})$ is a partially ordered set (of course not a totally ordered set, for not every pair in it belongs to the order relation). As a matter of fact it is a partially ordered set of a highly structured kind. First of all, it is a complete lattice: this means that every family of elements of $\mathcal{P}(\mathcal{H})$ admits a least upper bound and a greatest lower bound. Second, it satisfies a number of important regularity conditions: among others, the ones called "ortho-modularity", "atomicity", and the "covering property". We do not go into these properties now; we shall do that exhaustively in the following chapters.

Having recognized the ordered structure of propositions, we can ask how the notions of state and of physical quantity are related to it. Let us have a brief look at this problem.

If D is any density operator on $\mathcal{H}$, then the function from $\mathcal{P}(\mathcal{H})$ into the real interval $[0,1]$ defined by

$$P \mapsto \mathrm{tr}(PD)$$

(i) maps the least and greatest elements of $\mathcal{P}(\mathcal{H})$ (that is, the null and the identity projectors) into the numbers 0 and 1 respectively;

(ii) is additive on orthogonal sequences, that is, if $\{P_i\}$ is a sequence of projectors such that $P_i P_j = 0$ when $i \neq j$, then $\mathrm{tr}((\Sigma_i P_i)D) = \Sigma_i \mathrm{tr}(P_i D)$.

These properties justify saying that the states are probability measures on the lattice $\mathcal{P}(\mathcal{H})$; furthermore, due to a deep theorem of Gleason (Section 11.2), we can even say that the set of all states can be fully identified with the set of all probability measures abstractly defined on $\mathcal{P}(\mathcal{H})$.

Consider now the connection between physical quantities and $\mathcal{P}(\mathcal{H})$. By the spectral theorem, every self-adjoint operator on $\mathcal{H}$ is completely determined by its spectral measure $E \mapsto P_A(E)$ $[E \in \mathcal{B}(\mathbb{R}), P_A(E) \in \mathcal{P}(\mathcal{H})]$; thus the physical quantities can be viewed as measures on the real line $\mathbb{R}$ taking values in $\mathcal{P}(\mathcal{H})$.

The fact that states and physical quantities can be defined in terms of $\mathcal{P}(\mathcal{H})$ suggests the following problem: Take from the outset a partially ordered set $\mathcal{L}$, to be physically interpreted as the set of "propositions" and having some of the lattice properties of $\mathcal{P}(\mathcal{H})$ but without any notion of Hilbert space. Consider the set $\mathcal{S}$ of all probability measures on $\mathcal{L}$, and the set $\mathcal{O}$ of all functions from $\mathcal{B}(\mathbb{R})$ into $\mathcal{L}$ that have the formal properties of spectral measures. Then is it possible to determine a Hilbert space $\mathcal{H}$ such that $\mathcal{L}$ is identified with $\mathcal{P}(\mathcal{H})$, $\mathcal{S}$ with the set of all density operators on $\mathcal{H}$, and $\mathcal{O}$ with the set of all self-adjoint operators on $\mathcal{H}$? Briefly, the question is: to what extent is the Hilbert-space description of quantum systems coded into the ordered structure of propositions? The answer to this question constitutes an approach to quantum mechanics that has been the focus of considerable research since the pioneering paper "The logic of quantum mechanics" by Birkhoff and von Neumann.[1] This approach will take a large part of the rest of this volume.

9.2 The Closure-Space Structure of Pure States

In the Hilbert-space formulation of quantum mechanics pure states are represented by one-dimensional subspaces of $\mathcal{H}$, and the structure of the linear space carried by $\mathcal{H}$ manifests itself in the notion of superposition of pure states. Given a collection M of pure states (that is, a collection M of one-dimensional subspaces of $\mathcal{H}$), consider the smallest subspace $\mathfrak{M}$ of $\mathcal{H}$ containing all the elements of M, and denote by $\overline{M}$ the set of all one-dimensional subspaces in $\mathfrak{M}$. The distinction $\mathfrak{M}$ versus $\overline{M}$, though pedantic, emphasizes the fact that we are not interested in the whole structure of $\mathfrak{M}$, but only in the collection of one-dimensional subspaces it contains. From the physical point of view, $\overline{M}$ is the collection of pure states obtained by adding to M all pure states that can be expressed as superpositions of states in M.

Clearly

$$M \subseteq \overline{M},$$

and if M' denotes another collection of pure states,

$$M \subseteq \overline{M}' \quad \text{implies} \quad \overline{M} \subseteq \overline{M}'.$$

These two facts qualify the mapping $M \mapsto \overline{M}$ as a "closure relation" in the (set of all subsets of the) set of all pure states. A set endowed with a closure relation is called a closure space. So we can summarize the above by saying that, from the Hilbert-space description of a quantum system, a closure-space structure emerges for the set of all pure states. Of course, we can expect that the underlying Hilbert space makes this closure relation a rather specialized one.

In Chapter 17 we shall examine a number of mathematical facts about closure spaces and we shall see connections between closure spaces and ordered structures; it will become interesting to ask which part of the edifice of quantum mechanics can be motivated by assuming from the outset that the set of pure states, viewed as primitive entities, is equipped with a closure-space structure. Roughly speaking, this amounts to asking to what extent an abstract notion of superposition of pure states (that is, a notion of superposition not referring to an underlying vector-space structure) can give rise to the usual formalism of quantum mechanics. Such a question characterizes what might be called a closure-space approach to quantum mechanics; this approach has not received so much attention in the literature as the approach based on the ordered structure of propositions, but nevertheless provides an abstract and attractive format for future research.

9.3 The Transition-Probability Structure of Pure States

The Hilbert-space description of quantum systems embodies the notion of a scalar product between any two vectors, and hence the notion of a function from the pairs of pure states into a number field ($\mathbb{C}$). If we consider the squared modulus of the scalar product, we have a function from the pairs of pure states into the real interval $[0, 1]$. Following von Neumann* (see also Section 2.5), we call this function the *transition probability* between two pure states.

The underlying Hilbert-space structure ensures the following properties of the transition probability function:

   (i) $|(\varphi, \psi)|^2 = 1$ if and only if φ and ψ represent the same pure state;
   (ii) $|(\varphi, \psi)|^2 = |(\psi, \varphi)|^2$ for every φ, ψ;
   (iii) if $\{\varphi_i\}$ is an orthonormal basis in $\mathcal{H}$, then $\Sigma_i |(\varphi_i, \psi)|^2 = 1$ for every unit vector ψ.

We summarize these properties by saying that the set of pure states of a quantum system forms a *transition-probability space*.

Now, the notion of transition-probability space can be given in more general terms, without the need of a Hilbert space as background. In fact, we can start from a set of pure states thought of as primitive entities, we can endow this set with a function from the pairs of pure states into the real interval $[0, 1]$, and we can express the content of conditions (i)–(iii) in an autonomous form without reference to a Hilbert-space structure. The mathematical object so emerging has interesting aspects: we shall be concerned with it in Chapter 18. We shall find connections with ordered structures, and we shall point out those features of the Hilbert-space description of quantum systems that can be recovered if one assumes from the outset that the set of pure states has the structure of a transition-probability space.

---

*J. von Neumann, *Continuous Geometries with a Transition Probability*, unpublished manuscript, 1937.

9.4 The Convex Structure of States

In the Hilbert-space formulation of quantum mechanics the states are represented by the density operators on $\mathcal{H}$. As already remarked in Sections 2.1 and 2.3, the set of all density operators is a convex set, and the pure states are geometrically characterized as the extreme points of this convex set. As a concrete example of this situation the reader is referred to Section 4.2, where the convex structure of states of the spin-$\frac{1}{2}$ system was examined. This example showed that many features of the spin-$\frac{1}{2}$ system are embodied in the geometrical structure of the convex set of states and, more specifically, in the shape of the boundary of this convex set.

All this is a hint to explore the inverse problem, taking the states of the physical system as primitive notions and assuming for them the structure of a convex set, whose extreme points define the pure states. Then the problem becomes one of ascertaining how much of the description of quantum systems is embodied in the geometrical structure of the convex set of the states. The possibility of approaching quantum mechanics from this direction will be considered in Chapter 19; in particular we shall examine connections between convex structures and the typical ordered structures that underlie quantum mechanics.

9.5 The Algebra of Physical Quantities

The bounded physical quantities of a quantum system, being represented by the bounded self-adjoint operators on $\mathcal{H}$, generate a distinguished algebraic structure, the algebra $\mathbb{B}(\mathcal{H})$ of all bounded operators on $\mathcal{H}$. $\mathbb{B}(\mathcal{H})$ is an involutive and noncommutative algebra. Every state with density operator D induces on $\mathbb{B}(\mathcal{H})$ a complex-valued linear function $B \mapsto \mathrm{tr}(BD)$ which, in particular, is normalized $[\mathrm{tr}(ID) = 1]$ and positive $[\mathrm{tr}(BB^*(D)) \geqslant 0]$.

Under the name of "algebraic approach to quantum mechanics" are grouped a number of slightly different versions of a same idea. These use as basic ingredient a non-commutative involutive algebra, abstractly defined without any reference to Hilbert space, whose self-adjoint elements are interpreted as the bounded physical quantities, while the states are defined as normalized, positive, linear functions on this algebra. The algebraic approach to quantum mechanics has proved successful not only in motivating the Hilbert-space description of quantum systems with a finite number of degrees of freedom, but also, and particularly, in the domain of quantum field theory and quantum statistical mechanics.[2]

The analysis of such an approach falls outside the scope of this volume for two reasons at least: first, it uses, as a starting point, a highly structured mathematical framework that contains, in a sense, the more elementary mathematical structures referred to in the preceding subsections; second, it is not a central tool for the understanding of the logic of nonrelativistic quantum mechanics, but rather is relevant in broadening the area of application of quantum-mechanical theories.

References

1. G. Birkhoff and J. von Neumann, *Ann. Math.* 37 (1936) 823.
2. G. G. Emch, *Algebraic Methods in Statistical Mechanics and Quantum Field Theory*, Wiley, New York, 1972.

PART II

Basic Structures in the Description of Quantum Systems

The aim of this Part is to discuss simple mathematical structures that occur in the description of quantum systems. Chapter 9 will be taken as a model. The approach particularly emphasized is the probabilistic one where states are probability measures on the propositions. The main mathematical tools to be used are the theory of orthomodular posets and of probability measures on them. The exposition is essentially self-contained; some technical results are collected and proved in Appendix C. There are papers that are not explicitly mentioned in the References but nevertheless have been used in some way; they are collected in the form of additional bibliographies to single chapters or to groups of related chapters.

CHAPTER 10

The Typical Mathematical Structure of Propositions: Orthomodular AC Lattices

10.1 Some Remarks and Basic Definitions

As anticipated in the Preface, the typical mathematical structure attached to the set of propositions of a physical system is an ordered structure. In the Hilbert-space formulation of quantum mechanics, where propositions are identified with projection operators in the Hilbert space $\mathcal{H}$, the emerging ordered structure is $\mathcal{P}(\mathcal{H})$; it has been introduced in Section 9.1, and it will be further examined in Section 10.3 of this chapter. We shall see that $\mathcal{P}(\mathcal{H})$ belongs to a rather specialized family of ordered structures: in technical words, the family of complete, separable, orthomodular, AC lattices, where AC is an abbreviation for "atomistic with the covering property". In following chapters we shall examine from different points of view, and without assuming from the outset a Hilbert-space framework, how the occurrence of this family of ordered structures is physically motivated; in Part III we shall see that once a structure of this kind is established for the set of propositions, the Hilbert-space formulation of quantum mechanics is almost in view.

In the present chapter we are only committed to outline what an orthomodular AC lattice is; to give some examples, among them $\mathcal{P}(\mathcal{H})$ itself; and to point out some continuity between the theory of von Neumann algebras and the theory of orthomodular lattices. As a general reference on lattice theory we refer to Maeda and Maeda's book.[1]

Now we proceed with some basic definitions.

ENCYCLOPEDIA OF MATHEMATICS and Its Applications, Gian-Carlo Rota (ed.). Vol. 15: E. G. Beltrametti and G. Cassinelli, The Logic of Quantum Mechanics

ISBN 0-201-13514-0

The word *poset* is an abbreviation for "partially ordered set" and thus denotes a set $\mathcal{L}$ with a reflexive antisymmetric transitive relation (that is, an order relation), for which the symbol $\leqslant$ is customary. With some abuse of notation we let $\mathcal{L}$ denote the poset itself, giving up the more correct notation $(\mathcal{L}, \leqslant)$.

If $\mathcal{L}$ is a poset and X is a collection of elements of $\mathcal{L}$, we say that c ($\in \mathcal{L}$) is the *meet* of the elements of X if: (i) $c \leqslant a$ for every $a \in X$, (ii) $b \leqslant a$ for every $a \in X$ implies $b \leqslant c$. Intuitively, c is the largest element of $\mathcal{L}$ smaller than each element of X. We say that c ($\in \mathcal{L}$) is the *join* of the elements of X if: (i) $c \geqslant a$ for every $a \in X$, (ii) $b \geqslant a$ for every $a \in X$ implies $b \geqslant c$. Intuitively c is the smallest element of $\mathcal{L}$ greater than each element of X. It is elementary to prove that the meet and the join of any subset X of $\mathcal{L}$, if they exist, are unique. They are denoted, respectively, by $\bigwedge_X a$ and by $\bigvee_X a$, or by $\bigwedge a_i$ and by $\bigvee a_i$ if X is written in the form $\{a_i\}$, or simply by $a \wedge b$ and $a \vee b$ if one has to do with just two elements. Alternative names for meet and join are "greatest lower bound" and "least upper bound", or "infimum" and "supremum".

It may happen that a poset has a least element and a greatest element: they will be denoted by 0 and 1 respectively; of course they may also be thought of as the meet and join of all the elements of the poset. Hereinafter, by the word poset we shall always mean a poset with 0 and 1, for we shall generally encounter posets of this sort.

The next relevant notion is that of orthogonal complement (orthocomplement). An *orthocomplementation* in a poset $\mathcal{L}$ is a mapping $a \mapsto a^\perp$ of $\mathcal{L}$ onto itself such that

(i) $a^{\perp\perp} = a$,
(ii) $a \leqslant b$ implies $b^\perp \leqslant a^\perp$,
(iii) $a \wedge a^\perp = 0$, $a \vee a^\perp = 1$.

A poset with an orthocomplementation is called *orthocomplemented*; $a^\perp$ is the *orthocomplement* of a.

The orthocomplementation makes the meet and join, when they exist, not independent. In fact, the so-called De Morgan laws hold:

$$\left(\bigwedge a_i \right)^\perp = \bigvee a_i^\perp, \qquad \left(\bigvee a_i \right)^\perp = \bigwedge a_i^\perp .$$

The proof of these equalities is a straightforward consequence of the fact that orthocomplementation reverses the order relation, according to (ii) above. It may be noticed in particular that the previous definition of orthocomplementation is redundant; in fact the two statements in (iii) follow one from the other, since

$$0^\perp = 1, \qquad 1^\perp = 0.$$

In an orthocomplemented poset the notion of disjointness, or orthogonality, can be introduced. Two elements a, b of $\mathfrak{L}$ are said to be *disjoint* (*orthogonal*), and we write $a \perp b$, when $a \leqslant b^\perp$ (or, equivalently, when $b \leqslant a^\perp$). Notice that if $b \perp a_i$, i running over some index set, then $b \perp (\wedge a_i)$ as well as $b \perp (\vee a_i)$.

A subset of $\mathfrak{L}$ formed by pairwise disjoint elements will be simply called a disjoint (or orthogonal) subset. $\mathfrak{L}$ is called *separable* if every disjoint subset of $\mathfrak{L}$ is at most countable.

Let us now approach the notion of orthomodularity. A poset $\mathfrak{L}$ is called σ-*orthocomplete* when it is orthocomplemented and there exists in $\mathfrak{L}$ the join of every countable orthogonal subset of $\mathfrak{L}$. If $a \leqslant b$, we write, as usual, $b-a$ in place of $b \wedge a^\perp$ (whose existence follows from $a \perp b^\perp$); if $a \perp b$, we write $a+b$ in place of $a \vee b$ (similarly, if $\{a_i\}$ is an orthogonal sequence, we write $+_i a_i$ instead of $\vee_i a_i$). Notice that $a \leqslant b$ implies $b-a \perp a$, and hence the existence of $a+(b-a)$. A poset $\mathfrak{L}$ is called *orthomodular* when it is σ-orthocomplete and, in addition, satisfies the condition

$$\text{if} \quad a \leqslant b \quad \text{then} \quad b = a + (b-a), \tag{10.1.1}$$

which is usually referred to as the orthomodular identity. This is the usual definition, but it contains some redundancy. In fact (10.1.1) could be replaced by: if $a \leqslant b$, then there exists $c \in \mathfrak{L}$ such that $b = c + a$. It is then an easy theorem[1] that $c = b - a$.

Now we come to the notion of lattice. A lattice is a poset in which meet and join always exist. To be more precise about the word "always", one speaks of a *complete lattice* $\mathfrak{L}$ when the meet and join of any subset of $\mathfrak{L}$ exist, of a σ-*lattice* when the existence of meet and join is guaranteed for countable subsets, and of a *lattice* when it is guaranteed for finite subsets. When we are not especially interested in these distinctions, we shall simply say "lattice", even for complete lattices.

Of course, the extension to lattices of the previously given notions of orthocomplementation and orthomodularity is straightforward. Let us remark, in particular, that $a \perp b$ in an orthocomplemented lattice implies $a \wedge b = 0$, a fact that motivates viewing $\perp$ as a disjointness relation.

Another definition will be useful: that of distributive triple. We say that the elements a, b, c of a lattice form a *distributive triple* if the equalities

$$a \wedge (b \vee c) = (a \wedge b) \vee (a \wedge c), \quad a \vee (b \wedge c) = (a \vee b) \wedge (a \vee c)$$

hold, together with the other four equalities obtained by cyclical permutation of a, b, c.

A lattice is called distributive if and only if every triple is distributive. A distributive lattice is obviously orthomodular. Orthocomplemented distributive lattices are generally called Boolean algebras.

A lattice is called modular if and only if $a \leqslant b$ implies that the triple (a, b, c) is distributive for every c in the lattice. Thus, distributive lattices are modular, but the converse is in general not the case.

It is immediate that the definition of orthomodularity given above can be read (for lattices) as follows: a lattice is orthomodular if and only if it is orthocomplemented and $a \leqslant b$ implies that $(a, b, a^{\perp})$ is a distributive triple. This makes clear that, within orthocomplemented lattices,

$$\text{distributivity} \Rightarrow \text{modularity} \Rightarrow \text{orthomodularity}.$$

Anticipating conclusions to be drawn in the sequel, we can say that distributivity is a natural framework for the description of classical mechanics, and orthomodularity for quantum mechanics, though modularity was suspected, for a time, to be a good candidate for the passage from the classical to the quantum case.

As has already been said, the main orthomodular lattices considered in the description of quantum systems are atomic and have the covering property. Thus we proceed with the pertinent definitions.

An element p of a lattice is called an *atom* if the only element it majorizes is 0: symbolically, if $0 \leqslant a \leqslant p$ implies either $a = 0$ or $a = p$. Atoms are reminiscent of points in the geometrical sense, so we shall often use "point" in place of "atom". A lattice is called *atomic* when every nonzero element majorizes at least one atom.

When an orthomodular lattice is atomic, then it is atomic in a strong way: in fact, every nonzero element is the joint of (all) the atoms it contains. The proof of this statement is straightforward when the lattice is complete: let b be the join of all the atoms contained in a ($\neq 0$), and suppose $a \neq b$ and hence $a > b$, so that $a - b \neq 0$; then the atomicity of the lattice ensures that there exists an atom in $a - b$—a contradiction, since this atom would be contained in a but not in b. When every element of a lattice is the join of the atoms it contains, the lattice is called *atomistic*. Thus we can say that if an orthomodular lattice is atomic, then it is also atomistic.

We come now to Birkhoff's covering property. In a lattice we say that a *covers* b if $a > b$ and if the relation $a \geqslant c \geqslant b$ implies either $c = a$ or $c = b$—in other words, if a majorizes b in a minimal way (no intermediate element between a and b). An atomic lattice $\mathcal{L}$ has the *covering property* if for every a in $\mathcal{L}$ and every atom p such that $a \wedge p = 0$, the element $a \vee p$ covers a. When $\mathcal{L}$ is orthomodular, and thus atomistic if atomic, the covering property asserts that the join of any element a with an atom not contained in a covers a.

There are other equivalent ways of expressing the covering property in atomic orthomodular lattices. We mention a couple of them.[1] $\mathcal{L}$ has the covering property if, for every $a \in \mathcal{L}$ and for every two atoms p and q not contained in a, $p \leqslant a \vee q$ implies $q \leqslant a \vee p$. We have also: $\mathcal{L}$ has the covering property if, for every $a \in \mathcal{L}$ and every atom p not contained in $a^{\perp}$, the element $a \wedge (p \vee a^{\perp})$, often called the Sasaki projection of p onto a, is an atom.

10.2 Examples

Due to the fact that, in orthocomplemented lattices, distributivity and modularity are particular cases of orthomodularity, any example of a Boolean algebra or of an orthocomplemented modular lattice is also an example of an orthomodular lattice. So, for instance, the set of all subsets of a set Ω is an orthomodular lattice: it is a Boolean algebra in which the ordering is set inclusion, the orthocomplement is set complementation relative to Ω, and the meet and join are set intersection and union respectively.

Now we consider more genuine examples of orthomodular posets and lattices. Consider first the set of all subspaces of a three-dimensional real Euclidean space V: the elements of this set are the origin 0, the lines through the origin (extending to infinity in both directions), the planes through the origin, and the whole of V. With the ordering relation defined by set inclusion we get a lattice, the origin *0* and the whole of V serving as least and greatest element. The meet of two (different) lines is the origin, while their join is the plane they span; the meet of a plane and a line not lying in that plane is the origin, while their join is V; and so on. This lattice becomes orthocomplemented if we adopt as orthocomplement of *0* the whole of V (and vice versa), and as orthocomplement of a line the plane orthogonal to that line (and vice versa). The orthomodularity condition

$$a \leqslant b \quad \Rightarrow \quad b = a + (b - a)$$

is also satisfied; if $a = 0$ or $b = V$ the check is trivial, and if a is a line in the plane b, then $b - a$ is a line in b perpendicular to a, so that the orthomodular identity states the obvious fact that a plane is spanned by two perpendicular lines lying in it. One can easily check that this lattice satisfies also the stronger modular law ($a \leqslant b \Rightarrow (a, b, c)$ a distributive triple). This lattice is obviously atomic, its atoms being the lines, and has the covering property.

The representation of posets and lattices by means of the so called Hasse diagrams, in which lines rise from smaller to greater elements, is a useful tool for displaying the structure of simple cases. Consider, for instance the diagrams in Figure 10.1. The first one is a poset but not a lattice (the join of a and b does not exist), and it is not orthocomplemented. The second is a lattice but not an orthomodular lattice (in fact, $a \leqslant b^{\perp}$ but $b^{\perp} \neq a \vee (b^{\perp} \wedge a^{\perp}) = a$). It is interesting to note that any orthocomplemented lattice contains the second as a sub-orthocomplemented lattice if and only if it is not orthomodular.

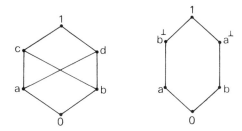

Figure 10.1.   A non-orthocomplemented poset and a non-orthomodular lattice.

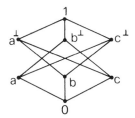

Figure 10.2.   A distributive orthocomplemented lattice.

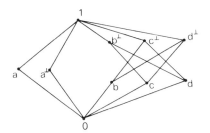

Figure 10.3.   A genuine orthomodular lattice.

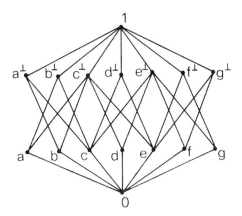

Figure 10.4.   Dilworth's lattice D_{16}.

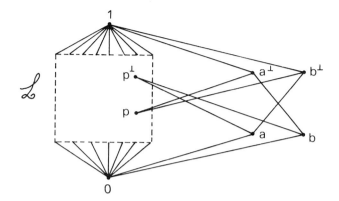

Figure 10.5. Pasting of orthomodular lattices.

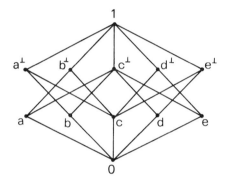

Figure 10.6. G_{12}.

Consider now the diagram in Figure 10.2. It is easily seen to be not only orthomodular but also distributive (hence also modular). It is atomic, with atoms a, b, c, and has the covering property, as is easily verified. To obtain an orthomodular but nonmodular lattice one needs to add one more element with its orthocomplement, as in Figure 10.3. This diagram corresponds to the smallest nonmodular orthomodular lattice. It is an atomic lattice, with atoms $a, a^{\perp}, b, c, d$, but it has not the covering property, as is immediately recognized by noticing that b is not covered by $b \vee a = 1$.

Another example of an orthomodular but nonmodular lattice is provided by the diagram in Figure 10.4. It is called Dilworth's lattice, after its discoverer, and often denoted D_{16} (it has 16 elements). It is obviously atomic, with atoms a, b, c, d, e, f, g, but again, it has not the covering property, a fact which follows from the remark that f is not covered by $f \vee a = 1$.

As the last two examples indicate, the invention of genuine finite orthomodular (i.e. nonmodular) lattices is not a trivial task. Recipes for producing a wide variety of them have been discovered by Greechie,[2] who has found methods of getting new orthomodular lattices by pasting together two or more orthomodular lattices. We do not enter into Greechie's technique, but only mention two special cases of his pasting or "amalgamation". The first is called the "horizontal sum" and was known earlier.[3] The orthomodular lattices to be pasted together are placed side by side; then the greatest elements are tied together, and the same is done for the least elements, while in between, the lattices are left untouched. The orthomodular lattice so constructed is modular if and only if, in each of the lattices to be pasted, the maximal chains are formed by just three elements, two of them being the extremal elements 0 and 1. The lattice of Figure 10.3 is a horizontal sum of two Boolean lattices. Another kind of pasting is illustrated by Figure 10.5: if $\mathfrak{L}$ is an orthomodular lattice, then the amalgamation of $\mathfrak{L}$ with the four elements $a, b, a^\perp, b^\perp$ is a new orthomodular lattice if and only if p is an atom of $\mathfrak{L}$. The lattice of Figure 10.6 is an explicit example of an amalgamation of this kind: it is denoted by G_{12}.

Greechie's examples are particularly good representatives of orthomodular lattices, for they seem to be free from any additional, inessential structure; S. Holland[4] calls them "wild", in contrast with the "nice" projection lattices (considered in the sequel), which in general are rich in structures besides orthomodularity. Thus Greechie's examples provide us with a selective filter for conjectures on orthomodular lattices; if a conjecture passes this filter, then it is likely to be a theorem of orthomodular lattice theory.

Let us add one more example of a finite orthomodular poset. Let S be a set formed by an even number of objects, say eight, and write $S = \{1, 2, 3, 4, 5, 6, 7, 8\}$. The set $\mathfrak{L}$ of all subsets of S formed by an even number of elements is a poset with respect to the ordering relation defined by set-theoretic inclusion. Clearly, it is not a lattice (for instance, there is no join of $\{1, 2\}$ and $\{2, 3\}$). After introducing in $\mathfrak{L}$ the natural orthocomplementation defined by set-theoretic complementation (relative to S), it is immediately verified that $\mathfrak{L}$ is orthomodular.

After these examples, we want to call attention to the remarkable fact that each example of an involutive ring with identity carries a concomitant example of an ordered structure. Indeed, we have that

$$\text{the projections of an involutive ring with identity} \atop \text{form an orthomodular poset.} \qquad (10.2.1)$$

The proof of this statement is a simple check recalling a few definitions. An *involutive ring* (*-ring for short) is a ring R together with a mapping $x \mapsto x^*$ of R onto itself such that

$$(x+y)^* = x^* + y^*, \qquad (xy)^* = y^* x^*, \qquad x^{**} = x.$$

A *projection* in an involutive ring R is an element e satisfying

$$ee=e^*=e;$$

in words, the projections are the idempotent self-adjoint elements. In the set of the projections of R the relation $ef=e$ is easily seen to be reflexive, antisymmetric, and transitive; it is thus an ordering relation:

$$e\leqslant f \quad \Leftrightarrow \quad ef=e,$$

(notice that $ef=e$ is equivalent to $fe=e$). Thus, the set of the projections of an involutive ring is a poset. This poset becomes orthocomplemented if the ring has the identity (denoted by 1); in fact, the mapping

$$e\mapsto e^{\perp}=1-e,$$

where the minus sign here refers to the ring operation, provides the orthocomplementation, as is immediately verified. Notice that the orthogonality relation $e\perp f$ now reads $e(1-f)=e$, that is, $ef=0$. We have, by simple check,

$$e\vee f=e+f \qquad \text{whenever} \quad e\perp f,$$

where the plus sign refers to the ring operation. To conclude the proof of the statement (10.2.1), we are left with the orthomodular identity

$$e\leqslant f \quad \Rightarrow \quad f=e\vee(f\wedge e^{\perp}),$$

which, inasmuch as $e\perp(f\wedge e^{\perp})=(f^{\perp}\vee e)^{\perp}$ and $e\perp f^{\perp}$, reads $ef=e \Rightarrow$ $f=e\vee(f^{\perp}+e)^{\perp}=e+1-(1-f+e)=f$, a tautology.

Notice that, should the ring be supposed commutative, the set of its projections would become a distributive lattice with meet and join given by $e\wedge f=ef$, $e\vee f=e+f-ef$, according to the classical result of Marshall Stone[5] on the analogy between formal logic and Boolean algebras.

We come now to an explicit example of a lattice arising as the projection lattice of an involutive ring. Consider the set of two-by-two matrices with complex entries, and look at it as a ring with respect to matrix addition and row-by-column multiplication. The usual hermitian-conjugate (adjoint) operation makes it an involutive ring. Thus, the statement (10.2.1) applies: its projections (i.e., the hermitian idempotent matrices) form an orthomodular poset. Explicitly, the nontrivial projections have the form (as easily verified)

$$\frac{1}{2}\begin{pmatrix} 1+r_3 & r_1-ir_2 \\ r_1+ir_2 & 1-r_3 \end{pmatrix},$$

where r_1, r_2, r_3 are real numbers such that $r_1^2+r_2^2+r_3^2=1$. We denote this matrix by $\sigma(r_1,r_2,r_3)$. Besides these projections there are the two trivial projections

$$0=\begin{pmatrix} 0 & 0 \\ 0 & 0 \end{pmatrix} \quad\text{and}\quad 1=\begin{pmatrix} 1 & 0 \\ 0 & 1 \end{pmatrix}.$$

Recalling the definition of ordering for the projections of an involutive ring, we see that, for every r_1, r_2, r_3,

$$0\leqslant\sigma(r_1,r_2,r_3)\leqslant 1,$$

while no ordering relation exists between two σ's. Clearly, the meet of any two σ's is 0, and the join in 1 (if they are different). Thus we have a lattice, not only a poset. We define orthocomplementation in the standard way (by means of ring operations):

$$\sigma(r_1,r_2,r_3)^{\perp}=1-\sigma(r_1,r_2,r_3)=\sigma(r_1,r_2,r_3).$$

The lattice structure so determined can be depicted as a continuous line between 0 and 1, with every point connected with 0 and 1, as in Figure 10.7. The orthomodularity, secured by (10.2.1), is immediate by direct inspection: it is also evident that this lattice is atomic (all the nontrivial elements are atoms) and has the covering property.

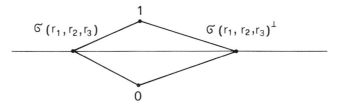

Figure 10.7. The polarization lattice of a spin-$\frac{1}{2}$ system.

The explicit example just considered is a projection lattice originating from a special kind of involutive ring, namely, from a von Neumann algebra. In Section 10.4 we shall return to the orthomodular lattices arising from von Neumann algebras.

By comparison with Section 4.2, the reader will recognize that the lattice just considered can also be viewed as the lattice of projectors on the Hilbert space $\mathbb{C}^2$, the Hilbert space of the spin-$\frac{1}{2}$ system; the σ's correspond to the polarizations, and the lattice of Figure 10.7 can be called the polarization lattice for spin $\frac{1}{2}$.

For a recent overview of orthomodular-lattice theory, see Reference 6.

10.3 The Lattice of Projectors on a Complex Hilbert Space

Projectors on a Hilbert space correspond one-to-one to closed subspaces, so that the title of this section could be expressed, in geometrical terms, as "the lattice of closed subspaces of a complex Hilbert space". The object under discussion has been already introduced in Section 9.1. We denote it by $\mathcal{P}(\mathcal{H})$, and we shall initially refer to its geometrical description: $\mathcal{H}$ is a complex, separable Hilbert space, and $\mathcal{P}(\mathcal{H})$ is the set of all closed (in the strong topology of $\mathcal{H}$) subspaces of $\mathcal{H}$. Let us examine its structure. We shall use a number of properties of Hilbert-space theory: as a general reference the reader may consult Reed and Simon's book.[7]

10.3.1 *Ordering*

$\mathcal{P}(\mathcal{H})$ is partially ordered by set-theoretic inclusion: if $\mathfrak{M}, \mathfrak{N}$ are closed subspaces, we write $\mathfrak{M} \leqslant \mathfrak{N}$ when $\mathfrak{M}$ is contained, as a set, in $\mathfrak{N}$.

10.3.2 *Lattice Structure*

The set-theoretic intersection of closed subspaces is a closed subspace, and, of course, it is precisely the meet with respect to the chosen ordering. Less obvious is the join. If $\mathfrak{M}$ and $\mathfrak{N}$ are closed subspaces, $\mathfrak{M} \cup \mathfrak{N}$ is not a subspace while their linear sum $\mathfrak{M} + \mathfrak{N}$ is not, in general, a closed subspace:* we must go to the closure (in the strong topology of $\mathcal{H}$) of $\mathfrak{M} + \mathfrak{N}$, which has indeed the properties of the join; thus

$$\mathfrak{M} \vee \mathfrak{N} = \overline{\mathfrak{M} + \mathfrak{N}}.$$

In some particular cases this formula can be simplified. For example, if $\mathfrak{M}$ or $\mathfrak{N}$ is finite-dimensional, then $\mathfrak{M} + \mathfrak{N}$ is closed and $\mathfrak{M} \vee \mathfrak{N} = \mathfrak{M} + \mathfrak{N}$; if $\mathfrak{M}$ and $\mathfrak{N}$ are orthogonal, in the sense that every vector of $\mathfrak{M}$ is orthogonal

---

*If $\mathcal{H}$ is infinite-dimension, $\mathfrak{M} + \mathfrak{N}$ need not be closed,[8] a fact connected with the nonmodularity of $\mathcal{P}(\mathcal{H})$, as we shall see later.

to every vector of $\mathfrak{N}$, the direct sum $\mathfrak{M} \oplus \mathfrak{N}$ is closed and $\mathfrak{M} \vee \mathfrak{N} = \mathfrak{M} \oplus \mathfrak{N}$.

The definition of meet and join can be generalized to an arbitrary family $\{\mathfrak{M}_i\}$ of closed subspaces: $\mathcal{P}(\mathcal{H})$ is thus a complete lattice with meet and join given by

$$\bigvee \mathfrak{M}_i = \bigcap \mathfrak{M}_i, \tag{10.3.1}$$

$$\bigvee \mathfrak{M}_i = \overline{\sum \mathfrak{M}_i}. \tag{10.3.2}$$

The whole space $\mathcal{H}$ and the set consisting of the vector 0 are, respectively, the greatest and the least elements of $\mathcal{P}(\mathcal{H})$.

10.3.3 *Orthocomplementation*

If $\mathfrak{M}$ is a closed subspace, define

$$\mathfrak{M}^{\perp} = \{\varphi \in \mathcal{H} : (\varphi, \psi) = 0 \text{ for every } \psi \in \mathfrak{M}\}$$

[where $(\cdot, \cdot)$ is the scalar product in $\mathcal{H}$]. It is an elementary property of the geometry of Hilbert spaces that for every closed subspace $\mathfrak{M}$, $\mathfrak{M}^{\perp}$ is also a closed subspace. It is easy to verify that the map $\mathfrak{M} \mapsto \mathfrak{M}^{\perp}$ is an ortho-complementation in the lattice $\mathcal{P}(\mathcal{H})$; in fact

(i) $\mathfrak{M} \wedge \mathfrak{M}^{\perp} = \mathfrak{M} \cap \mathfrak{M}^{\perp} = \{0\}$, $\mathfrak{M} \vee \mathfrak{M}^{\perp} = \mathfrak{M} \oplus \mathfrak{M}^{\perp} = \mathcal{H}$;
(ii) $\mathfrak{M} \leqslant \mathfrak{N} \Rightarrow \mathfrak{N}^{\perp} \leqslant \mathfrak{M}^{\perp}$;
(iii) $\mathfrak{M} = (\mathfrak{M}^{\perp})^{\perp}$.

Hence $\mathcal{P}(\mathcal{H})$ becomes an orthocomplemented lattice. Since two subspaces are lattice-theoretically disjoint in $\mathcal{P}(\mathcal{H})$ if and only if they are orthogonal in $\mathcal{H}$, we can say that if $\mathfrak{M}_1, \ldots, \mathfrak{M}_k$ is any finite disjoint sequence of $\mathcal{P}(\mathcal{H})$, then the join of the $\mathfrak{M}_i$'s is simply the algebraic direct sum:

$$\mathfrak{M}_1 \vee \cdots \vee \mathfrak{M}_k = \mathfrak{M}_1 \oplus \ldots \oplus \mathfrak{M}_k.$$

Should the sequence of the $\mathfrak{M}_i$'s be not finite, their algebraic direct sum would not be closed, and the join would become the Hilbert direct sum

$$\bigvee \mathfrak{M}_i - \overline{\oplus \mathfrak{M}_i}.$$

10.3.4 *Orthomodularity*

If $\mathfrak{M}, \mathfrak{N}$ are elements of $\mathcal{P}(\mathcal{H})$ and $\mathfrak{M} \leqslant \mathfrak{N}$, put

$$\mathfrak{L} = \{\varphi \in \mathfrak{N} : (\varphi, \psi) = 0 \text{ for every } \psi \in \mathfrak{M}\};$$

we have $\mathfrak{L} \in \mathcal{P}(\mathcal{H})$, $\mathfrak{L} \perp \mathfrak{M}$, and $\mathfrak{N} = \mathfrak{M} \oplus \mathfrak{L} = \mathfrak{M} \vee \mathfrak{L}$. This shows that $\mathcal{P}(\mathcal{H})$ is orthomodular.

10.3.5 *Atomicity*

The one-dimensional subspaces of $\mathcal{H}$ are obviously the atoms of $\mathcal{P}(\mathcal{H})$. It is also obvious that $\mathcal{P}(\mathcal{H})$ is atomic and atomistic.

10.3.6 *Covering Property*

Let $\mathcal{M} \in \mathcal{P}(\mathcal{H})$, and assume that there exists $\mathcal{N} \in \mathcal{P}(\mathcal{H})$ such that $\mathcal{M} < \mathcal{N} < \mathcal{M} + \hat{\varphi}$ (strict inequalities) for some $\varphi \notin \mathcal{M}$. Then there would be a nonzero vector ψ such that $\psi \in \mathcal{N}$ and $\psi \notin \mathcal{M}$, which would imply $\mathcal{N} \geqslant \mathcal{M} + \hat{\psi}$, and hence $\mathcal{M} + \hat{\psi} < \mathcal{M} + \hat{\varphi}$. This is absurd, and we conclude that $\mathcal{P}(\mathcal{H})$ has the covering property.

10.3.7 *Separability*

Let $\{\mathcal{M}_i\}$ be an orthogonal set of closed subspaces of $\mathcal{H}$. Choose in $\mathcal{M}_i$ a unit vector ψ_i; the set $\{\psi_i\}$ is an orthonormal set of vectors, which can be extended to an orthonormal basis of $\mathcal{H}$. If $\mathcal{H}$ is separable, then every orthonormal basis of $\mathcal{H}$ is at most countable, and then $\{\psi_i\}$ is a subset of a countable set; hence it is at most countable. Thus $\mathcal{P}(\mathcal{H})$ is separable.

10.3.8 *Lack of Modularity*

$\mathcal{P}(\mathcal{H})$ is orthomodular, but if the Hilbert space $\mathcal{H}$ is not finite-dimensional $\mathcal{P}(\mathcal{H})$ cannot be modular. This fact is a consequence of the existence of nonclosed algebraic sums of closed subspaces, as the following arguments shows. Choose two closed subspaces $\mathcal{M}$ and $\mathcal{N}$ whose algebraic sum $\mathcal{M} + \mathcal{N}$ is not closed; then $\mathcal{M} \vee \mathcal{N} \supset \mathcal{M} + \mathcal{N}$ and there exists in $\mathcal{M} \vee \mathcal{N}$ a nonzero vector φ not belonging to $\mathcal{M} + \mathcal{N}$. Consider the closed subspace $\mathcal{Q} = \mathcal{M} + \hat{\varphi}$; we want to prove that $\mathcal{N} \wedge \mathcal{Q} = \mathcal{N} \wedge \mathcal{M}$. Obviously, $\mathcal{N} \wedge \mathcal{Q} \geqslant \mathcal{N} \wedge \mathcal{M}$. If $\psi \in \mathcal{N} \wedge \mathcal{Q}$ then $\psi \in \mathcal{Q}$ and we can write $\psi = \chi + \lambda \varphi$, where $\chi \in \mathcal{M}$ and λ is a complex number. Hence $\lambda \varphi = \psi - \chi \in \mathcal{M} + \mathcal{N}$, but φ is not contained in $\mathcal{M} + \mathcal{N}$, so that $\lambda = 0$, and $\psi = \chi \in \mathcal{M}$. Since $\psi \in \mathcal{N} \wedge \mathcal{Q}$, we have also $\psi \in \mathcal{N}$ and hence $\psi \in \mathcal{N} \wedge \mathcal{M}$; hence $\mathcal{N} \wedge \mathcal{Q} \leqslant \mathcal{N} \wedge \mathcal{M}$. This shows that $\mathcal{N} \wedge \mathcal{Q} = \mathcal{N} \wedge \mathcal{M}$. If $\mathcal{P}(\mathcal{H})$ were modular, we would have, since $\mathcal{M} \leqslant \mathcal{Q}$,

$$(\mathcal{M} \vee \mathcal{N}) \wedge \mathcal{Q} = \mathcal{M} \vee (\mathcal{N} \wedge \mathcal{Q});$$

but $\mathcal{N} \wedge \mathcal{Q} = \mathcal{N} \wedge \mathcal{M}$, so that the right-hand side is simply $\mathcal{M}$, which does not contain φ, while $\varphi \in (\mathcal{M} \vee \mathcal{N}) \wedge \mathcal{Q}$. Thus $\mathcal{P}(\mathcal{H})$ cannot be modular. In fact, $\mathcal{P}(\mathcal{H})$ is modular if and only if $\mathcal{H}$ is finite-dimensional.

10.3.9 $\mathcal{P}(\mathcal{H})$ *as a Projection Lattice*

To a projection operator there corresponds the closed subspace that is its range; conversely, given a closed subspace $\mathcal{M}$, every element φ of $\mathcal{H}$ can be uniquely written as a sum $\varphi = \psi_1 + \psi_2$ with $\psi_1 \in \mathcal{M}$ and $\psi_2 \in \mathcal{M}^\perp$ (this is the content of the so-called projection theorem in Hilbert spaces), and the linear operator $P^{\mathfrak{M}}$ defined by $P^{\mathfrak{M}}\varphi=\psi_1$ is a projector whose range is $\mathfrak{M}$. Then the lattice structure of $\mathcal{P}(\mathcal{H})$ can be rephrased in terms of projection operators: we write $P^{\mathfrak{M}}\leqslant P^{\mathfrak{N}}$ when $\mathfrak{M}\leqslant\mathfrak{N}$, and we have $\bigvee P^{\mathfrak{M}_i}=P^{\vee\mathfrak{M}_i}$, $\bigwedge P^{\mathfrak{M}_i}=P^{\wedge\mathfrak{M}_i}$, $(P^{\mathfrak{M}})^{\perp}=P^{\mathfrak{M}^{\perp}}$.

It is interesting to note that some lattice operations in $\mathcal{P}(\mathcal{H})$ can be expressed in terms of the algebraic structure of the linear bounded operators on $\mathcal{H}$. The order relation can be equivalently characterized by "$P^{\mathfrak{M}}\leqslant P^{\mathfrak{N}}$ whenever $P^{\mathfrak{M}}P^{\mathfrak{N}}=P^{\mathfrak{M}}$". (Notice that if $P^{\mathfrak{M}}<P^{\mathfrak{N}}$, then $P^{\mathfrak{M}}P^{\mathfrak{N}}=P^{\mathfrak{N}}P^{\mathfrak{M}}$). The product of two projectors is a projector if and only if they commute in the sense of bounded operators on $\mathcal{H}$; in that case

$$P^{\mathfrak{M}}\wedge P^{\mathfrak{N}}=P^{\mathfrak{M}}P^{\mathfrak{N}} \quad\text{and}\quad P^{\mathfrak{M}}\vee P^{\mathfrak{N}}=P^{\mathfrak{M}}+P^{\mathfrak{N}}-P^{\mathfrak{M}}P^{\mathfrak{N}}.$$

$$(10.3.3)$$

Also the orthocomplementation and the disjointness relation can be characterized algebraically: $(P^{\mathfrak{M}})^{\perp}=I-P^{\mathfrak{M}}$ (I is the identity operator), and $P^{\mathfrak{M}}\perp P^{\mathfrak{N}}$ is equivalent to $P^{\mathfrak{M}}\leqslant I-P^{\mathfrak{N}}$, hence to $P^{\mathfrak{M}}=P^{\mathfrak{M}}(I-P^{\mathfrak{N}})$, and hence to $P^{\mathfrak{M}}P^{\mathfrak{N}}=0$, in which case $P^{\mathfrak{M}}\vee P^{\mathfrak{N}}=P^{\mathfrak{M}}+P^{\mathfrak{N}}$. This expression for the join can be extended, with some care, to any disjoint sequence $\{P^{\mathfrak{M}_i}\}$ of projectors: indeed, we can write $\bigvee P^{\mathfrak{M}_i}=\Sigma P^{\mathfrak{M}_i}$ provided the sum is interpreted, when infinitely many terms occur, in the sense of the strong operator topology, i.e.,

$$\left\| \sum_i P^{\mathfrak{M}_i} - \sum_{i=1}^{k} P^{\mathfrak{M}_i} \right\| \to 0 \quad \text{for } k\to\infty.$$

It can be shown[7] that the meet of two noncommuting projections is given by

$$P^{\mathfrak{M}}\wedge P^{\mathfrak{N}} = \underset{k\to\infty}{\text{s-lim}}\,(P^{\mathfrak{M}}P^{\mathfrak{N}})^k, \qquad (10.3.4)$$

where, as indicated, the limit on the right-hand side must be evaluated in the strong operator topology. When $P^{\mathfrak{M}}$ and $P^{\mathfrak{N}}$ commute, this expression reduces to the one given in (10.3.3). Notice also that, in terms of projection operators, the orthomodularity of $\mathcal{P}(\mathcal{H})$ reduces to the obvious assertion: if $P^{\mathfrak{M}}\leqslant P^{\mathfrak{N}}$, then $P^{\mathfrak{M}}$ and $P^{\mathfrak{N}}-P^{\mathfrak{M}}$ are orthogonal, and hence $P^{\mathfrak{M}}\vee(P^{\mathfrak{N}}-P^{\mathfrak{M}})=P^{\mathfrak{M}}+P^{\mathfrak{N}}-P^{\mathfrak{M}}=P^{\mathfrak{N}}$.

Let us remark that $\mathcal{P}(\mathcal{H})$ can even be viewed as a lattice of projections of an involutive ring, in the sense discussed in the last section. In fact the set $\mathbb{B}(\mathcal{H})$ of all bounded operators on $\mathcal{H}$ has the structure of an involutive ring, with involution given by the adjoint, and $\mathcal{P}(\mathcal{H})$ is precisely the set of self-adjoint idempotent elements of $\mathbb{B}(\mathcal{H})$. Of course $\mathbb{B}(\mathcal{H})$ is a highly specialized involutive ring: it is a von Neumann algebra.

10.3.10 *Remark*

The importance of $\mathcal{P}(\mathcal{H})$ is also outlined by the following result, due to
Mackey and Kakutani.[9] Let V be a separable, complex, infinite-dimensional
Banach space, and let $\mathcal{L}(V)$ denote the complete lattice of all closed
subspaces of V, ordered by set-theoretic inclusion. If $\mathcal{L}(V)$ is orthocomplemented, then (i) it is possible to define a scalar product in V making it an
Hilbert space, (ii) the topology of V as a Hilbert space is the original one,
and (iii) the assumed orthocomplementation coincides with the one induced
by the scalar product. In a few words, if the structure $\mathcal{L}(V)$, with V a
Banach space, is implemented by an orthocomplementation, then we end up
with the $\mathcal{P}(\mathcal{H})$ structure.

10.4 From von Neumann Algebras to Orthomodular Lattices

Recall that von Neumann algebra N is a *-closed subalgebra of $\mathbb{B}(\mathcal{H})$
containing the identity and such that it equals its double commutant:[†]

$$N=N''.$$

Of course, $\mathbb{B}(\mathcal{H})$ itself is a von Neumann algebra.

One of the most useful and typical features of a von Neumann algebra is
that the set of its projections (self-adjoint idempotent operators) is so rich as
to specify uniquely the algebra itself; more precisely, if $\mathcal{P}(N)$ stands for the
set of the projections of the von Neumann algebra N, then N is the smallest
von Neumann algebra containing $\mathcal{P}(N)$.

A von Neumann algebra can, of course, be viewed as an involutive ring
with identity, so that the statement (10.2.1) applies, leading to the conclusion that its projections form an orthomodular poset. Actually they form
not only a poset but even a lattice, and hence an orthomodular lattice.

To better appreciate the fact that von Neumann algebras have to be put
in the family tree of orthomodular lattices, one has to realize that projection
lattices of von Neumann algebras can be quite different from one another;
they provide a wide variety of lattices, the main common attribute being
precisely orthomodularity. The reader is referred to the famous Murray–
von Neumann classification[10] of "factors" (the von Neumann algebras
whose centers consist only of the multiples of the identity): the projection
lattices of the so-called type-I factors (the ones occurring in standard
quantum mechanics) are atomic and have the covering property, while the
projection lattices of type-II and of type-III factors do not contain atoms.

The theory of von Neumann algebras with their projection lattices, and
the theory of orthomodular lattices, can be viewed as extreme points of a

---

[†] We are using the same notation as in Section 3.2; *-closed means that if $A \in N$, then also
the adjoint A^* belongs to N.

historical sequence: the main step between them is the invention of von Neumann lattices, or, according to the older (perhaps better known) terminology, the continuous geometries.[11] Using an image of Samuel Holland,[4] von Neumann algebra theory is the mother theory, von Neumann lattice theory the first-born son, and orthomodular lattice theory the second son. Continuous geometries are modular lattices: this fits with the fact that the search for a generalization of distributive lattices, also demanded by the 1936 paper of Birkhoff and von Neumann with its indication of the distributive law as suspect in a logical account of quantum mechanics, naturally rested for a certain time on the replacement of the distributive law by the weaker modular law. Only later was it realized that modularity was too stringent for the needs of quantum theory and that a further relaxation—that is, to orthomodularity—was necessary. Nevertheless the theory of continuous geometries has been a decisive step toward the theory of orthomodular lattices. Contrary to the theory of projection lattices of von Neumann algebras, the theory of continuous geometries lies in a purely lattice-theoretic framework: in the former case the objects one is dealing with are operators on a Hilbert space, and a lot of definitions and properties are inherited from their operator algebra; in the latter case one is dealing with elements of abstract lattices. The theory of continuous geometries, with its completeness and its rich supply of results, has historically assumed the role of stimulating further lattice-theoretic generalizations and of pointing out, coherently with indications coming from physics, the next natural step: the theory of orthomodular lattices.

References

1. F. Maeda and S. Maeda, *Theory of Symmetric Lattices*, Springer, Berlin, Heidelberg, New York, 1970.
2. R. Greechie, *J. Combinatorial Theory* 4 (1968) 210.
3. M. D. MacLaren, *Trans. Amer. Math. Soc.* 114 (1965) 401.
4. S. S. Holland, in *Trends in Lattice Theory*, J.C. Abbott, ed., Van Nostrand–Reinhold, New York, 1970; reprinted in *The Logico-algebraic Approach to Quantum Mechanics*, Vol. 1, C. A. Hooker, ed., Reidel, Dordrecht, 1975.
5. M. Stone, *Trans. Amer. Math. Soc.* 40 (1936) 37.
6. G. Kalmbach, *Omolattices*, Academic Press, New York, in press.
7. M. Reed and B. Simon, *Methods of Modern Mathematical Physics*, Vol. I, Academic Press, New York, 1972.
8. P. R. Halmos, *Introduction to Hilbert Space and the Theory of Spectral Multiplicity*, Chelsea, New York, 1957.
9. G. W. Mackey and S. Kakutani, *Bull. Amer. Math. Soc.* 52 (1946) 727.
10. F. J. Murray and J. von Neumann, *Ann. of Math.* 37 (1936) 116.
11. J. von Neumann, *Continuous Geometry*, Princeton University Press, Princeton, N.J., 1960.

CHAPTER 11

Probability Measures on Orthomodular Posets and Lattices

11.1 Basic Definitions; Convexity Properties

The simplest class of physical quantities of a quantum system, that is, the class of two-valued physical quantities or propositions, is naturally endowed with the structure of an orthomodular poset; the candidates to represent the states of the physical system then become the probability measures on this poset. All this has been partially anticipated in the last two chapters and will be justified in future chapters. Hence our interest in the mathematical digression presented in this chapter. Though the set of propositions is often assumed to be not only a poset but also a lattice, we shall here adopt a more flexible framework in which the lattice structure will not be systematically assumed.

Given an orthomodular poset $\mathcal{L}$, we say that a real-valued function α on $\mathcal{L}$ is a *probability measure* on $\mathcal{L}$ if

(i) $0 \leqslant \alpha(a) \leqslant 1$ for all $a \in \mathcal{L}$, $\alpha(0)=0$, and $\alpha(1)=1$;
(ii) α is countably additive (or σ-additive), that is, for every countable orthogonal sequence $\langle a_i \rangle$ in $\mathcal{L}$ the series $\sum_i \alpha(a_i)$ converges and

$$\alpha(+_i a_i) = \sum_i \alpha(a_i).$$

Notice that the definition of probability measure on $\mathcal{L}$ does not make use of the orthomodular structure: it only uses the σ-orthocompleteness of $\mathcal{L}$.

A first property of probability measures is that they are monotone functions, namely, if $a \leqslant b$ in $\mathcal{L}$, then $\alpha(a) \leqslant \alpha(b)$: in fact we have $a \perp b^{\perp}$

ENCYCLOPEDIA OF MATHEMATICS and Its Applications, Gian-Carlo Rota (ed.).
Vol. 15: E. G. Beltrametti and G. Cassinelli, The Logic of Quantum Mechanics

ISBN 0-201-13514-0

and hence $1 \geqslant \alpha(a+b^{\perp})=\alpha(a)+1-\alpha(b)$. Of course (i) entails $\alpha(a^{\perp})=1-\alpha(a)$.

Let us now proceed to mention some convexity properties of probability measures (we refer to Appendix C for the discussion of continuity properties). Given two probability measures on $\mathcal{L}$, say α_1 and α_2, and given two positive numbers w_1, w_2 such that $w_1+w_2=1$, it is immediate to verify that the real-valued function $w_1\alpha_1+w_2\alpha_2$ defined by

$$(w_1\alpha_2+w_2\alpha_2)(a)=w_1\alpha_1(a)+w_2\alpha_2(a) \qquad \text{for all} \quad a \in \mathcal{L} \quad (11.1.1)$$

is a new probability measure on $\mathcal{L}$, called the *convex combination*, or mixture, of α_1 and α_2 with *weights* w_1, w_2. The notion of convex combination can be extended to any (countably) infinite sequence $\langle \alpha_i \rangle$ of probability measures: if $\langle w_i \rangle$ is an infinite sequence of positive numbers such that $\sum_i w_i=1$, the real-valued function $\sum_i w_i\alpha_i$ defined by

$$\left(\sum_i w_i\alpha_i\right)(a)=\sum_i w_i\alpha_i(a) \qquad \text{for all} \quad a \in \mathcal{L} \quad (11.1.2)$$

is a probability measure on $\mathcal{L}$.

Thus, the set of all probability measures on an orthomodular poset $\mathcal{L}$ is σ-convex, that is, it is closed not only under convex combination of a finite number of probability measures but also under convex combination of countably many probability measures. The probability measures that cannot be written as convex combinations are called *pure*. In the nomenclature of convex sets, the probability measures that are pure can be called the extremal points of $\mathcal{S}$.

Let us now introduce certain distinguished families of probability measures: they will be frequently used in the sequel, and their consideration can clarify the geometrical structure of the convex set of probability measures.* If $\mathcal{S}$ is a (nonempty) set of probability measures on $\mathcal{L}$, and X is any subset of $\mathcal{L}$, define the following subsets of $\mathcal{S}$:

$$\mathcal{S}_1(X)=\{\alpha \in \mathcal{S} : \alpha(a)=1 \text{ for all } a \in X\}, \qquad (11.1.3)$$

$$\mathcal{S}_0(X)=\{\alpha \in \mathcal{L} : \alpha(a)=0 \text{ for all } a \in X\}. \qquad (11.1.4)$$

In case X contains just one element, say a, we shall write $\mathcal{S}_1(a)$ and $\mathcal{S}_0(a)$ instead of the more consistent but pedantic $\mathcal{S}_1(\{a\})$ and $\mathcal{S}_0(\{a\})$. If $\mathcal{S}$ is σ-convex (which is guaranteed if it is the set of all the probability measures on $\mathcal{L}$), then it is easily verified that $\mathcal{S}_1(X)$ and $\mathcal{S}_0(X)$ are closed under

---

*In Chapter 19 the convex structure of states will be examined in some detail. The probability measures now considered form a particular realization of the convex structure there examined.

convex combination in the following strong sense:

THEOREM 11.1.1. *A sequence* $\alpha_1, \alpha_2, \ldots$ *of probability measures belongs to* $\mathcal{S}_1(X) [\mathcal{S}_0(X)]$ *if and only if there is a convex combination of them belonging to* $\mathcal{S}_1(X) [\mathcal{S}_0(X)]$.

Of course, when a convex combination of $\alpha_1, \alpha_2, \ldots$, belongs to $\mathcal{S}_1(X)$ [or $\mathcal{S}_0(X)$], then every convex combination of them belongs to $\mathcal{S}_1(X)$ [or $\mathcal{S}_0(X)$].

Let us also introduce a sort of dual of $\mathcal{S}_1(X)$ and $\mathcal{S}_0(X)$. For any subset S of $\mathcal{S}$, define

$$\mathcal{L}_1(S) = \{ a \in \mathcal{L}: \alpha(a) = 1 \text{ for all } \alpha \in S \}, \tag{11.1.5}$$

$$\mathcal{L}_0(S) = \{ a \in \mathcal{L}: \alpha(a) = 0 \text{ for all } \alpha \in S \}; \tag{11.1.6}$$

in case S contains just one element α, we shall simply write $\mathcal{L}_1(\alpha)$ and $\mathcal{L}_0(\alpha)$. Clearly, if $\Sigma_i w_i \alpha_i$ is a convex combination of the sequence $\alpha_1, \alpha_2, \ldots$, then

$$\mathcal{L}_1(\{\alpha_1, \alpha_2, \ldots\}) = \mathcal{L}_1\left(\sum_i w_i \alpha_i\right),$$

$$\mathcal{L}_0(\{\alpha_1, \alpha_2, \ldots\}) = \mathcal{L}_0\left(\sum_i w_i \alpha_i\right). \tag{11.1.7}$$

11.2 Examples; Gleason's Theorem

In the preceding section we have defined what a probability measure on an orthomodular lattice is, but we have not proved its existence: thus one may question whether the generalization of measure theory we are pursuing is meaningful. The answer to this question reveals two important facts: the first is that the theory is meaningful, in the sense that there exist important examples of posets and lattices that are genuinely orthomodular (i.e., they have the orthomodular property but not other properties of which orthomodularity is a weakening) and admit probability measures; the second is that the existence of probability measures on an orthomodular poset or lattice, otherwise arbitrary, is not a trivial fact, for the existence of probability measures is not ensured by orthomodularity alone. Examples of orthomodular posets and lattices that do not admit probability measures have been exhibited by Greechie.[1]

To Maczynski and Traczyk[2] is due a complete characterization of the orthomodular posets that admit "enough" probability measures. "Enough" here means that the orthomodular poset $\mathcal{L}$ admits a set $\mathcal{S}$ of probability measures rich enough to be order determining on $\mathcal{L}$, that is,

$$\alpha(a) \leqslant \alpha(b) \text{ for all } \alpha \in \mathcal{S} \quad \text{implies} \quad a \leqslant b.$$

These authors have shown that this property is met if and only if $\bar{\mathcal{L}}$ can be represented as a set $\bar{\mathcal{L}}$ of functions from $\mathcal{S}$ into $[0,1]$ such that: (i) $\bar{\mathcal{L}}$ contains the zero function; (ii) if $f \in \bar{\mathcal{L}}$ then $1 - f \in \bar{\mathcal{L}}$; (iii) if $f, g, h \in \bar{\mathcal{L}}$ and $f + g \leqslant 1$, $g + h \leqslant 1, f + h \leqslant 1$, then $f + g + h \in \bar{\mathcal{L}}$.

Now we come to three examples of orthomodular lattices admitting probability measures. The first refers to a finite lattice; the last two provide the standard models of classical statistical mechanics and of quantum mechanics.

Example G_{12}. Consider the lattice G_{12} shown in Figure 10.6. A probability measure α on G_{12} is entirely determined by the values $\alpha(b), \alpha(c), \alpha(d)$, for we have $\alpha(a) = 1 - \alpha(b) - \alpha(c)$ and $\alpha(e) = 1 - \alpha(c) - \alpha(d)$. As is easily checked, the convex set $\mathcal{S}$ of all probability measures on G_{12} is in one-to-one correspondence with the set of points belonging to the pyramid of Figure 11.1, whose vertices will be recognized as the pure states.[3]

Example C. Let Ω be a set (the phase space of a classical system), and Σ a Boolean σ-algebra of subset of Ω (ordered by set inclusion). Of course Σ is, in particular, an orthomodular complete lattice. The ordinary probability measures on Σ are, obviously, also probability measures in the sense of the previous section. Among them there is a special class: the probability measures concentrated at a point of Ω. Recall that a probability measure α

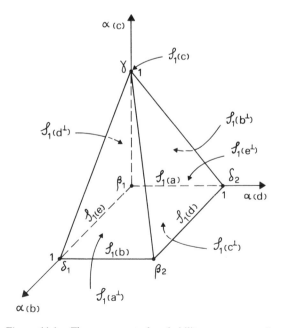

Figure 11.1. The convex set of probability measures on G_{12}.

on Σ is called *concentrated* at the point $\omega \in \Omega$ if $\alpha(a)=1$ whenever $\omega \in a$, while $\alpha(a)=0$ whenever $\omega \notin a$.

Example Q. Here $\mathcal{L}$ is $\mathcal{P}(\mathcal{H})$, the lattice examined in Section 10.3 viewed as a projection lattice. For any given density operator D the real-valued function α_D on $\mathcal{P}(\mathcal{H})$ defined by

$$\alpha_D(P)=\operatorname{tr}(DP), \qquad P \in \mathcal{P}(\mathcal{H}), \tag{11.2.1}$$

is a probability measure on $\mathcal{P}(\mathcal{H})$: the conditions $0 \leqslant \alpha_D(P) \leqslant 1$, $\alpha_D(0)=0$, $\alpha_D(1)=1$ are obvious, while the σ-additivity reads

$$\operatorname{tr}\left(D \sum_i P_i \right) = \sum_i \operatorname{tr}(DP_i)$$

for every orthogonal sequence $P_1, P_2, \ldots \in \mathcal{P}(\mathcal{H})$,

which is a known fact of Hilbert-space theory (see Appendix A). The natural question now arises whether every probability measure on $\mathcal{P}(\mathcal{H})$ must be of the form (11.2.1) for some density operator D. The affirmative answer was assumed by von Neumann, conjectured by Mackey,[4] and proved by Gleason[5] whose name is now attached to the following theorem:

THEOREM 11.2.1. *If the dimension of $\mathcal{H}$ is not less than 3, then every probability measure on $\mathcal{P}(\mathcal{H})$ arises from a density operator in $\mathcal{H}$, according to* (11.2.1).

The complete proof of this theorem is long and nontrivial; we refer, besides the original paper, to the book of Varadarajan.[6]

Thanks to Gleason's theorem, whenever $\dim \mathcal{H} \geqslant 3$ the set $\mathcal{S}(\mathcal{H})$ of all probability measures on $\mathcal{P}(\mathcal{H})$ is in one-to-one correspondence with the set of all density operators on $\mathcal{H}$. Both sets are convex [the convexity of $\mathcal{S}(\mathcal{H})$ comes from the preceding section, while the convexity of the set of density operators has been discussed in Section 2.1], and the question arises whether the correspondence preserves the convex structures. The answer is clearly affirmative if we take into account the formula (2.1.1) of Chapter 2. The probability measures on $\mathcal{P}(\mathcal{H})$ are thus in all respects the same thing as the density operators on $\mathcal{H}$ (provided $\dim \mathcal{H} \geqslant 3$).

11.3 Ordering Properties

Here we consider relations between poset-theoretic properties of $\mathcal{L}$ and properties of probability measures on it; to this purpose, $\mathcal{L}$ is only required to be a σ-orthocomplete poset. We recapitulate some nomenclature, only partially anticipated in previous sections.

A set $\mathcal{S}$ of probability measures on $\mathcal{L}$ is said to be

separating if $\alpha(a)=\alpha(b)$ for all $\alpha\in\mathcal{S}$ implies $a=b$;

sufficient if, for every $a\in\mathcal{L}$, $a\neq0$, there exists $\alpha\in\mathcal{S}$ such that $\alpha(a)=1$;

ordering if $\alpha(a)\leq\alpha(b)$ for all $\alpha\in\mathcal{S}$ implies $a\leq b$;

strongly ordering if $\mathcal{S}_1(a)\subseteq\mathcal{S}_1(b)$ implies $a\leq b$.

We list a number of facts:

(i) *If $\mathcal{S}$ is separating, then $\mathcal{L}$ is orthomodular.* Indeed, if $a\leq b$, then $a+(b-a)$ exists in $\mathcal{L}$ and for every $\alpha\in\mathcal{S}$, $\alpha(a+(b-a))=\alpha(a)+\alpha(b-a)=\alpha(a)+\alpha(b)-\alpha(a)=\alpha(b)$, so that $b=a+(b-a)$.

(ii) *If $\mathcal{S}$ is strongly ordering, then it is also ordering.* Indeed, $\alpha(a)\leq\alpha(b)$ for all $\alpha\in\mathcal{S}$ implies in particular $\mathcal{S}_1(a)\subseteq\mathcal{S}_1(b)$, and hence $a\leq b$ by the hypothesis that $\mathcal{S}$ is strongly ordering. The nonequivalence of the ordering and strongly ordering properties can be exhibited by counterexamples: Greechie[7] has given an orthomodular poset having a set of probability measures that is sufficient and ordering but not strongly ordering (this example can be generalized to an orthomodular lattice).

(iii) *If $\mathcal{S}$ is strongly ordering, then it is sufficient.* Indeed, if $a\neq0$, we must exclude $\mathcal{S}_1(a)=\varnothing$, for the hypothesis that $\mathcal{S}$ is strongly ordering would then imply $a=0$. Notice that, should $\mathcal{S}$ be supposed ordering, the sufficiency would not follow: Greechie and Miller[8] have given an example of an orthomodular poset admitting an ordering set of probability measures that is not sufficient.

(iv) *$\mathcal{S}$ is strongly ordering if and only if it is sufficient and for any nonorthogonal pair, $a,b\in\mathcal{L}$ there is an element α of $\mathcal{S}$ such that $\alpha(a)=1$ and $\alpha(b)\neq0$.* Indeed, suppose first $\mathcal{S}$ to be strongly ordering; then it is sufficient by (iii), and if $a\nleq b^\perp$ we must have $\mathcal{S}_1(a)\nsubseteq\mathcal{S}_1(b^\perp)=\mathcal{S}_0(b)$; hence there exists $\alpha\in\mathcal{S}_1(a)$ such that $\alpha(b)\neq0$. Conversely, let $\mathcal{S}_1(a)\subseteq\mathcal{S}_1(b)$ for some $a,b\in\mathcal{L}$, and suppose, if possible, that $a\nleq b$, so that $a\not\perp b^\perp$; then there exists $\alpha\in\mathcal{S}_1(a)$ such that $\alpha(b^\perp)\neq0$, i.e. $\alpha(b)\neq1$, which contradicts $\mathcal{S}_1(a)\subseteq\mathcal{S}_1(b)$.

(v) *If $\mathcal{S}$ is strongly ordering, then the following statements are equivalent*: (1) $a\leq b$, (2), $\alpha(a)\leq\alpha(b)$ *for every* $\alpha\in\mathcal{S}$, (3) $\mathcal{S}_1(a)\subseteq\mathcal{S}_1(b)$. Indeed, (1) implies (2) because probability measures are monotone; (2) obviously implies (3); (3) implies (1) because $\mathcal{S}$ is strongly ordering.

Let us add a few remarks on the examples introduced in the previous section.

Example G_{12} (continued). As is apparent from Figure 11.1, the set of all probability measures on G_{12} is strongly ordering.

Example C (continued). A set $\mathcal{S}$ of probability measures containing the ones concentrated at points of Ω is strongly ordering on Σ. In fact, suppose $\mathcal{S}_1(a) \subseteq \mathcal{S}_1(b)$; if we had $a \not\leq b$, then there would be a point $\omega \in \Omega$ contained in a but not in b, and the probability measure concentrated at ω would contradict the premise.

Example Q (continued). The set $\tilde{\mathcal{S}}(\mathcal{H})$ of probability measures on $\mathcal{P}(\mathcal{H})$ defined by density operators [if dim $\mathcal{H} \geq 3$, it coincides with the set $\mathcal{S}(\mathcal{H})$ of all probability measures, by Gleason's theorem] is strongly ordering on $\mathcal{P}(\mathcal{H})$. Suppose $\mathcal{S}_1(P_1) \subseteq \mathcal{S}_1(P_2)$ for some $P_1, P_2 \in \mathcal{P}(\mathcal{H})$. If $P_1 \not\leq P_2$, there would exist at least one unit vector ψ of $\mathcal{H}$ contained in the range of P_1 but not contained in the range of P_2, and the projector P^ψ would satisfy $\text{tr}(P^\psi P_1) = 1$, $\text{tr}(P^\psi P_2) \neq 1$, thus contradicting the premise $\mathcal{S}_1(P_1) \subseteq \mathcal{S}_1(P_2)$.

11.4 Regularity Properties

The measure theory we have outlined so far is rather poor. Not only is the very existence of probability measures for arbitrary orthomodular posets not guaranteed, but also some basic properties of measures on classical measure spaces do not hold here. Therefore, to attain a decent theory, one has to adopt some additional requirements. They will be examined in this section.

We first need a definition:

DEFINITION 11.4.1. Let α be a probability measure on the orthomodular poset $\mathcal{L}$. We say that $a \in \mathcal{L}$ is the *support* of α when $\alpha(b) = 0$ if and only if $a \perp b$.

Equivalently, we might say, with reference to the subsets of $\mathcal{L}$ defined by 11.1.7:

DEFINITION 11.4.2. a is the support of α if $\mathcal{L}_0(\alpha) = \{b \in \mathcal{L}: a \perp b\}$.

Or we might say:

DEFINITION 11.4.3. a is the support of α if $\mathcal{L}_1(\alpha) = \{b \in \mathcal{L}: a \leq b\}$.

The support of α, when it exists, is unique (see in particular Definition 11.4.3) and will be denoted by $s(\alpha)$.

Before going on let us return to

Example Q (continued). Consider again the set $\tilde{\mathcal{S}}(\mathcal{H})$ of probability measures on $\mathcal{P}(\mathcal{H})$ arising from density operators. If the density operator is a one-dimensional projector P^ψ, then it is precisely the support of the associated probability measure. If the density operator D is not one-dimensional, let $D=\sum_i w_i P^{\hat{\psi}_i}$ be the canonical decomposition of D into the orthonormal set $\langle \psi_i \rangle$; then (see Exercises 1–3) the support of D is $\sum_i P^{\hat{\psi}_i}$. Conversely, if P is any nonzero element of $\mathcal{P}(\mathcal{H})$, then it is the support of every density operator of the form $D=\sum_i w_i P^{\hat{\psi}_i}$ where $\langle \psi_i \rangle$ is an orthonormal basis in the range of P.

The existence of supports is related to the ordering properties of $\mathcal{S}$ and to the poset structure of $\mathcal{L}$. In fact we have the following two important results (see Appendix C):

THEOREM 11.4.1. *If every (nonzero) element of the orthomodular poset $\mathcal{L}$ is the support of some element of a set $\mathcal{S}$ of probability measures on $\mathcal{L}$, then $\mathcal{S}$ is strongly ordering on $\mathcal{L}$.*

THEOREM 11.4.2. *Let $\mathcal{S}$ be a convex set of probability measures on an orthomodular poset $\mathcal{L}$. If every element of $\mathcal{S}$ has support in $\mathcal{L}$, and if every (nonzero) element of $\mathcal{L}$ is the support of some element of $\mathcal{S}$, then $\mathcal{L}$ is a lattice; if $\mathcal{S}$ is σ-convex, then $\mathcal{L}$ is σ-complete (and if in addition $\mathcal{L}$ is separable, then it is complete).*

We want to stress that these results by no means have a converse: there exist examples of orthomodular lattices with a strongly ordering convex set of probability measures such that not every element of $\mathcal{S}$ has support in $\mathcal{L}$. This is the case with

Example G_{12} (continued). Take for instance the probability measure β_2 of Figure 11.1; we have $\mathcal{L}_1(\beta_2) = \{b, d, c^\perp, a^\perp, e^\perp\}$, and this set has no minimal element (see Figure 10.6), so that β_2 has no support (refer to Definition 11.4.3). Similarly, β_1, δ_1, and δ_2 have no support, while c is the support of γ. Notice however that each nonzero element of G_{12} is the support of some probability measure: for instance b is support of the probability measure represented by points on the edge $\delta_1\beta_2$ (vertices excluded), and $b^\perp$ is support of the ones on the face $\beta_1\delta_2\gamma$ (vertices and edges excluded).

Now we come to an important regularity condition for probability measures: let us pick it up from our classical and quantum examples.

Example C (continued). Given $a, b \in \Sigma$ $(a, b \neq 0)$, suppose $\alpha(a) = \alpha(b) = 1$ for some probability measure α. From the distributivity law we have $a = (a \wedge b) + (a \wedge b^\perp)$; hence $\alpha(a) = \alpha(a \wedge b) + \alpha(a \wedge b^\perp) = \alpha(a \wedge b)$, since by hypothesis $\alpha(b^\perp) = 0$, so that $\alpha(a \wedge b^\perp) = 0$. Summing up, $\alpha(a) = \alpha(b) = 1$ entails $\alpha(a \wedge b) = 1$.

Example Q (continued). Given $P_1, P_2 \in \mathcal{P}(\mathcal{H})$ $(P_1, P_2 \neq 0)$, suppose $\text{tr}(DP_1) = \text{tr}(DP_2) = 1$ for some density operator D. This means $DP_1 = DP_2$

$= D$ (see Exercise 2). Now consider the product $D(P_1 \wedge P_2) = D$ ($\text{s-lim}_{n \to \infty}(P_1 P_2)^n$), and use the continuity of multiplication in the strong operator topology, thus getting $D(P_1 \wedge P_2) = \text{s-lim}_{n \to \infty} D(P_1 P_2)^n = D$. Summing up, $\text{tr}(DP_1) = \text{tr}(DP_2) = 1$ entails $\text{tr}(D(P_1 \wedge P_2)) = 1$.

Thus we are led to consider, for a general orthomodular lattice $\mathfrak{L}$ endowed with a set $\mathfrak{S}$ of probability measures, the following regularity condition:

$$\alpha(a) = \alpha(b) = 1 \quad \text{implies} \quad \alpha(a \wedge b) = 1, \tag{11.4.1}$$

or the equivalent one

$$\alpha(a) = \alpha(b) = 0 \quad \text{implies} \quad \alpha(a \vee b) = 0. \tag{11.4.1'}$$

The inobviousness of this condition is outlined by known counterexamples. Rüttimann[9] has proved the following nice result: if $\mathfrak{L}$ is a finite orthomodular lattice, then it is distributive if and only if the set $\mathfrak{S}$ of all its probability measures is sufficient and satisfies (11.4.1). This result points to a rich collection of finite counterexamples. A counterexample referring to an infinite lattice will be encountered in Chapter 25: it will be based on the projection lattice of a two-dimensional Hilbert space (for which Gleason's theorem does not apply).

It is sometimes useful to consider a condition like (11.4.1) just for a poset, rather than a lattice: then we rewrite it in the form

$$\begin{gathered} \text{if } \alpha(a) = \alpha(b) = 1, \\ \text{then there exists } c \in \mathfrak{L} \text{ such that } c \leqslant a, b \text{ and } \alpha(c) = 1, \end{gathered} \tag{11.4.2}$$

which does not explicitly require the existence of meets in $\mathfrak{L}$. Of course, an analogous transcription can be made for (11.4.1').

This condition is related to ordering properties of $\mathfrak{S}$ and to the existence of supports. We summarize these relations in the following statements (further details are in Appendix C).

THEOREM 11.4.3. *If $\mathfrak{L}$ is a separable orthomodular poset and $\mathfrak{S}$ is a σ-convex set of probability measures on $\mathfrak{L}$, then $\mathfrak{S}$ is sufficient and (11.4.2) holds true if and only if every α in $\mathfrak{S}$ has support in $\mathfrak{L}$ and every nonzero element of $\mathfrak{L}$ is support of some $\alpha \in \mathfrak{S}$.*

THEOREM 11.4.4. *If $\mathfrak{L}$ is an orthomodular lattice and $\mathfrak{S}$ is a sufficient set of probability measures on $\mathfrak{L}$ that satisfies (11.4.2) [equivalently, (11.4.1)], then $\tilde{\mathfrak{S}}$ is strongly ordering on $\mathfrak{L}$.*

11.5 The Space of Probability Measures

In a set $\mathcal{S}$ of probability measures on an orthomodular poset $\mathcal{L}$ one can recognize much of the ordered structure of $\mathcal{L}$. Indeed, we shall see in this section that under certain hypotheses there is a distinguished family of subsets of $\mathcal{S}$ that mirrors the structure of $\mathcal{L}$.

With reference to our previous notation (11.1.3) and (11.1.5), consider the following definition:

DEFINITION 11.5.1. $\alpha \in \mathcal{S}$ is called a *superposition* of the elements of the subset S of $\mathcal{S}$ when $\mathcal{L}_1(S)$ is nonempty and $\alpha \in \mathcal{S}_1(\mathcal{L}_1(S))$.

Notice that this notion of superposition includes that of convex combination; of course it is interesting only insofar as there are superpositions that are not convex combinations, namely superpositions that are pure. The use of the word "superposition", the same word which, referring to states, is familiar in quantum mechanics, is not accidental: as soon as we recognize probability measures as representatives of states, we shall see that Definition 11.5.1 matches the quantum-mechanical notion of superposition. This will be discussed in next Example Q and in Section 14.8.

Let us introduce the abbreviation

$$\bar{S} = \mathcal{S}_1(\mathcal{L}_1(\dot{S})),$$

and observe that for every $S \subseteq \mathcal{S}$,

$$S \subseteq \bar{S}, \tag{11.5.1}$$

and that for $S, T \subseteq \mathcal{S}$,

$$S \subseteq \bar{T} \quad \text{implies} \quad \bar{S} \subseteq \bar{T}. \tag{11.5.2}$$

These properties qualify the function $S \mapsto \bar{S}$ as a "closure relation" in $\mathcal{S}$,[10] thus motivating for $\bar{S}$ the name of closure of S. Whenever $S = \bar{S}$, that is, whenever S contains all superpositions of its elements, we shall say that S is closed. Closed subsets of $\mathcal{S}$ will be denoted by $M, N, \ldots$, and the set they form by $\mathcal{K}$. Now we have the following facts: $\mathcal{K}$ is a complete lattice with respect to the ordering defined by set inclusion; meet and join are given by

$$M \wedge N = M \cap N, \quad M \vee N = \overline{M \cup N}, \quad M, N \in \mathcal{K}; \tag{11.5.3}$$

and the empty set and the whole of $\mathcal{S}$ are the least and the greatest elements of $\mathcal{K}$. This, as we shall see in Section 17.2, simply follows from the fact that the function $S \mapsto \bar{S}$ is a closure relation and holds true irrespective of the particular form of this function.

We come now to the relation between $\mathcal{L}$ and $\mathcal{K}$. After noticing that for every $a \in \mathcal{L}$ the set $\mathcal{S}_1(a)$ is certainly closed, we focus our attention on the mapping

$$a \mapsto \mathcal{S}_1(a), \tag{11.5.4}$$

which maps $\mathcal{L}$ into $\mathcal{K}$, and which is order-preserving, since probability measures are monotone. As shown in Appendix C, we have the following results: if $\mathcal{L}$ is a complete lattice, $\mathcal{S}$ is a sufficient σ-convex set of probability measures, and every element of $\mathcal{S}$ has support in $\mathcal{L}$, then the mapping (11.5.4) is a bijection between $\mathcal{L}$ and $\mathcal{K}$, and it preserves orders with its inverse. Moreover we can introduce in $\mathcal{S}$ an orthogonality relation by saying that α is orthogonal to β and writing $\alpha \perp \beta$ when there exists $a \in \mathcal{L}$ such that $\alpha(a) = 0$ and $\beta(a) = 1$. For every $M \in \mathcal{K}$ we define

$$M^{\perp} = \{ \alpha \in \mathcal{S} : \alpha \perp \beta \text{ for all } \beta \in M \}. \tag{11.5.5}$$

It is then possible to show that $M \mapsto M^{\perp}$ is an orthocomplementation in $\mathcal{K}$ preserved by the mapping (11.5.4) and its inverse, in the sense that $a^{\perp} \mapsto \mathcal{S}_1(a^{\perp}) = (\mathcal{S}_1(a))^{\perp}$.

It is worth remarking that under the hypotheses stated above, the notion of superposition (see Definition 11.5.1) takes the following form: α is superposition of the elements of S when its support $s(\alpha)$ is majorized by the join of the supports of the elements of S; formally,

$$s(\alpha) \leqslant \bigvee (s(\beta) : \beta \in S).$$

The inverse of the mapping (11.5.4) then takes the explicit form (we are still referring to Appendix C): given $M \in \mathcal{K}$, its image in $\mathcal{L}$ is $\bigvee (s(\beta) : \beta \in M)$.

Let us end this section by outlining how the bijection between $\mathcal{L}$ and $\mathcal{K}$ works in the previous examples.

Example Q (continued). For probability measures coming from density operators the notion of superposition appears as follows. Given a set S of density operators, consider all eigenvectors of all its elements and the closed subspace of $\mathcal{K}$ they span. A density operator D is a superposition of the elements of S when its eigenvectors belong to the foregoing subspace. In view of this characterization, the function (11.5.4) now maps $P \in \mathcal{P}(\mathcal{K})$ into the set of all density operators whose eigenvalues lie in the range of P. The bijection between $\mathcal{L}$ and $\mathcal{K}$ simply reduces to the fact that $\mathcal{P}(\mathcal{K})$ can be viewed either as a projection lattice or as a lattice of the closed subspaces of $\mathcal{K}$.

Example G_{12} (continued). As already noticed, not every probability measure has support in G_{12}, so we can expect that the isomorphism between

$\mathcal{L}$ and $\mathcal{K}$ will not work in this example. This is indeed the case, as is seen by noticing that the edges of the pyramid of Figure 11.1 that have a vertex in γ are closed under superposition but do not arise from elements of G_{12} via the mapping (11.5.4).

11.6 Pure Probability Measures and Atomicity

There is a deep link between the existence of atoms of $\mathcal{L}$ and the existence of pure probability measures, namely, elements of $\mathcal{S}$ that cannot be written as convex combinations of other (different) elements of $\mathcal{S}$. To comment on this fact is the purpose of this section. It is perhaps useful to see at once what the connection looks like in our familiar examples.

Example C (continued). Suppose for simplicity that Σ coincides with the set of all subsets of Ω. The probability measures concentrated at the points of Ω are pure, and conversely, every pure probability measure on Σ is concentrated at a point of Ω. The proof of this we leave as an exercise. On the other hand the subsets formed by just one point of Ω are the atoms of Σ, so that there is indeed a one-to-one correspondence between the pure probability measures and the atoms of Σ. Notice also that, restricting ourselves to pure probability measures, no superposition can exist, so that every subset of pure probability measures is closed and the lattice $\mathcal{K}$ of last section is trivially isomorphic to Σ.

Example $\tilde{Q}$ (continued). We have already noticed in Section 2.1 that an element of $\tilde{\mathcal{S}}(\mathcal{K})$ is pure if and only if it is associated with a projector onto a one-dimensional subspace of $\mathcal{K}$: explicitly, a pure probability measure has the form $P \mapsto (\psi, P\psi)$, where ψ is a unit vector. But the one-dimensional projectors obviously represent the atoms of $\mathcal{P}(\mathcal{K})$, so that also in this example there is a one-to-one correspondence between pure probability measures and atoms. This correspondence is the restriction to the pure probability measures of the mapping from $\tilde{\mathcal{S}}(\mathcal{K})$ onto $\mathcal{P}(\mathcal{K})$ provided by the support.

Example G_{12} (continued). The lattice G_{12} is atomic and has five atoms: a, b, c, d, e (see Figure 10.6). It is easily checked that the set of probability measures represented by Figure 11.1 has five pure elements: the vertices $\beta_1, \beta_2, \delta_1, \delta_2, \gamma$ of the pyramid. Of course we can invent a one-to-one correspondence between the atoms and the pure probability measures; however, this correspondence cannot be given by the support function, for we know (see Section 11.4) that $\beta_1, \beta_2, \delta_1, \delta_2$ have no support in G_{12}.

The question suggested by the foregoing examples is thus the following: under which conditions does the support function provide a one-to-one correspondence between the atoms of $\mathcal{L}$ and the pure elements of $\mathcal{S}$? Let us stress that we are not looking for a generic one-to-one correspondence: we shall see in Section 14.8 that the correspondence of physical interest is indeed the one given by the support.

The answer to this question goes as follows (we refer to Appendix C):

(i) Let $\mathcal{L}$ be an orthomodular poset and $\mathcal{S}$ a σ-convex set of probability measures containing a nonempty set $\mathcal{S}^P$ of pure probability measures such that for every $\alpha \in \mathcal{S}^P$ the set $\mathcal{L}_1(\alpha)$ is nonempty. Then the statement

$$\mathcal{L}_1(\alpha) \subseteq \mathcal{L}_1(\beta) \text{ implies } \alpha = \beta, \qquad \alpha \in \mathcal{S}^P, \quad \beta \in \mathcal{S} \qquad (11.6.1)$$

is equivalent to the statement

$$\begin{array}{l} \text{if } \alpha \in \mathcal{S}^P, \\ \text{the singleton } \{\alpha\} \text{ is closed under superposition.} \end{array} \qquad (11.6.2)$$

(ii) Let $\mathcal{L}$ be an orthomodular poset and $\mathcal{S}$ a σ-convex set of probability measures such that every element of $\mathcal{S}$ has support in $\mathcal{L}$ and every nonzero element of $\mathcal{L}$ is the support of some element of $\mathcal{S}$. If $\mathcal{S}$ contains a sufficient subset $\mathcal{S}^P$ of pure elements, then each of the statements (11.6.1) and (11.6.2) is equivalent to the following statement:

(A) $\mathcal{L}$ is atomic, and the restriction to $\mathcal{S}^P$ of the support function defined on $\mathcal{S}$ determines a one-to-one correspondence between the set of atoms of $\mathcal{L}$ and $\mathcal{S}^P$.

11.7 Covering Property and Probability Measures

The covering property has been introduced in Section 10.1 as a lattice-theoretic property. Let us outline a property of probability measures on $\mathcal{L}$ that corresponds to the covering property of $\mathcal{L}$. We refer to the following characterization of the covering property:[11] if p and q are atoms of $\mathcal{L}$ not contained in $a \in \mathcal{L}$, then $p \leqslant a \vee q$ implies $q \leqslant a \vee p$.

Let us take into account only a set $\mathcal{S}^P$ of pure probability measures on $\mathcal{L}$ and assume all the premises that lead to the isomorphism between $\mathcal{L}$ and the lattice of closed (under superposition) subsets of $\mathcal{S}^P$ (see Section 11.5) as well as all the premises that ensure the one-to-one correspondence between the atoms of $\mathcal{L}$ and $\mathcal{S}^P$ given by the support function (see Section 11.6). Then the covering property can be rephrased as follows:

(CP) Let α and β be pure probability measures, not contained into some closed (under superposition) subset M of $\mathcal{S}^P$. If α is superposition of β and of the elements of M, then β is superposition of α and of the elements of M.

We shall see in Section 14.8 that, once the probability measures are interpreted as states of a quantum system, such a characterization of the covering property will translate a physically quite natural statement.

Exercises

1. Let $D=\sum_i w_i P^{\hat{\psi}_i}$ be a decomposition of D into an orthonormal set. Though in case some w_i is repeated there are other decompositions into orthonormal bases, show that $\sum_i P^{\hat{\psi}_i}$ does not depend upon the choice of the basis.

[*Hint.* if $w_1 = w_2$, the vectors ψ_1, ψ_2 are not uniquely determined, but if we take any other pair, say φ_1, φ_2, of orthonormal vectors in the subspace spanned by $\{\psi_1, \psi_2\}$, we have $P^{\hat{\psi}_1} + P^{\hat{\psi}_2} = P^{\hat{\varphi}_1} + P^{\hat{\varphi}_2}$.]

2. Prove that, for every $P \in \mathcal{P}(\mathcal{H})$,

    (a) $\operatorname{tr}(DP) = 0$ if and only if $DP = 0$,
    (b) $\operatorname{tr}(DP) = 1$ if and only if $DP = D$.
[*Hint.* $\operatorname{tr}(DP) = \operatorname{tr}((\sqrt{DP})^*(\sqrt{DP}))$ and $(\sqrt{DP})^*(\sqrt{DP}) \geqslant 0$; hence $\operatorname{tr}(DP) = 0$ implies $\sqrt{DP} = 0$; hence (a). (b) follows from (a) by observing that $\operatorname{tr}(DP) = 1$ means $\operatorname{tr}(D(I - P)) = 0$.]

3. Show that $(\sum_i P^{\hat{\psi}_i})P = 0$ if and only if $P^{\hat{\psi}_i}P = 0$ for all i.
[*Hint.* use the continuity of multiplication in the strong operator topology.]

References

1. R. J. Greechie, *J. Combinatorial Theory* 10A (1971) 119.
2. M. J. Maczynski and T. Traczyk, *Bull. Acad. Polonaise des Sciences* (Série sc. math., astr. et phys.) 21 (1973) 3.
3. G. T. Rüttimann, *Noncommutative Measure Theory*, Habilitationsschrift, Universität Bern, 1979.
4. G. W. Mackey, *Mathematical Foundations of Quantum Mechanics*, Princeton University Press, Princeton, N.J., 1955.
5. A. M. Gleason, *J. Math. Mech.* 6 (1957) 885.
6. V. S. Varadarajan, *Geometry of Quantum Theory*, Vol. I, Van Nostrand, Princeton, N.J., 1968.
7. R. J. Greechie, *Caribbean J. Math.* 1 (1969) 15.
8. R. J. Greechie and F. R. Miller, Kansas State University Technical Report No. 14, 1970.
9. G. T. Rüttimann, *J. Math. Phys.* 18 (1977) 189.
10. G. Birkhoff, *Lattice Theory*, 2nd ed., Amer. Math. Soc. Colloq. Publ., Vol. 25, Amer. Math. Soc., Providence, R.I., 1948.
11. E. A. Schreiner, *Pacific J. Math.* 19 (1966) 519.

CHAPTER 12

Characterization of Commutativity

12.1 Commutativity in Orthomodular Posets

The occurrence of noncommutative algebraic structure marks a branching point between classical and quantum mechanics. Thus it is worthwhile to follow the notion of commutativity into the foundations of quantum theory.

The familiar notion of commutativity makes reference to the existence of a binary operation (multiplication). The usual formulation of quantum mechanics uses commutativity in the context of the algebra of operators on a Hilbert space: the statement that two operators commute is a statement of the invariance of their product under the exchange of the factors. However, there is a place for the notion of commutativity even in absence of binary operation; the purpose of this section is precisely to review how that notion can be transplanted within the mathematical structure of orthomodular posets.

Let $\mathcal{L}$ be an orthomodular poset, and let $a, b \in \mathcal{L}$. We say that a and b commute, and write $(a, b)C$, if there are pairwise orthogonal elements a_1, b_1, c such that $a = a_1 + c$ and $b = b_1 + c$. This relation is symmetric: $(a, b)C$ entails $(b, a)C$.

As examples of commuting elements we can mention: (i) two orthogonal elements of $\mathcal{L}$, in particular any element and its orthocomplement; (ii) two ordered elements—in fact, if $a \leqslant b$, we can use the orthomodularity to deduce that $(a, b)C$. These examples are expected for every acceptable definition of commutativity. Nonetheless, at first sight the adopted definition looks nothing like any intuitive idea of commutativity; thus, before examining its properties, it seems useful to check whether the restriction of

ENCYCLOPEDIA OF MATHEMATICS and Its Applications, Gian-Carlo Rota (ed.).
Vol. 15: E. G. Beltrametti and G. Cassinelli, The Logic of Quantum Mechanics

ISBN 0-201-13514-0

the poset $\mathcal{L}$ to some familiar situation causes this abstract notion of commutativity to match more familiar ideas. For this purpose, choose for $\mathcal{L}$ the poset of projections of an involutive ring. We have seen in Section 10.2 that the projections of an involutive ring form precisely an orthomodular poset, with ordering relation given by $e \leqslant f$ whenever $ef = e$. In this poset we have at our disposal the "natural" notion of commutativity, the one expressed in terms of ring multiplication: two projections e, f commute if $ef = fe$. Now the question is: what is the relation between this "natural" notion of commutativity (ring commutativity) and the one making reference only to the poset structure? The answer is that they are completely equivalent, as the reader can easily verify (Exercise 1). Recall that the von Neumann algebra $\mathbb{B}(\mathcal{H})$ of all bounded operators on the Hilbert space $\mathcal{H}$ is an example of involutive ring, and its projection lattice is our familiar $\mathcal{P}(\mathcal{H})$; hence in $\mathcal{P}(\mathcal{H})$ the abstract notion of commutativity, expressed in terms of the poset structure, is equivalent to the one expressed in terms of operator multiplication.

We proceed now to enumerate some preliminary properties (see Exercises 2–4).

THEOREM 12.1.1. *If $a, b \in \mathcal{L}$ and $(a, b)C$, then the join and meet of any two among $a, b, a^{\perp}, b^{\perp}$ exist in $\mathcal{L}$.*

THEOREM 12.1.2. *If $(a, b)C$, the pairwise orthogonal elements a_1, b_1, c that make $a = a_1 + c, b = b_1 + c$ are uniquely determined as $a_1 = a \wedge b^{\perp}, b_1 = b \wedge a^{\perp}, c = a \wedge b$.*

Thus we can rephrase the definition of commutativity by saying that (see Exercise 5)

THEOREM 12.1.3. *$(a, b)C$ if and only if $(a \wedge b)$ and $(a \wedge b^{\perp})$ exist and $a = (a \wedge b^{\perp}) + (a \wedge b)$.*

Another relevant fact, implicit in what is said above, is that

THEOREM 12.1.4. *If $(a, b)C$, then any two of $a, b, a^{\perp}, b^{\perp}$ commute.*

12.2 Commutativity and Boolean Algebras

We need first some nomenclature. We say that $\mathcal{B}$ is a *Boolean subalgebra* of the orthomodular poset $\mathcal{L}$ if: (i) $\mathcal{B}$ is a subset of $\mathcal{L}$ ordered by restriction of the order relation of $\mathcal{L}$; (ii) the join of any two elements of $\mathcal{B}$ exists in $\mathcal{L}$ and $\mathcal{B}$ is closed under the join operation of $\mathcal{L}$; (iii) $\mathcal{B}$ is closed under the orthocomplementation of $\mathcal{L}$; (iv) $\mathcal{B}$ is distributive [notice that items (i)–(iii) imply that $\mathcal{B}$ is an orthocomplemented lattice]. We say that $\mathcal{B}$ is the Boolean subalgebra of $\mathcal{L}$ *generated by* the elements a, b of $\mathcal{L}$ if it is the smallest Boolean subalgebra of $\mathcal{L}$ containing a and b.

Now we have the following fact[1] (see Exercises 6–8):

THEOREM 12.2.1. *In an orthomodular poset $\mathcal{L}$, two elements commute if and only if they generate a Boolean subalgebra of $\mathcal{L}$.*

This result raises a natural question: Is the number two really essential? Could we replace two by three, or any other number, still getting a true statement? The answer is no. More precisely, the answer is no if we have to do just with orthomodular posets; it becomes yes if we have to do with something more, namely with orthomodular lattices. This situation became clear after Ramsay[1] and Pool[2] gave an explicit example of an orthomodular poset where three pairwise commuting elements do not generate a Boolean subalgebra or a lattice. The typical gap that can prevent the construction of the Boolean subalgebra is the fact that in an orthomodular poset, the condition that a, b, c are pairwise commuting is, in general, not sufficient to imply $(a, b \vee c)C$. In this context the following statement, due to Varadarajan[3] (see also Pool,[2] Ramsay,[1] Gudder,[4] Guz[5]), has to be recalled:

THEOREM 12.2.2. *Let $a, b_1, b_2, \ldots$ be elements of the orthomodular poset $\mathcal{L}$. If $(a, b_i)C$ for every i, and if $\vee b_i$ and $\vee(a \wedge b_i)$ exist, then $(a, \vee b_i)C$, and one has $a \wedge (\vee b_i) = \vee(a \wedge b_i)$.*

In the light of this result we see that if a, b, c are pairwise commuting, the conclusion $(a, b \vee c)C$ comes from the existence of $(a \wedge b) \vee (a \wedge c)$. Of course the existence of this object is secured when $\mathcal{L}$ is a lattice, and in this case a, b, c generate a Boolean subalgebra of $\mathcal{L}$.[4,5] As a matter of fact, requiring that a, b, c generate a Boolean subalgebra is equivalent to requiring that they satisfy $(a, b \vee c)C$.

12.3 Commutativity and Distributivity

Let us definitely assume, in the rest of this chapter, that $\mathcal{L}$ is an orthomodular lattice, rather than only an orthomodular poset. We dwell a little more on the distributive properties connected with commutativity. Recall that according to Theorem 12.2.2 if $a \in \mathcal{L}$ commutes with all the members of some subset $\{b_i\}$ of $\mathcal{L}$,* then one has the generalized distributivity relations

$$a \wedge (\vee b_i) = \vee (a \wedge b_i),$$
$$a \vee (\wedge b_i) = \wedge (a \vee b_i),$$

---

*If $\mathcal{L}$ is a complete lattice, then $\{b_i\}$ is any subset; if $\mathcal{L}$ is σ-complete, then $\{b_i\}$ has to be countable; if $\mathcal{L}$ is just a lattice, then $\{b_i\}$ has to be finite.

where the latter equality follows from the former thanks to the orthocomplementation in $\mathcal{L}$. Foulis[6] and Holland[7] discovered that when the subset $\{b_i\}$ contains just two elements, a much stronger result holds true:

THEOREM 12.3.1. *If, in an orthomodular lattice, one of the elements a, b, c commutes with the other two, then (a, b, c) is a distributive triple.*

This theorem is of great practical importance, and by no means expected, since the three elements are not required to be pairwise commuting [we might have $(a, b)C$ and $(a, c)C$ without having $(b, c)C$], so that the triple (a, b, c) does not, in general, generate a Boolean algebra.

Notice, as a particular case, that if a and b commute, then the triple $a, b, a^\perp$ is distributive, so that $a \wedge (b \vee a^\perp) = a \wedge b$. Conversely, if this equality holds true, we have $(a, b)C$ for $(a \wedge b^\perp) + (a \wedge b) = (a \wedge b^\perp) + [a - (a \wedge b^\perp)] = a$, where the last step simply expresses the orthomodular identity for the ordered elements $a, (a \wedge b^\perp)$. Thus we have:

THEOREM 12.3.2. *In an orthomodular lattice $(a, b)C$ if and only if $a \wedge b = a \wedge (b \vee a^\perp)$.*

A subset S of $\mathcal{L}$ is called a Foulis-Holland set if whenever $a, b, c \in$ S are distinct then one of them commutes with the other two. Greechie[8] has proved that Foulis-Holland sets generate distributive sublattices of $\mathcal{L}$.

12.4 The Center of an Orthomodular Lattice and the Notion of Irreducibility

The notion of commutativity in orthomodular lattices (posets) allows the introduction of the "commutant": if X is a nonempty subset of the orthomodular lattice (poset) $\mathcal{L}$, the set X' formed by all elements of $\mathcal{L}$ that commute with every element of X is called the *commutant* of X. Notice that there are remarkable analogies between this prime operation and the centralizer operation in von Neumann algebras.

The commutant $\mathcal{L}'$ of the whole of $\mathcal{L}$ is called the *center* of $\mathcal{L}$: we shall prefer for it the more familiar notation $Z(\mathcal{L})$. The center $Z(\mathcal{L})$ is a nonempty set, for the least and the greatest element of $\mathcal{L}$ certainly belong to it, since they commute with every a in $\mathcal{L}$. When 0 and 1 are the only elements of $Z(\mathcal{L})$, $\mathcal{L}$ is called *irreducible*.

The commutant X' of any nonempty subset of $\mathcal{L}$ is an orthomodular sublattice of $\mathcal{L}$.[9] However, where the commutant $Z(\mathcal{L})$ of the whole of $\mathcal{L}$ is concerned, we have a much stronger result:

THEOREM 12.4.1. *The center of an orthomodular lattice $\mathcal{L}$ is a Boolean subalgebra of $\mathcal{L}$.*

The proof is an immediate consequence of what was said in Section 12.2.

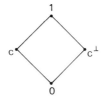

Figure 12.1. Center of G_{12}.

From the remark that $Z(\mathcal{L})$ contains the greatest element of $\mathcal{L}$ it follows that every element of $\mathcal{L}$ can be majorized by a central element; given $a \in \mathcal{L}$, the least element in $Z(\mathcal{L})$ that majorizes a is called the *central cover* of a and denoted $e(a)$ (see Exercise 9). The following facts are worthy of mention: [10]

$$e\left(\bigvee_i a_i \right) = \bigvee_i e(a_i), \qquad a_i \in \mathcal{L},$$

$$e(z \wedge a) = z \wedge e(a), \qquad z \in Z(\mathcal{L}), \quad a \in \mathcal{L}. \tag{12.4.1}$$

We end this section mentioning a few examples.

Example 1. Consider the finite lattice D_{16} (Figure 10.4). Its center consists of the elements $0, 1$ alone. Thus D_{16} is an example of irreducible orthomodular lattice.

Example 2. Consider the finite lattice G_{12} (see Figure 10.6). Its center is the Boolean algebra shown in Figure 12.1, so that G_{12} is not irreducible.

Example 3. Consider our familiar projection lattice $\mathcal{P}(\mathcal{H})$. We have $\mathbb{B}(\mathcal{H}) = \mathcal{P}(\mathcal{H})''$, and we know that $\mathbb{B}(\mathcal{H})$ is expressed as a commutant by $\mathbb{B}(\mathcal{H}) = \{\lambda I\}'$, where $\{\lambda I\}$ denotes the set of all multiples of the identity. Therefore $\mathcal{P}(\mathcal{H})' = \{\lambda I\}$. The only projections in $\mathcal{P}(\mathcal{H})'$ are thus zero and the identity: $\mathcal{P}(\mathcal{H})$ has trivial center, thus being irreducible.

Example 4. If $\mathbf{A}$ is any von Neumann algebra on $\mathcal{H}$, consider the orthomodular lattice $\mathcal{P}(\mathbf{A})$ of its projections. $\mathbf{A}' \cap \mathbf{A}$ is the center of $\mathbf{A}$, while the center of $\mathcal{P}(\mathbf{A})$ takes the form $\mathbf{A}' \cap \mathbf{A} \cap \mathcal{P}(\mathbf{A})$. We see that the projection lattice of a von Neumann algebra is not necessarily irreducible.

12.5 Morphisms and Direct Sums

The notions of morphisms and direct sums are here briefly reviewed in the framework of orthomodular lattices, though of course they may be given in a more general context. We confine ourselves to a collection of definitions and facts to be used in the sequel, omitting proofs, and without claim of systematicity.

Let $\mathfrak{L}_1, \mathfrak{L}_2$ be orthomodular σ-complete lattices.* A mapping h of $\mathfrak{L}_1$ onto $\mathfrak{L}_2$ is called a *morphism* if it preserves the join operation and orthogonality — symbolically, if for every (countable) subset $\{a_i\}$ of $\mathfrak{L}_1$,

$$h\left(\bigvee_i a_i \right) = \bigvee_i h(a_i)$$

and

$$a \perp b \quad \text{implies} \quad h(a) \perp h(b).$$

The following facts are then consequences (Exercise 10):

THEOREM 12.5.1. $h(0)=0$ (*where the first 0 refers to $\mathfrak{L}_1$ and the second to $\mathfrak{L}_2$*).

THEOREM 12.5.2. *h is monotone: $a \leqslant b$ (in $\mathfrak{L}_1$) implies $h(a) \leqslant h(b)$.*

THEOREM 12.5.3. *$h(a^\perp) = h(a)^\perp \wedge h(1)$, for every $a \in \mathfrak{L}_1$.*

THEOREM 12.5.4. *$h(a \wedge b) = h(a) \wedge h(b)$.*

THEOREM 12.5.5. *h is injective if and only if $h(a)=0$ implies $a=0$.*

THEOREM 12.5.6. *if h is injective, then it is order-preserving in both ways, that is, $a \leqslant b$ if and only if $h(a) \leqslant h(b)$.*

When the morphism is bijective (one-to-one), it is generally called an isomorphism. Notice that if $\mathfrak{L}_1$ and $\mathfrak{L}_2$ are isomorphic, then the greatest element of $\mathfrak{L}_1$ is mapped into the greatest element of $\mathfrak{L}_2$. When $\mathfrak{L}_1$ coincides with $\mathfrak{L}_2$, morphisms are called endomorphisms, while isomorphisms are called automorphisms.

Let us now introduce the notion of segment of an orthomodular lattice $\mathfrak{L}$. If $b \in \mathfrak{L}$, we denote by $\mathfrak{L}[0, b]$ the set of all elements of $\mathfrak{L}$ which are $\leqslant b$, and call it the *segment* from 0 to b. Clearly $\mathfrak{L}[0, b]$ is itself a lattice, with the ordering inherited by restriction from $\mathfrak{L}$, and becomes orthomodular if equipped with the so-called relative orthocomplementation $a \mapsto a^r$ defined by $a^r = a^\perp \wedge b$ (see Exercise 11). The canonical inclusion of $\mathfrak{L}[0, b]$ in $\mathfrak{L}$ is an example of morphism. If we consider a segment of $\mathfrak{L}$ having a central element as greatest element, (that is, a segment of the form $\mathfrak{L}[0, z]$ with

---

*In the rest of this volume the notion of morphism will be always considered in the context of σ-complete lattices. What we call a morphism coincides with what is often called a σ-morphism.

$z \in Z(\pounds)$), then it is easily seen that the mapping

$$a \mapsto a \wedge z$$

provides a morphism from $\pounds$ onto $\pounds[0, z]$.

Segments of this last kind—with a central element as greatest element—are of particular importance for the study of structural properties of orthomodular lattices. The following is a crucial fact:[10]

THEOREM 12.5.7. *Let $\pounds$ be an orthomodular lattice, and let $\langle z_i \rangle$ be a sequence of central elements, all different from $0, 1$, such that $\bigvee_i z_i = 1$, $z_i \wedge z_j = 0$ when $i \neq j$. Then every a in $\pounds$ can be written in the form $a = \bigvee_i a_i$ with $a_i \in \pounds[0, z_i]$, and this expression is unique, since $a_i = a \wedge z_i$.*

We paraphrase this result by saying that whenever $\langle z_i \rangle$ is a pairwise disjoint sequence in $Z(\pounds)$ not containing 0 and 1 and whose join is 1, we have a one-to-one correspondence between the elements of $\pounds$ and the sequences formed by picking up one element from each $\pounds[0, z_i]$: given $a \in \pounds$, the associated sequence is $\langle a \wedge z_i \rangle$; given the sequence $\langle a_i \rangle$ with $a_i \in \pounds[0, z_i]$, the associated element of $\pounds$ is $\bigvee_i a_i$. All this we express by saying that $\pounds$ is the *direct sum* of the segments $\pounds[0, z_i]$ and we write

$$\pounds = \biguplus \pounds[0, z_i].$$

Notice that if $\pounds$ is irreducible, then it does not admit a (nontrivial) direct-sum decomposition.

What is said above suggests comparing the notion of direct sum with another lattice-theoretic notion: that of *direct product*. Let $\pounds_i$, i running over some index set $\mathcal{I}$, be a family of lattices, and consider the sequences obtained by picking one element from each $\pounds_i$; write $\langle a_i \rangle$ for the generic sequence, and write $\prod \pounds_i$ for the set of all such sequences. $\prod \pounds_i$ inherits a natural lattice structure once the following definitions of ordering, meet, and join are adopted:

$$\langle a_i \rangle \leqslant \langle b_i \rangle \quad \text{whenever} \quad a_i \leqslant b_i \, (\text{in } \pounds_i) \text{ for every } i \in \mathcal{I},$$

$$\langle a_i \rangle \wedge \langle b_i \rangle = \langle a_i \wedge b_i \rangle, \quad \langle a_i \rangle \vee \langle b_i \rangle = \langle a_i \vee b_i \rangle.$$

If the $\pounds_i$'s are orthocomplemented, we make $\prod \pounds_i$ orthocomplemented by defining $\langle a_i \rangle^\perp = \langle a_i^\perp \rangle$, and if the $\pounds_i$'s are orthomodular then also $\prod \pounds_i$, the direct product of the $\pounds_i$'s, is orthomodular. Now we have the following fact: if $\pounds$ is an orthomodular lattice that admits the direct-sum expression $\pounds = \biguplus \pounds[0, z_i]$, then $\pounds$ is isomorphic to the direct product $\prod \pounds[0, z_i]$. Therefore, in our framework the direct sum is essentially the same as the direct product. To visualize the foregoing facts we briefly consider some examples.

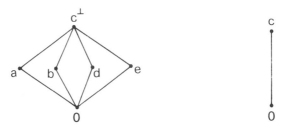

Figure 12.2. The irreducible components of G_{12}.

Example 1. Take the finite lattice G_{12} (see Figure 10.6), whose center is shown in Figure 12.1. G_{12} is the direct sum of the irreducible segments shown in Figure 12.2, where b and e are the relative complements of a and d, respectively (and vice versa); we can equivalently express G_{12} as direct product of these segments. The reader is referred to Exercise 12 for a more explicit check.

Example 2. Consider the projection lattice $\mathcal{P}(\mathbf{A})$ of a von Neumann algebra on $\mathcal{H}$. It is a known fact of the theory of von Neumann algebras that there exist in $Z(\mathcal{P}(\mathbf{A}))$ three pairwise orthogonal projectors $P_{\mathrm{I}}, P_{\mathrm{II}}, P_{\mathrm{III}}$ whose sum is the identity. Then $\mathcal{P}(\mathbf{A})$ has the direct-sum decomposition

$$\mathcal{P}(\mathbf{A}) = \mathcal{P}(\mathbf{A}_{P_I}) \cup \mathcal{P}(\mathbf{A}_{P_{II}}) \cup \mathcal{P}(\mathbf{A}_{P_{III}})$$

where $\mathbf{A}_{P_\mathrm{I}}, \mathbf{A}_{P_\mathrm{II}}, \mathbf{A}_{P_\mathrm{III}}$ are the reduced von Neumann algebras.[11] Notice that the segments of this direct sum need not be irreducible.

Example 3. Consider a separable Hilbert space $\mathcal{H}$ and in it a sequence $\langle \mathcal{M}_i \rangle$ of closed, pairwise orthogonal subspaces such that $\bigoplus_i \mathcal{M}_i = \mathcal{H}$. Let $P^{\mathcal{M}_i}$ be the projector onto $\mathcal{M}_i$, and write $\mathbf{A}$ for the von Neumann algebra $\{P^{\mathcal{M}_i}\}'$. The projection lattice $\mathcal{P}(\mathbf{A})$ admits the following characterization (see Exercise 13): $P \in \mathcal{P}(\mathbf{A})$ if and only if $P = \bigvee_i(P \wedge P^{\mathcal{M}_i})$, where the join and meet operations are those of the projection lattice $\mathcal{P}(\mathcal{H})$. Notice that each closed subspace $\mathcal{M}_i$ can itself be viewed as a Hilbert subspace, and the corresponding projection lattice $\mathcal{P}(\mathcal{M}_i)$ can also be interpreted as the segment $\mathcal{L}[0, P^{\mathcal{M}_i}]$ of $\mathcal{P}(\mathbf{A})$. Then we easily recognize the direct-sum structure

$$\mathcal{P}(\mathbf{A}) = \bigcup \mathcal{L}[0, P^{\mathcal{M}_i}].$$

In order to view $\mathcal{P}(\mathbf{A})$ as a direct product, notice that whenever we have a sequence $\langle \mathcal{H}_i \rangle$ of Hilbert spaces and we consider the new Hilbert space $\mathcal{H} = \bigoplus_i \mathcal{H}_i$, then each $\mathcal{H}_i$ can be canonically identified with a closed subspace $\mathcal{M}_i$ of $\mathcal{H}$ and the sequence $\langle \mathcal{M}_i \rangle$ has the properties mentioned before.

12.6 Atomicity, Covering Property, and Center Characterization

The proposition lattice of a quantum system is usually assumed to be not only orthomodular (and complete), but also atomic (hence atomistic) and with the covering property. Thus it is interesting to review how the notion of center and of irreducibility can be further characterized for a lattice of this kind.

A first relevant fact reads as follows (see Exercise 14):

THEOREM 12.6.1. *If the orthomodular complete lattice $\mathcal{L}$ is atomic, then also its center $Z(\mathcal{L})$ is atomic, and the atoms* of $Z(\mathcal{L})$ are the central covers of the atoms of $\mathcal{L}$.*

Now, the atoms of $Z(\mathcal{L})$ form a pairwise disjoint sequence having 1 as join [in fact $Z(\mathcal{L})$ is atomistic]; therefore, on account of Theorem 12.5.7 we conclude:

THEOREM 12.6.2. *Any orthomodular atomic lattice $\mathcal{L}$ is the direct sum of all the segments of the form $\mathcal{L}[0, z_i]$ with z_i an atom of $Z(\mathcal{L})$.*

The segments $\mathcal{L}[0, z_i]$ with z_i an atom of $Z(\mathcal{L})$ are irreducible, and conversely, if a segment $\mathcal{L}[0, z]$ with $z \in Z(\mathcal{L})$ is irreducible, then z is an atom of $Z(\mathcal{L})$.[10] Thus the direct sum referred to in Theorem 12.6.2 is the only one that decomposes $\mathcal{L}$ into irreducible summands.

Let us proceed with some consequences of the covering property of $\mathcal{L}$. We need first a definition: we say that two distinct atoms p, q of $\mathcal{L}$ belong to the relation $\approx$, and we write $p \approx q$, when there exists a third atom r, distinct from p and q, such that $r < p \vee q$. It is immediately seen, thanks to the covering property, that $p \approx q$ is fully equivalent to the existence of an atom r, distinct from p and q, such that

$$p \vee r = q \vee r = p \vee q. \tag{12.6.1}$$

(This property is reminiscent of the notion of perspectivity as introduced in the context of projection lattices of von Neumann algebras, where two elements are called perspective whenever they admit a common complement, in which case they are unitarily equivalent in the von Neumann algebra, as Fillmore[12] has shown.)

The relation $\approx$ defined above is transitive: if p, q, r are distinct atoms of $\mathcal{L}$, then $p \approx q$ and $q \approx r$ implies $p \approx r$. This follows from the covering property.[10] We have also the following fact (see Exercise 15):

$$\text{if} \quad p \approx q \quad \text{then} \quad e(p) = e(q). \tag{12.6.2}$$

---

*Of course, by "atom of $Z(\mathcal{L})$" we mean an atom with respect to the lattice (Boolean algebra) structure of $Z(\mathcal{L})$.

Thus we see that the relation $\approx$, viewed as an equivalence relation, partitions the set of all atoms of $\mathcal{L}$ into a family of equivalence classes that is in one-to-one correspondence with the family of all atoms of the Boolean algebra $Z(\mathcal{L})$, that is, with the family of the central covers of the atoms of $\mathcal{L}$. Moreover, owing to Theorem 12.6.2 we have the following fact:

THEOREM 12.6.3. *An orthomodular atomic lattice with the covering property is irreducible if and only if the join of any two atoms contains a third atom.*

Exercises

1. Let $\mathcal{R}$ be an involutive ring with identity, and $\mathcal{P}(\mathcal{R})$ its projection lattice. Then in $\mathcal{P}(\mathcal{R})$ we have $(e, f)C$ if and only if $ef = fe$.

[*Hint.* If $e = e_1 + g$, $f = f_1 + g$ with e_1, f_1, g pairwise orthogonal, that is, $e_1 f_1 = e_1 g = f_1 g = 0$, then check that $ef = fe$. Conversely, if $ef = fe$, choose $e_1 = e - ef$, $f_1 = f - ef$, $g = ef$.]

2. If $a = a_1 + c$, $b = b_1 + c$, then show that $a \vee b$ and $a \wedge b$ exist, that a_1, b_1, c are uniquely determined, and that $c = a \wedge b$.

[*Hint.* Write $d = a_1 + b_1 + c$, and observe that $d = a \vee b$. Using orthomodularity, write $d = a + (d - a) = b + (d - b)$; then compare with $d = a + b_1 = b + a_1$ to get $b_1 = d - a$, $a_1 = d - b$. Deduce $a^\perp = d^\perp + b_1$, $b^\perp = d^\perp + a_1$, and conclude that $(a^\perp, b^\perp)C$, so that $a^\perp \vee b^\perp$ exists, and hence also $a \wedge b$ exists. Since $d = (a_1 + b_1) + [d - (a_1 + b_1)]$, it follows $c = d - (a_1 + b_1) = (a^\perp \vee b^\perp)^\perp = a \wedge b$.]

3. If $(a, b)C$, show that $(a, b^\perp)C$ and $(a^\perp, b)C$.

[*Hint.* From the previous exercise $a = a_1 + c$, $b^\perp = a_1 + d^\perp$; hence $(a, b^\perp)C$, and similarly $(a^\perp, b)C$. It follows that $(a \wedge b^\perp)$ and $(a^\perp \wedge b)$ also exist.]

4. If $a = a_1 + c$, $b = b_1 + c$, then $a_1 = a \wedge b^\perp$, $b_1 = b \wedge a^\perp$.

[*Hint.* From Exercise 2, $a_1 = d - b = (a \vee b) \wedge b^\perp$; hence $a_1 \geqslant a \wedge b^\perp$. But also $a_1 \leqslant b^\perp$ and $a_1 \leqslant a$; hence $a_1 \leqslant a \wedge b^\perp$. Similarly for b_1.]

5. If $a = (a \wedge b^\perp) + (a \wedge b)$, then $(a, b)C$ and $b = (b \wedge a^\perp) + (b \wedge a)$.

[*Hint.* From orthomodularity $b = (b \wedge a) + [b - (b \wedge a)]$, but $b - (b - a) \perp a \wedge b^\perp$; hence $(a, b)C$.]

6. Assume $(a, b)C$, and write $a = a_1 + c$, $b = b_1 + c$. If $a_1, b_1, c, 0$ are all different, we get the greatest Boolean subalgebra a and b can generate, and it has the form shown in Figure 12.3, where $d = (a_1 + b_1 + c)^\perp$.

[*Hint.* Recall (3.1.1), (3.1.2), (3.1.4) and use the orthomodular identity.]

7. If $a, b \in \mathcal{L}$ generate a Boolean subalgebra of $\mathcal{L}$, then $(a, b)C$.

[*Hint.* In the Boolean subalgebra take the orthogonal elements $a \wedge b^\perp$, $b \wedge a^\perp$, $a \wedge b$, and use distributivity to get $a = (a \wedge b) \vee (a \wedge b^\perp)$, $b = (b \wedge a) \vee (b \wedge a^\perp)$.]

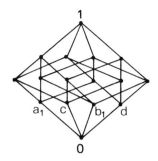

Figure 12.3. Boolean algebra generated by two commuting elements.

8. Verify that the commuting elements a, b of the lattices D_{16} and G_{12} shown, respectively, in Figures 10.4 and 10.6 generate the Boolean subalgebra given by Figure 10.2, while the noncommuting elements a, e generate nondistributive lattices.

9. Check the uniqueness of the central cover.
[*Hint.* If $a \in Z(\mathcal{L})$ then $a = e(a)$. If $a \notin Z(\mathcal{L})$, suppose z_1, z_2 are two central covers of a; then $z_1 \wedge z_2 \in Z(\mathcal{L})$ and $z_1 \wedge z_2 \geqslant a$, whence a contradiction unless $z_1 = z_2$.]

10. Verify that a morphism h of $\mathcal{L}_1$ on $\mathcal{L}_2$ fulfills Theorems 12.5.1–12.5.6.

11. Prove that the segment $\mathcal{L}[0, b]$ is an orthomodular lattice with respect to the relative orthocomplementation $a \mapsto a^r = a^\perp \wedge b$.
[*Hint.* First prove that $a \mapsto a^r$ is an orthocomplementation: (i) $a^{rr} = (a \vee b^\perp) \wedge b = a$ because $(a, b, b^\perp)$ is a distributive triple; (ii) $a_1 \leqslant a_2$ entails $a_2^\perp \leqslant a_1^\perp$ and hence $a_2^r \leqslant a_1^r$; (iii) $a \vee a^r = 1$ by the orthomodularity of $\mathcal{L}$. To prove orthomodularity notice that for $a_1, a_2 \in \mathcal{L}[0, b]$ we have $a_1 \vee (a_2 \wedge a_1^r) = a_1 \vee (a_2 \wedge a_1^\perp)$.]

12. Show that G_{12} is the direct sum as well as the direct product of the segments of Figure 12.2.
[*Hint.* Call the two segments $\mathcal{L}_1$ and $\mathcal{L}_2$, and write explicitly the correspondence between the elements of $\mathcal{L}$ and the sequences $\langle x_1, x_2 \rangle$, $x_1 \in \mathcal{L}_1$, $x_2 \in \mathcal{L}_2$. Build up the direct product of $\mathcal{L}_1$ and $\mathcal{L}_2$, and check that the mapping $\langle x_1, x_2 \rangle \mapsto x_1 \vee x_2$ is an isomorphism between the direct product and the direct sum.]

13. If $\langle \mathfrak{M}_i \rangle$ is an orthogonal sequence of subspaces of a separable Hilbert space $\mathcal{H}$ and $\bigoplus_i \mathfrak{M}_i = \mathcal{H}$, then the projection lattice $\mathcal{P}(\mathbf{A})$ of the von Neumann algebra $\{P^{\mathfrak{M}_i}\}'$ admits the following characterization: $P \in \mathcal{P}(\mathbf{A})$ if and only if $P = \bigvee_i (P \wedge P^{\mathfrak{M}_i})$, where $\vee$ and $\wedge$ are the operations in $\mathcal{P}(\mathcal{H})$.
[*Hint.* If $P \in \mathcal{P}(\mathbf{A})$ then $(P, P^{\mathfrak{M}_i})C$; hence $P = P \wedge (\bigvee_i P^{\mathfrak{M}_i}) = \bigvee_i (P \wedge P^{\mathfrak{M}_i})$ by Theorem 12.2.2. If $P = \bigvee_i (P \wedge P^{\mathfrak{M}_i})$, notice that $P \wedge P^{\mathfrak{M}_i} \perp P \wedge$

$P^{\mathfrak{M}_j}$ $(i \neq j)$, so that P leaves invariant each $\mathfrak{M}_i$. Hence $(P, P^{\mathfrak{M}_i})C$ and $P \in \mathcal{P}(\mathbf{A})$.]

14. Prove that p is an atom of $\mathfrak{L}$ if and only if its central cover $e(p)$ is an atom of $Z(\mathfrak{L})$.

[*Hint.* Suppose p to be atom of $\mathfrak{L}$. If $0 < z \le e(p)$ with $z \in Z(\mathfrak{L})$, use (12.4.1) to write $z = z \wedge e(p) = e(z \wedge p)$, hence $z \wedge p \neq 0$, hence $p \le z$, and hence $z = e(p)$. Suppose z is an atom of $Z(\mathfrak{L})$; take an atom p of $\mathfrak{L}$ contained in z, and exclude the strict inequality $e(p) < z$.]

15. Prove (12.6.2).

[*Hint.* Show that every central element z majorizing p majorizes q; if r is a third atom in $p \vee q$, notice that $q = q \wedge (p \vee r) \le q \wedge (z \vee r) = z$.]

16. For $a, b \in \mathfrak{L}[0, b], (a, b)C$ in $\mathfrak{L}$ if and only if $(a, b)C$ in $\mathfrak{L}[0, b]$.

References

1. A. Ramsay, *J. Math. Mech.* 15 (1966) 227.
2. J. C. T. Pool, Ph.D. Thesis, State University of Iowa, 1963.
3. V. S. Varadarajan, *Comm. Pure and Appl. Math.* 15 (1962) 189.
4. S. P. Gudder, in *Probabilistic Methods in Applied Mathematics*, Vol. 2, A. T. Bharucha-Reid, ed., Academic Press, New York, 1970.
5. W. Guz, *Rep. Math. Phys.* 2 (1971) 53.
6. D. J. Foulis, *Portugal Math.* 21 (1962) 65.
7. S. S. Holland, Jr., *Trans. Amer. Math. Soc.* 108 (1963) 66.
8. R. J. Greechie, *Proc. Amer. Math. Soc.* 67 (1977) 17 and 76 (1979) 216.
9. S. S. Holland, Jr. in *Trends in Lattice Theory*, J. C. Abbott, ed., Van Nostrand–Reinhold, New York, 1970; reprinted in *The Logico-algebraic Approach to Quantum Mechanics*, Vol. 1, C. A. Hooker, ed., Reidel, Dordrecht, 1975.
10. F. Maeda and S. Maeda, *Theory of Symmetric Lattices*, Springer, Berlin, Heidelberg, New York, 1970.
11. J. Dixmier, *Les Algèbres d'Opérateurs dans l'Espace Hilbertien*, Gauthiers-Villars, Paris, 1957.
12. P. A. Fillmore, *Proc. Amer. Math. Soc.* 16 (1965) 383.

CHAPTER 13

States and Propositions of a Physical System

13.1 Yes-No Experiments

The first five sections of this chapter are intended to provide some intuitive physical content to the notions of state and proposition of a quantum system. We deliberately avoid any deductive attitude, and we also avoid formalizations: the ingredients we shall start with (yes-no experiments and preparations) are so primitive and unstructured that the construction of quantum mechanics out of them is certainly not at hand. What we are going to see is simply that there is natural physical motivation for the occurrence at the very basis of every statistical theory—quantum mechanics in particular—of an ordered structure of propositions, of a convex structure of states, and of a relationship between these structures that makes the states behave as probability measures on propositions. From Section 13.6 on, we shall sketch a more formalized model, based on Mackey's approach.

By a yes-no experiment is meant an experiment that makes use of a measuring instrument having just two outcomes, which without loss of generality we can agree to label "yes" and "no". Of course, when talking of a yes-no experiment we always imagine that we have specified the physical system to which it pertains. We also imagine that the interaction between the physical system and the measuring instrument starts at a certain time, and after a finite time interval either the "yes" or the "no" outcome appears; that instructions are given specifying how to let the measuring instrument interact with the physical system; and that the notion of system preparation before interaction with the instrument has an unambiguous meaning. This last notion is understood as the list of all maneuvers and

ENCYCLOPEDIA OF MATHEMATICS and Its Applications, Gian-Carlo Rota (ed.). Vol. 15: E. G. Beltrametti and G. Cassinelli, The Logic of Quantum Mechanics

ISBN 0-201-13514-0

manipulations that have been used to isolate the physical system under discussion and to bring it into contact with the instrument employed in the yes-no experiment we are dealing with. The preparation of the system thus embodies all information about the interactions suffered by the system during its past, up to the instant in which the interaction with the instrument begins. Here we are ignoring information about the system after the execution of the yes-no experiment, because the way in which the physical system is affected by the experiment reflects the particular realization of the measuring instrument rather than essential properties of the system itself. We leave open the realistic possibility that the act of measurement can even destroy the physical system.*

We write $\mathcal{E}$ for the collection of all yes-no experiments pertaining to the physical system, and we write Π for the set of all preparations of the physical system. Of course the sets $\mathcal{E}$ and Π depend on, and actually represent, the knowledge reached of the physical system: an improvement of this knowledge may result in the development of new yes-no experiments and of new preparation procedures.

In this context the description of the physical system is equivalent to the knowledge of the probability of the "yes" outcome (or of the "no" outcome) of every yes-no experiment for every preparation of the system, and hence to the knowledge of a function from the cartesian product $\mathcal{E} \times \Pi$ into the real interval $[0, 1]$ (though, from the empirical point of view, recourse to the rationals would be sufficient, for only frequencies of events are measured). The framework consisting of the sets $\mathcal{E}$ and Π and of the foregoing function from $\mathcal{E} \times \Pi$ into $[0, 1]$ is indeed a very general one and encompasses all known physical theories. Notice that the restriction to dichotomic experiments is not a serious loss of generality: intuitively, the information provided by an experiment that uses a measuring instrument with more than two outcomes can also be provided by a suitable collection of yes-no experiments.

13.2 States

In a specific physical theory, the sets $\mathcal{E}$ and Π might contain some redundancy, so that we might abandon the consideration of all yes-no experiments and all preparation procedures in favor of physically exhaustive representatives of $\mathcal{E}$ and Π. This is indeed the case when we want to approach the structure of quantum mechanics.

---

* In theoretical analyses of quantum phenomena (we refer to Sections 8.1 and 16.2) it is often useful to think of yes-no experiments that have a further property: at least in case of a yes outcome, the physical system emerges from the measuring instrument, becoming again spatially separated from it, so that the experiment itself gives rise to a new preparation of the system, a preparation that is usually different from the one before the experiment.

In this section we focus attention on the possible redundancy of the set Π. Since $\mathcal{E}$ is intended to be the set of all conceivable yes-no experiments on our physical system, it becomes natural to think that in order to view two preparations as significantly different there must be at least one yes-no experiment whose yes outcome is produced with different probabilities by the two preparations. Thus we are led to recognize two preparation procedures $\pi_1, \pi_2 \in \Pi$ as physically equivalent, and we write $\pi_1 \sim \pi_2$, if the yes-no experiments are not able to distinguish between them, in other words, if for every element of $\mathcal{E}$ the probability of a yes outcome relative to the preparation π_1 is the same as the one relative to the preparation π_2. The relation $\sim$ is obviously reflexive, symmetric, and transitive: it is an equivalence relation, which partitions Π into equivalence classes. We call them the states of the system, and write $\mathcal{S}$ for the set they form ($\mathcal{S} = \Pi/\sim$).

Let us remark that the above-stipulated equivalence of preparations is not a trivial one. Consider for instance the following two preparations of a spin-$\frac{1}{2}$ particle: (1) a beam polarized along the positive direction of a z-axis is mixed with another beam of equal intensity but polarized along the negative direction of the z-axis, and then a particle is isolated without privileging any spin direction; (2) the same as (1) but with "z-axis" changed to "x-axis". Clearly the two preparations are, from a certain point of view, distinct, for they use different experimental devices (different polarizers), but we must take as an empirical fact that a yes-no experiment able to distinguish which one of the two preparations is being used has never been invented. Thus we put these two preparations into the same equivalence class, and we say that they provide the same state (recall also the discussion in Sections 2.4 and 4.2).

We shall use Greek letters $\alpha, \beta, \ldots$ to denote the elements of $\mathcal{S}$, the states. Clearly, every state determines and can be viewed as a function from the set $\mathcal{E}$ of all the yes-no experiments into the real interval $[0, 1]$. If $\alpha \in \mathcal{S}$ and $e \in \mathcal{E}$, we write $\alpha(e)$ for the value of α at e, so that the number $\alpha(e)$ gives the probability of yes outcome of e when the initial state of the system is α. A notion of convexity of the set $\mathcal{S}$ of states is now at hand. We say that the state α is a convex combination, or a mixture, of the states α_1 and α_2 if there exist two positive numbers w_1, w_2 with $w_1 + w_2 = 1$ such that

$$\alpha(e) = w_1\alpha_1(e) + w_2\alpha_2(e) \qquad \text{for every } e \in \mathcal{E}. \qquad (13.2.1)$$

In this case we write $\alpha = w_1\alpha_1 + w_2\alpha_2$, and w_1, w_2 are called the weights of the mixture. The notion of convex combination can be extended to arbitrary numbers of states: α is said to be mixture of $\alpha_1, \alpha_2, \ldots$ with weights $w_1, w_2, \ldots$ ($w_1 + w_2 + \cdots = 1$) if, for every $e \in \mathcal{E}$, the sum $\sum_i w_i\alpha_i(e)$ exists and equals $\alpha(e)$. If a state cannot be written as a convex combination of other states, then it is called pure. Thus we see that the distinction between pure and nonpure states is a very primitive one, and can be expressed without need of any mathematically structured background.

Now, it is an empirical fact that, given any two states, say α_1, α_2, it is possible to prepare the physical system in the mixed state $w_1\alpha_1 + w_2\alpha_2$ for arbitrary choice of the weights w_1, w_2. This gives physical content to the assertion that $\mathbb{S}$ is a convex set, relative to the convex combination defined in (13.2.1). Pure states appear as extremal points of the convex set: they form the "boundary" of $\mathbb{S}$.

Remark. Formally, we might endow $\mathbb{S}$ with an ordering relation by defining $\alpha \leqslant \beta$ whenever $\alpha(e) \leqslant \beta(e)$ for every $e \in \mathcal{E}$ (with some abuse of notation we use $\leqslant$ both for the ordering of $\mathbb{S}$ and for that of real numbers). However, we must take as an empirical fact that this order relation is physically void, in the sense that it is impossible to find two states belonging to it. Note that, interchanging the yes with the no outcome of e, we get a new yes-no experiment, say e', and $\alpha(e) \leqslant \beta(e)$ implies $\beta(e') \leqslant \alpha(e')$.

13.3 Propositions

Clearly, yes-no experiments have to do with the intuitive idea of "proposition", as it has been advanced in the Preface. Nevertheless, the notion of yes-no experiment is still too generic to be identified with a useful notion of proposition. This fact will be discussed in this and in the next two sections. As already remarked, we are now giving up the description of the physical system after a yes-no experiment has been performed on it: this agrees with the fact that quantum mechanics is committed to predicting outcomes of experiments but not the actual situation of the physical system after the experiment, for this situation depends upon the particular instruments used, and the choice of the instruments does not belong to the prescriptions of the theory. Therefore, it becomes natural to recognize in the set $\mathcal{E}$ of all yes-no experiments a level of redundancy, quite analogous to the one we have recognized in the set Π of all preparation procedures. Two yes-no experiments $e_1, e_2 \in \mathcal{E}$ are considered as physically equivalent, and we write $e_1 \simeq e_2$, if there is no state (or no preparation) that assigns them different probabilities of a yes outcome. $\simeq$ is obviously an equivalence relation, and it partitions $\mathcal{E}$ into equivalence classes. We call these equivalence classes the propositions of the system, though some further sharpening of this notion will appear necessary in the sequel. We use $a, b, c, \ldots$ to

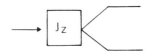

Figure 13.1.   Scheme of a Stern-Gerlach device.

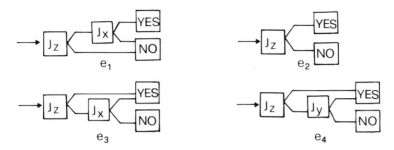

Figure 13.2. Yes-no experiments.

denote propositions, and we write $\mathcal{L}$ for the set they form: hence $\mathcal{L}$ is the quotient set $\mathcal{E}/\simeq$.

To visualize what we said above, consider again a spin-$\frac{1}{2}$ particle and denote by Figure 13.1 the Stern-Gerlach experiment* that separates the positive and negative polarizations along a z-axis (the two lines on the right are to represent the two possible trajectories of the emerging particle). Then imagine the yes-no experiments depicted in Figure 13.2, where x, y, z are supposed to be orthogonal axes. Clearly, each initial state of the system assigns to the device e_1 a yes-outcome probability smaller than that assigned to e_2, and the latter is smaller than that assigned to e_3; thus, according to our present definition of proposition, e_1, e_2, and e_3 are representatives of different propositions. On the other hand we must take as an empirical fact that nobody has, up to now, invented a state (a preparation) that assigns to e_3 and e_4 different yes-outcome probabilities, despite of the fact that different Stern-Gerlach devices are used there: thus e_3 and e_4 have to be regarded as representatives of the same proposition.

Return now to the set $\mathcal{L}$ of propositions, and notice that, by definition, each state assigns to every proposition a definite yes-outcome probability; thus every state determines, and can be viewed as, a function from $\mathcal{L}$ into the real interval $[0,1]$. Conversely, every proposition determines, and can be viewed as, a function from $\mathcal{S}$ into $[0,1]$: this function represents the yes-outcome probability that the various states assign to the proposition. To emphasize what is the function and what is the variable, we shall write $\alpha(a)$ for the function from $\mathcal{L}$ into $[0,1]$ determined by $\alpha \in \mathcal{S}$, and $a(\alpha)$ for the function from $\mathcal{S}$ into $[0,1]$ determined by $a \in \mathcal{L}$ [of course, for fixed α and a, $\alpha(a)$ and $a(\alpha)$ are the same number].[†]

---

* Stern-Gerlach experiments are described in any textbook on quantum mechanics. We only recall that, due to the interaction between the magnetic momentum of the particle and an inhomogenuous magnetic field cylindrically symmetric along a z-axis, the particle is forced into one of two spatially separated trajectories. Particles on one of these trajectories are polarized along the positive direction of the z-axis; particles on the other trajectory are polarized along the opposite direction.

† When this distinction appears pedantic, we shall drop $a(\alpha)$ in favor of $\alpha(a)$.

Viewing propositions as real functions on $\mathfrak{S}$, it is natural to say that the proposition a is less than b if the yes-outcome probability of a is systematically (i.e. for every state) smaller than that of b. Formally,

$$a \leqslant b \quad \text{whenever} \quad a(\alpha) \leqslant b(\alpha) \quad \text{for every } \alpha \in \mathfrak{S}, \qquad (13.3.1)$$

where, with some abuse of notation, we have used the same symbol to denote the order relation in $\mathfrak{L}$ and the order of real numbers. That (13.3.1) really defines in $\mathfrak{L}$ an order relation (that is, a reflexive, transitive, and antisymmetric relation) follows at once from the ordering of real numbers.

This ordering is physically nonvoid, in the sense that there are elements of $\mathfrak{L}$ that belong to the order relation. Think, e.g., of propositions a, b that test whether the physical system is, respectively, in a volume V or in a volume W containing V: clearly $a \leqslant b$; or let a, b be propositions testing whether the energy of the system is, respectively, in the interval E or in the interval F containing E: again $a \leqslant b$; or let a, b, c be the propositions, relative to a spin-$\frac{1}{2}$ system, represented respectively by the yes-no experiments e_1, e_2, and e_3 of Figure 13.2: we have $a \leqslant b \leqslant c$.

It will be useful to include in the set $\mathfrak{L}$ also two trivial propositions: the one represented by the yes-no experiment whose outcome is always (i.e. for every state) no and the one represented by the yes-no experiment whose outcome is always yes. We denote the former by 0 and the latter by 1: they are the least and the greatest elements of $\mathfrak{L}$.

Remark. Formally, we might rephrase for $\mathfrak{L}$ a notion of convex combination analogous to the one given for $\mathfrak{S}$ in the previous section: we might say that $a \in \mathfrak{L}$ is a mixture of $a_1, a_2 \in \mathfrak{L}$ with weights w_1, w_2 if, for every $\alpha \in \mathfrak{S}$, $a(\alpha) = w_1 a_1(\alpha) + w_2 a_2(\alpha)$. Though a notion of this kind has sometimes been judged useful,[1] it is generally considered to be unnecessary, in the sense that there is no real evidence of a need to take into account mixtures of propositions.

13.4 On the Ordering of Propositions

The ordering among propositions expressed by (13.3.1) is not the favorite in logical approaches to quantum mechanics. When a logical calculus is imposed on the propositions of a physical system,* it is usually preferred to postulate that a proposition a is less than a proposition b whenever the truth of a entails the truth of b. To give to this criterion a clear meaning we have to specify what "truth of a" means. The simplest choice appears to be the one of saying that a is true in the state α if $\alpha(a) = 1$, that is, if α causes *with certainty* the yes outcome of a. Then, defining the *certainly-yes domain*, or

---

* The reader is referred to Chapter 20.

"truth set", of a as

$$S_1(a) = \{\alpha \in S : \alpha(a) = 1\},$$

we come to the following candidate for the order relation among propositions:

$$a \preccurlyeq b \quad \text{whenever} \quad S_1(a) \subseteq S_1(b) \tag{13.4.1}$$

In other words, we say that $a \preccurlyeq b$ if every state that makes certain the yes outcome of a also makes certain the yes outcome of b. (This is reminiscent of the ordering introduced by Jauch[2] and Piron,[3] though they do not make use of the concept of state as a primitive concept.)

But is the $\preccurlyeq$ relation really an ordering? It is clearly reflexive and transitive, but the antisymmetry requirement reads

$$\text{if} \quad S_1(a) = S_1(b) \quad \text{then} \quad a = b, \tag{13.4.2}$$

and this property is generally not satisfied in the set of propositions we have defined in last section. Indeed, from the premise that a and b have the same certainly-yes domain, we cannot infer that any other state will assign to a and b the same probability of a yes outcome (for this is the meaning of $a = b$). We can easily imagine counterexamples. Take for instance a spin-$\frac{1}{2}$ particle and the yes-no experiments of Figure 13.2: the certainly-yes domain of (the propositions represented by) e_1 is empty, like that of the trivial proposition 0, but the yes-outcome probability of e_1 is not zero except for the spin-down state; similarly, the certainly-yes domains of e_2 and e_3 coincide, being formed by the spin-up state alone, but any other state assigns to them different probabilities of yes outcome (for instance, the spin-down state gives probability 0 to e_2 but probability $\frac{1}{2}$ to e_3).

Thus, we see that in order to make $\preccurlyeq$ an order relation, we must make a selection among propositions. In each family of propositions having the same certainly-yes domain we must select just one, dropping all the others. In the foregoing example we would have to drop e_1 if we wanted to preserve the trivial proposition 0, and we would have to make a choice between e_2 and e_3 (needless to say, e_2 is more palatable and more economic than e_3).

Suppose that we have made a selection of propositions so that $\preccurlyeq$ becomes an ordering. At this stage the set of selected propositions would be endowed with two orderings: $\leqslant$ defined by (13.3.1) and $\preccurlyeq$ defined by (13.4.1), and in general these two relations will not coincide, for though $a \leqslant b$ implies $a \preccurlyeq b$, we have not that $a \preccurlyeq b$ implies $a \leqslant b$ (from $a \preccurlyeq b$ we can only exclude $b < a$). Now the question arises whether the selection of one element in each family of propositions having the same certainly-yes domain can be made in such a way that the two order relations coincide. Having in mind the Hilbert-space formulation of quantum mechanics, we are led to expect that for quantum systems this question should be answered affirmatively. Indeed, from Hilbert-space quantum mechanics only one ordered structure emerges: it is our familiar projection lattice $\mathcal{P}(\mathcal{H})$, where the two orderings reduce to the same thing.* We are not committed, in this chapter, to trying to characterize intrinsically the yes-no experiments to be selected so as to make $\leqslant$ equivalent to $\preccurlyeq$; something in this direction will be said in Chapter 16.

In the sequel we shall endow the set of propositions with a structure that goes beyond the simple poset structure discussed up to now. In particular we shall introduce a notion of orthocomplementation and the orthomodularity. At that stage our set $\mathcal{S}$ of states will qualify as a set of probability measures on propositions, and the order relations (13.3.1) and (13.4.1) will become the counterparts of what in Section 11.3 we have called, respectively, the ordering and the strongly ordering property of $\mathcal{S}$. Thus we refer to Section 11.3 for the mathematical intertwining between the two order relations; notice however that, while there the order of $\mathcal{L}$ was assumed to be autonomously given, in the present intuitive analysis we have used the states to define the order among propositions.[†]

13.5 Orthocomplementation

A natural requirement of completeness for the set of propositions is that to each proposition there correspond another one that constitutes its negation. This requirement is certainly a crucial one when a logical calculus is built on the propositions (we refer to Chapter 20, especially Section 20.2). If the familiar and basic properties of negation are translated into the language of the ordered structure of propositions, we see that, writing $a^{\perp}$ for the negation of a, the function $a \mapsto a^{\perp}$ has to behave as an orthocomplementation (see Section 10.1), that is,

$$a^{\perp\perp} = a, \tag{13.5.1}$$

$$a \leqslant b \quad \text{implies} \quad b^{\perp} \leqslant a^{\perp}, \tag{13.5.2}$$

$$a \wedge a^{\perp} = 0, \qquad a \vee a^{\perp} = 1. \tag{13.5.3}$$

As to the physical characterization of $a^{\perp}$, given a, it is intuitive to think of the following: take any yes-no experiment representing a, exchange the "yes" with the "no" label on the instrument display, and denote by $a^{\perp}$ the

---

* See Example Q of Section 11.3.

[†] This explains why, though in Section 11.3 we had that the strongly ordering property implies the ordering property, here we have that the order in the sense of (13.3.1) implies the order in the sense of (13.4.1).

proposition represented by the new yes-no experiment.* Equivalently we can say that $a^\perp$ is defined by

$$a^\perp(\alpha) = 1 - a(\alpha) \qquad \text{for all} \quad \alpha \in \mathbb{S}, \qquad (13.5.4)$$

or that the yes-outcome probability of $a^\perp$ equals (for every state) the no-outcome probability of a, and vice versa. Notice however that while this definition obviously satisfies (13.5.1) and (13.5.2), it generally fails to satisfy (13.5.3) unless some selection of privileged propositions is made—a situation similar to the one encountered in the last section. So, for instance, the proposition represented by e_3 of Figure 13.2 should be removed, for we would have $e_3 > e_3^\perp$, so that $e_3 \wedge e_3^\perp = e_3^\perp$ and $e_3 \vee e_3^\perp = e_3$. There are several prescriptions for selecting propositions so as to guarantee (13.5.3) also. One of them will be seen in next section; here we mention a simple one proposed by Garola:[4] Besides the trivial propositions 0 and 1, select the propositions a for which there exist a state α such that $a(\alpha) > \frac{1}{2}$ and a state β such that $a(\beta) < \frac{1}{2}$. In the family of propositions so selected (13.5.3) holds true; in fact, from $b \leq a$ and $b \leq a^\perp$ we get $b(\alpha) \leq \frac{1}{2}$ for every $\alpha \in \mathbb{S}$; hence $b = 0 = a \wedge a^\perp$, and similarly $a \vee a^\perp = 1$. Hence, in this family (13.5.4) defines an orthocomplement.

Up to now we have made reference to the ordering (13.3.1); if we go to the one expressed by (13.4.1) and we want to make the mapping $a \mapsto a^\perp$ defined by (13.5.4) an orthocomplementation relative to this ordering, then we are faced again with the need of restricting ourselves to some privileged collection of propositions. Notice for instance that (13.5.2) reads

$$\mathbb{S}_1(a) \subseteq \mathbb{S}_1(b) \quad \text{implies} \quad \mathbb{S}_0(b) \subseteq \mathbb{S}_0(a), \qquad (13.5.5)$$

where $\mathbb{S}_0(a)$ and $\mathbb{S}_0(b)$ stand for the certainly-no domains of a and b; explicitly

$$\mathbb{S}_0(a) = \{\alpha \in \mathbb{S} : \alpha(a) = 0\}.$$

This condition is obviously not satisfied by every yes-no experiment; for instance, if we return to Stern-Gerlach devices for a spin-$\frac{1}{2}$ system and consider the two yes-no experiments shown in Figure 13.3, we get a violation of (13.5.5). If we take, (as seems natural) e_2, then we must drop e_5.

Summing up, we have seen that starting from the set of all propositions, as defined in Section 13.3, we can easily motivate the order relation (13.3.1), but we do not reach any richer structures. If we want to do that—for

---

* Recall that in Section 13.3 a proposition was defined as an equivalence class of yes-no experiments relative to the equivalence relation $\simeq$; notice that the exchange of the "yes" and "no" labels maps an equivalence class into an equivalence class.

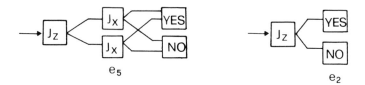

Figure 13.3. Two yes-no experiments which cause a violation of the orthocomplementation condition.

instance, if we want (13.3.1) and (13.4.1) to be equivalent orderings, or if we want to introduce orthocomplementation—then we must proceed to a sharper idea of proposition: not every yes-no experiment can be taken, but only a specialized class of them.

13.6 On Mackey's Approach

In this section we approach the notion of proposition on a more formalized level, taking from the outset, as primitive ingredients, states and physical quantities of the physical system; but of course no Hilbert space is in view at this stage. We shall mainly refer to Mackey's *Mathematical Foundations of Quantum Mechanics*,[5] but it is not our purpose to give a detailed exposition of this classical work. We shall also make particular reference to Maczynski,[6] especially for his quick and elegant way of introducing the orthomodular poset structure of propositions.

Let $\mathcal{O}$ be the set of all physical quantities of the physical system under consideration, and let $\mathcal{S}$ be the set of its states. The result of the measurement of a physical quantity will be assumed, as usual, to be expressible by a real number in proper units. The description of the physical system here considered is at fixed time, as in the preceding sections.

The starting point of a physical theory can be formulated as follows: if the system is known to be in the state $\alpha \in \mathcal{S}$, what is the probability that the measurement of the physical quantity $A \in \mathcal{O}$ gives a value contained in a set E of real numbers? We shall write $p(A, \alpha, E)$ for this probability function. Regularity conditions suggest restricting ourselves from the outset to the family $\mathcal{B}(\mathbb{R})$ of Borel sets, a restriction that fits in with the interpretation of E as a set containing measured values of physical quantities. Thus, we have to do with a probability function from $\mathcal{O} \times \mathcal{S} \times \mathcal{B}(\mathbb{R})$ into $[0, 1]$.

Consider now the ordered pairs $\langle A, E \rangle$ with $A \in \mathcal{O}$, $E \in \mathcal{B}(\mathbb{R})$. From the physical point of view, we may picture the pair $\langle A, E \rangle$ as the experimental device (Figure 13.4) that is obtained from the instrument used to measure the physical quantity A by attaching to the reading scale a window that isolates the numerical subset E; according as the pointer of the instrument does or does not appear within the window, one may answer yes or no, to

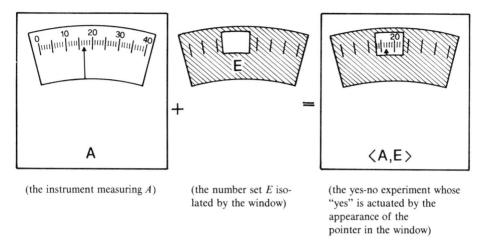

| (the instrument measuring A) | (the number set E isolated by the window) | (the yes-no experiment whose "yes" is actuated by the appearance of the pointer in the window) |

Figure 13.4. How to get a yes-no experiment.

the question "Is the measured result of A contained in E?" So we see that the pair $\langle A, E \rangle$ conforms to the notion of yes-no experiment as discussed in previous sections.

It may happen that two pairs, say $\langle A, E \rangle$ and $\langle B, F \rangle$ with $A, B \in \mathcal{O}$ and $E, F \in \mathcal{B}(\mathbb{R})$, are not separated by the probability function p, in the sense that

$$p(A, \alpha, E) = p(B, \alpha, F) \qquad \text{for every} \quad \alpha \in \mathcal{S};$$

then we say that the two pairs are *equivalent*. This equivalence relation determines a partition of the set of all pairs into equivalence classes: we denote $|A, E|$ the equivalence class containing the pair $\langle A, E \rangle$, and denote by $\mathcal{L}$ the set having as elements the equivalence classes. Symbolically,

$$\mathcal{L} = \{|A, E| : A \in \mathcal{O}, E \in \mathcal{B}(\mathbb{R})\}.$$

Clearly, the equivalence relation just considered paraphrases the one considered in Section 13.3: it is not accidental that we use the same symbol $\mathcal{L}$ as we did there. We again use for the elements of $\mathcal{L}$ the letters $a, b, \ldots$, and if $a = |A, E|$, we write

$$a(\alpha) = p(A, \alpha, E), \qquad \alpha \in \mathcal{S}.$$

Each element of $\mathcal{L}$ maps $\mathcal{S}$ into $[0, 1]$, and conversely, each element of $\mathcal{S}$

maps $\mathfrak{L}$ into $[0, 1]$; as in previous sections, we prefer to write $\alpha(a)$ when a is thought of as the variable.

According to the discussion of Section 13.3, $\mathfrak{S}$ orders $\mathfrak{L}$ via the natural ordering of real functions, that is, by the prescription (13.3.1). We adopt this ordering, thus making $\mathfrak{L}$ a poset. Paraphrasing what was said in Section 13.2, we have also that $\mathfrak{L}$ induces in $\mathfrak{S}$ a notion of convex combination and we adopt for $\mathfrak{S}$ the structure of a σ-convex set. Though at this stage, this is an assumption (Mackey's fourth axiom), it is well founded on empirical grounds.

For arbitrary $A \in \mathcal{O}$, consider the pairs $\langle A, \varnothing \rangle$ and $\langle A, \mathbb{R} \rangle$ ($\varnothing$ being the empty set). With reference to Figure 13.4, it is apparent that $\langle A, \varnothing \rangle$ is a trivial yes-no experiment, for the reading scale of A is entirely screened, no window being left open, so that the "no" outcome of $\langle A, \varnothing \rangle$ is certain no matter what the state of the system is. Also $\langle A, \mathbb{R} \rangle$ is a trivial yes-no experiment: no part of the reading scale of A is screened; whatever the position of the pointer is, it will fall in $\mathbb{R}$, so that the "yes" outcome of $\langle A, \mathbb{R} \rangle$ is certain for every state of the system. The corresponding equivalence classes $|A, \varnothing|$ and $|A, \mathbb{R}|$ will be denoted by 0 and 1 and are the least and the greatest elements of $\mathfrak{L}$, relative to the adopted ordering (13.3.1).

There is a natural notion of orthogonality in $\mathfrak{L}$: we say that a is orthogonal to b, and write $a \perp b$, whenever $a(\alpha) + b(\alpha) \leqslant 1$ for all $\alpha \in \mathfrak{S}$. Notice, in particular, that every state that causes with certainty the "yes" outcome of a will cause with certainty the "no" outcome of every proposition orthogonal to a. As an intuitive example of orthogonal propositions we might take $a = |A, E|, b = |A, F|$ with E disjoint from F, and A arbitrary (Figure 13.4 may help visualize the situation).

One of Mackey's crucial axioms, formulated in terms of $\mathfrak{L}$ and $\mathfrak{S}$,* is as follows:[7,8]

AXIOM 13.6.1. For every pairwise orthogonal sequence $a_1, a_2, \ldots \in \mathfrak{L}$, there exists $b \in \mathfrak{L}$ such that for every $\alpha \in \mathfrak{S}$, $b(\alpha) + a_1(\alpha) + a_2(\alpha) + \cdots = 1$.

This axiom has far-reaching consequences: it makes $\mathfrak{L}$ orthomodular, and it allows viewing $\mathfrak{S}$ as a set of probability measures on $\mathfrak{L}$. Let us outline the various steps.

A first fact is that for every $a \in \mathfrak{L}$ the function $1 - a$ (from $\mathfrak{S}$ into $[0, 1]$) still belongs to $\mathfrak{L}$; we write $a^\perp$ for this function and notice that it coincides with the one defined by (13.5.4). The physical interpretation of $a^\perp$ is quite natural if we think of the elements of $\mathfrak{L}$ in terms of physical quantities and real Borel sets. If $a = |A, E|$, then $a^\perp = |A, \mathbb{R} - E|$; referring again to Figure 13.4, we can say that the experimental device corresponding to $a^\perp$ is obtained by simply screening the window E and leaving unscreened the rest of the reading scale of A.

A second fact is that the mapping $a \mapsto a^\perp$ is an orthocomplementation in $\mathfrak{L}$. The items (13.5.1) and (13.5.2) are obvious. To prove (13.5.3) it must be

---

* In Mackey's book the axiomatization is done in terms of the triple $(\mathcal{O}, \mathfrak{S}, p)$.

shown that if $c \geqslant a$ and $c \geqslant a^\perp$, then $c = 1$ (similarly, if $c \leqslant a$ and $c \leqslant a^\perp$, then $c = 0$). In fact the premise is equivalent to saying that $c^\perp, a, a^\perp$ are pairwise orthogonal; hence Axiom 13.6.1 ensures the existence of $b \in \mathcal{L}$ such that for every $\alpha \in \mathcal{S}$, $b(\alpha) + c^\perp(\alpha) + a(\alpha) + a^\perp(\alpha) = 1$; hence $b(\alpha) = c^\perp(\alpha) = 0$; therefore $c = 1$. By the way, we have that the orthogonality relation $a \perp b$ defined above reduces to $a \leqslant b^\perp$, that is, to the familiar orthogonality in orthocomplemented lattices.

A third fact is that $\mathcal{L}$ is σ-orthocomplete. If $a_1, a_2, \ldots$ is a sequence of pairwise orthogonal elements, then the orthocomplement of the proposition b whose existence is ensured by Axiom 13.6.1 is precisely their join: $b^\perp = +_i a_i$. We leave as an exercise the check of this assertion. Notice that, for every α,

$$\alpha(b^\perp) = \alpha(+_i a_i) = \sum_i \alpha(a_i). \tag{13.6.1}$$

After these remarks we have at once that Axiom 13.6.1 makes $\mathcal{L}$ orthomodular. Take two ordered elements $a \leqslant b$; then $a \perp b^\perp$, so that there exists $c \in \mathcal{L}$ such that $c(\alpha) + a(\alpha) + b^\perp(\alpha) = 1$ for every α. Hence $b = a \vee c$ with $a \perp c$: the orthomodular identity. Also the fact that $\mathcal{S}$ becomes a set of probability measures on $\mathcal{L}$ is now almost obvious: we refer to the definition of probability measure given in Section 11.1, and we simply remark that (13.6.1) expresses the additivity on orthogonal sequences. Notice however that $\mathcal{S}$ is not, at this stage, proved to be the set of *all* the probability measures on $\mathcal{L}$ (should this be the case, the convexity of $\mathcal{S}$ would be a by-product). Notice also that, according to the terminology of Section 11.3, we have that the ordering of $\mathcal{L}$ is by construction such that $\mathcal{S}$ is an ordering set of probability measures on $\mathcal{L}$.

Summing up, Mackey's approach leads with some naturalness to the following structure: an orthomodular poset $\mathcal{L}$, also called the "logic" of the theory, whose elements adhere to the idea of "propositions", and an ordering, σ-convex set $\mathcal{S}$ of probability measures on $\mathcal{L}$, whose elements are interpreted as states of the physical system. Notice that this approach does not provide any special hint in favor of the equivalence (discussed in Section 13.4) between the orderings (13.3.1) and (13.4.1); in other words, this approach does not ensure for $\mathcal{S}$ the property of being not only ordering but also strongly ordering on $\mathcal{L}$.

13.7 Physical Quantities as Proposition-Valued Measures on the Reals; a Consistency Problem

In the last section we have seen how to reach a proposition-state structure, the $(\mathcal{L}, \mathcal{S})$ structure, starting from a triple $(\mathcal{O}, \mathcal{S}, p)$ with $\mathcal{O}$ denoting the set of physical quantities and p a probability function on $\mathcal{O} \times \mathcal{S} \times \mathcal{B}(\mathbb{R})$. A question naturally arises: Have we lost something in that itinerary? Has the $(\mathcal{L}, \mathcal{S})$ structure the same descriptive capacity as the

$(\mathcal{O}, \mathcal{S}, p)$ structure? Is the path reversible, so that the physical quantities and the p-function could be reconstructed starting from propositions? The answer is affirmative, as we are going to sketch.

Let $\mathcal{L}$ be an orthomodular poset and $\mathcal{S}$ an ordering, convex set of probability measures on it. We need first a definition. We say that A is a *proposition-valued measure* on the reals if it is a function from $\mathcal{B}(\mathbb{R})$ into $\mathcal{L}$ such that:

(i) $A(\varnothing)=0, A(\mathbb{R})=1$,
(ii) for every disjoint sequence $E_1, E_2,\ldots \in \mathcal{B}(\mathbb{R})$, $A(E_i)$ is orthogonal to $A(E_j)$ if $i \neq j$, and $A(E_1 \cup E_2 \cup \cdots)=A(E_1)+A(E_2)+\cdots$.

Then, by inspection of how propositions were derived (in last section) from physical quantities, it becomes natural to *define* the $\mathcal{L}$-valued measures on $\mathcal{B}(\mathbb{R})$ as the "physical quantities" of the physical system. This motivates the symbol A we have chosen, and hence the symbol $\mathcal{O}$ to designate the set of all $\mathcal{L}$-valued measures on $\mathcal{B}(\mathbb{R})$. The candidate for the p-function is now obvious: we *define*

$$p(A, \alpha, E)=\alpha(A(E)).$$

It can be proved that the triple $(\mathcal{O}, \mathcal{S}, p)$ so defined has all the properties appearing in Mackey's axiomatization.[9]

But there is still a consistency problem. Suppose we start from the $(\mathcal{L}, \mathcal{S})$ structure and construct from it the triple $(\mathcal{O}, \mathcal{S}, p)$. If, starting from the latter, we go back to a proposition-state structure, do we recover (up to isomorphism) the same pair $(\mathcal{L}, \mathcal{S})$ we started from? Or, similarly, suppose we take from the outset an $(\mathcal{O}, \mathcal{S}, p)$ structure and deduce from it the pair $(\mathcal{L}, \mathcal{S})$: if we go back, do we reach (up to isomorphism) the original $(\mathcal{O}, \mathcal{S}, p)$ structure? These questions can be answered affirmatively,[9] showing the full equivalence between a description based on physical quantities, states, and p-function and a description based on propositions and states.

The $(\mathcal{L}, \mathcal{S})$ structure discussed in this chapter is still quite general:* it contains as special cases classical mechanics, quantum mechanics, and possibly other mechanics. The properties $\mathcal{L}$ must be endowed with to correspond to standard quantum mechanics are epitomized by $\mathcal{P}(\mathcal{H})$, the lattice of projectors of a separable complex Hilbert space. Mackey[5] put forth an axiom that, in his words, seems entirely *ad hoc*. It reads: the partially ordered set of all propositions ("questions" was the term he used)

---

* The scheme we have adopted assumes from the beginning that the equivalence of yes-no experiments, which leads to the notion of proposition (see Section 13.3), can be derived from the set of states. For the defense of a different position, especially with a view to providing a more general format for a discussion of the hidden-variables issue, we refer to Cooke and Hilgevoord.[10] (From this reference we have derived the schemes of Figures 13.2 and 13.3.)

in quantum mechanics is isomorphic to the partially ordered set of all closed subspaces of a separable, infinite-dimensional complex Hilbert space. Of course, such an axiom provides a bridge to ordinary quantum mechanics, but one has the feeling that an axiom of this sort is rather outside the spirit of a logical analysis of quantum mechanics, where the accent is put on the search for axioms having physical naturalness and direct interpretation, and able to imply the mathematical edifice of quantum mechanics. Thus (quoting Mackey again) we are far from being forced to accept this axiom as logically inevitable: ideally, one would like to have a list of physically plausible assumptions from which one could deduce that axiom; short of this, one would like to have a list from which one could deduce a set of possibilities for the structure of $\mathcal{L}$, all but one of which could be shown to be inconsistent with suitably planned experiments. And Mackey added that at the moment (1963) such lists were not available. But other relevant contributions came in the sixties: we refer in particular to the "Geneva school"[11,12] and to the contributions of Zierler[13] and MacLaren.[14] As a result, the situation is now improved: the axiom under discussion can be partially circumvented (see in particular Chapter 21), and the list of physically plausible assumptions Mackey wanted has some definite entries. We allude in particular to the assumptions that $\mathcal{L}$ is a lattice (not only a poset), that it is atomic, and that it has the covering property. These items have been briefly commented on in preceding sections, and will be more extensively reviewed in the sequel.

References

1. R. Giles, *J. Math. Phys.* 11 (1970) 2139.
2. J. M. Jauch, *Foundations of Quantum Mechanics*, Addison-Wesley, Advanced Book Program, Reading, Mass., 1968.
3. C. Piron, *Foundations of Quantum Physics*, Benjamin, Advanced Book Program, Reading, Mass., 1976.
4. C. Garola, *Int. J. Theor. Phys.*, 19 (1980) 369.
5. G. W. Mackey, *Mathematical Foundations of Quantum Mechanics*, Benjamin, Advanced Book Program, Reading, Mass., 1963.
6. M. J. Maczynski, *Rep. Math. Phys.* 3 (1972) 209.
7. M. J. Maczynski, *Studia Math.* 47 (1973) 253.
8. M. J. Maczynski and T. Traczyk, *Bull. Acad. Polonaise des Sciences* (Série Sc. Math., Astr. et Phys.) 21 (1973) 3.
9. E. G. Beltrametti and G. Cassinelli, *Rivista Nuovo Cimento* 6 (1976) 321.
10. R. M. Cooke and J. Hilgevoord, in *Current Issues in Quantum Logic*, E. G. Beltrametti and B. van Fraassen, eds., Plenum, New York, 1980.
11. J. M. Jauch and C. Piron, *Helv. Phys. Acta* 36 (1963) 827.
12. C. Piron, *Helv. Phys. Acta* 42 (1969) 330.
13. N. Zierler, *Pacific J. Math.* 11 (1961) 1151.
14. M. D. MacLaren, *Pacific J. Math.* 14 (1964) 597.

CHAPTER 14

Quantum-Mechanical Features in Terms of the Logic of the Physical System

14.1 The Logic of a Quantum System

In the preceding chapter we have seen that the pair $(\mathcal{L}, \mathcal{S})$ is a general format for the probabilistic description of physical systems. The aim of this chapter is the discussion of a number of characteristic features of quantum systems in the $(\mathcal{L}, \mathcal{S})$ framework. Since the $(\mathcal{L}, \mathcal{S})$ structure is general enough to encompass classical systems, we shall also strive to outline the branching points between classical and quantum behavior.

We shall start with an $(\mathcal{L}, \mathcal{S})$ structure that, besides the properties ensured by Mackey's approach (see Section 13.6), possesses some minimal requirements of mathematical smoothness. In particular, we shall choose properties able to guarantee that $\mathcal{S}$ is both ordering and strongly ordering on $\mathcal{L}$ (see Section 11.3), a fact whose physical interest has been commented on in Section 13.4.

Specifically, let us assume for $\mathcal{L}$ and $\mathcal{S}$ the following requirements:

(i) $\mathcal{L}$ is a separable orthomodular poset;

(ii) $\mathcal{S}$ is a sufficient, σ-convex set of probability measures on $\mathcal{L}$ (see Sections 11.1 and 11.3);

(iii) if, for some $\alpha \in \mathcal{S}$ and some $a, b \in \mathcal{L}$, $\alpha(a) = \alpha(b) = 1$, then there exists $c \in \mathcal{L}$ such that $c \leqslant a$, $c \leqslant b$, and $\alpha(c) = 1$.

ENCYCLOPEDIA OF MATHEMATICS and Its Applications, Gian-Carlo Rota (ed.).
Vol. 15: E. G. Beltrametti and G. Cassinelli, The Logic of Quantum Mechanics

ISBN 0-201-13514-0

From Chapter 11 we know that these conditions are sufficient to ensure that:

    (I) $\mathcal{L}$ is a complete lattice;

    (II) the assertions (1) $a \leqslant b$, (2) $\alpha(a) \leqslant \alpha(b)$ for all $\alpha \in \mathbb{S}$, (3) $\mathbb{S}_1(a) \subseteq \mathbb{S}_1(b)$ are equivalent (thus $\mathbb{S}$ is ordering and strongly ordering on $\mathcal{L}$);

    (III) if $\{a_i\}$ is any set (countable or not) of elements of $\mathcal{L}$ such that $\alpha(a_i) = 1$ for some $\alpha \in \mathbb{S}$ and all i, then $\alpha(\bigwedge_i a_i) = 1$;

    (IV) every element of $\mathbb{S}$ has support in $\mathcal{L}$, and every nonzero element of $\mathcal{L}$ is the support of some $\alpha \in \mathbb{S}$;

    (V) there exists an isomorphism between $\mathcal{L}$ and the lattice $\mathcal{K}$ of closed (with respect to superposition) subsets of $\mathbb{S}$, given by $a \mapsto \mathbb{S}_1(a)$.

For the specific purposes of this chapter, mainly devoted to physical quantities, we add the technical assumption that for every $\mathcal{L}$-valued measure on the real line $E \mapsto x(E)$ there exists a physical quantity A such that $x(E) = |A, E|$ (see Sections 13.6 and 13.7).

Of course some of the hypotheses contained in (i)–(iii) could be slightly modified (and indeed, one can find in the literature modified versions of them) without significant change in the consequences (I)–(V). (We refer to Appendix C for a discussion of interrelations among the various facts.)

14.2 Observables

We shall focus attention on the class of physical quantities of the system, defined as proposition-valued measures in accordance with Section 13.7. Since it has become customary to use the term "observables associated with $\mathcal{L}$" for the $\mathcal{L}$-valued measures on the set $\mathcal{B}(\mathbb{R})$ of real Borel sets, we can say that the physical quantities are represented by the observables associated with the logic $\mathcal{L}$ of the physical system. This might appear to be a terminological tautology, for "observable" has a physical appeal (something that can be observed); but "observable" has taken an autonomous role in the mathematical language. To this role we shall refer later.

The *spectrum* $\sigma(x)$ of the observable x is defined as the smallest closed subset E of $\mathbb{R}$ such that $x(E) = 1$. Physically, it corresponds to the values that (the physical quantity associated with) x may attain. When $\sigma(x)$ is a bounded set we say that the observable x is bounded. The probability distribution of the observable x in the state α (of course α is thought of as a probability measure on $\mathcal{L}$) is defined by

$$E \mapsto \alpha(x(E)), \qquad E \in \mathcal{B}(\mathbb{R}). \qquad (14.2.1)$$

The expectation, or mean value, $\mathcal{E}(x, \alpha)$ of the observable x in the state α is similarly defined as the mean value of the distribution (14.2.1), i.e., as

$$\mathcal{E}(x,\alpha) = \int_{\mathbb{R}} \lambda\alpha(x(d\lambda)), \qquad (14.2.2)$$

provided the integral exists. The variance of x in the state α, denoted $\mathcal{V}(x,\alpha)$, is defined as

$$\mathcal{V}(x,\alpha) = \int_{\mathbb{R}} [\lambda - \mathcal{E}(x,\alpha)]^2 \alpha(x(d\lambda))$$

$$= \int_{\mathbb{R}} \lambda^2 \alpha(x(d\lambda)) - \left[\int_{\mathbb{R}} \lambda\alpha(x(d\lambda))\right]^2. \qquad (14.2.3)$$

As a simple example of an observable, let $\{a_i\}$ be any (at most countable) orthogonal sequence in $\mathcal{L}$ such that $+_i a_i = 1$ and $\langle \lambda_i \rangle$ a corresponding sequence of distinct real numbers, and define, for any $E \in \mathcal{B}(\mathbb{R})$, $x(E) = +_k a_k$, where the sum is over all k such that $\lambda_k \in E$; it is a simple check to verify that this x is indeed an observable associated with $\mathcal{L}$. In particular, given $a \in \mathcal{L}$, consider the orthogonal sequence $\{a, a^\perp\}$, and take two distinct real numbers λ_1, λ_2; the definition above now reads, explicitly,

$$\begin{array}{lll}
x(E) = 0 & \text{if neither } \lambda_1 \text{ nor } \lambda_2 \text{ is contained in } E, \\
x(E) = a & \text{if } \lambda_1 \text{ is contained in } E \text{ but } \lambda_2 \text{ is not,} \\
x(E) = a^\perp & \text{if } \lambda_2 \text{ is contained in } E \text{ but } \lambda_1 \text{ is not,} & (14.2.4) \\
x(E) = 1 & \text{if both } \lambda_1 \text{ and } \lambda_2 \text{ are contained in } E.
\end{array}$$

Thus we see that every element of $\mathcal{L}$ can be associated to an observable having a two-point spectrum. Conversely, if x has a two-point spectrum, say $\sigma(x) = \{\lambda_1, \lambda_2\}$, then, writing $x(\{\lambda_1\}) = a$, we have necessarily the pattern (14.2.4) above. All this confirms that the propositions of a physical system can be viewed as the physical quantities that have dichotomic outcome.

The two-point spectrum represents indeed the simplest nontrivial situation. In fact, an observable having a one-point spectrum has range $\{0, 1\}$ and represents a constant physical quantity whose value is just the number forming its spectrum.

Before going on let us see what the notion of observable looks like in the framework of the usual examples repeatedly considered in Chapter 11.

Example C (continued). $\mathcal{L}$ is a Boolean σ-algebra Σ of subsets of a set Ω (the phase space of a classical system), and the observables reduce to the familiar measurable real-valued functions on Ω. Indeed, it is a known fact[1] that to every observable x associated with Σ there corresponds a Σ-measurable real-valued function f on Ω such that $x(E) = f^{-1}(E)$ for all

$E \in \mathcal{B}(\mathbb{R})$. Then, if α is a probability measure on Σ, the probability distribution of x in α takes the form $\alpha(f^{-1}(E))$, and the expectation $\mathcal{E}(x, \alpha)$ can be written as[2]

$$\mathcal{E}(x, \alpha) = \int_{\mathbb{R}} \lambda \alpha(f^{-1}(d\lambda)) = \int_{\Omega} f(\omega) \alpha(d\omega).$$

In this case the observables are better called "random variables", according to the standard nomenclature of probability theory, and the last equality shows that $\mathcal{E}(x, \alpha)$ is just the usual mean value of a random variable. A corresponding remark holds true for variances.

Example Q (continued). The observables associated with $\mathcal{P}(\mathcal{H})$ are projection-valued measures, and by the spectral theorem they can be identified with the self-adjoint operators on $\mathcal{H}$ (see also Exercise 1). In this example, if the state is defined by a density operator, then the notions of spectrum, probability distribution, expectation, and variance coincide with the corresponding ones examined in Section 3.1.

14.3 Compatible Propositions and Compatible Observables

We come now to the notion of compatible observables. In Chapter 12 we have examined the notion of commutativity in orthomodular posets, or lattices; when dealing with the lattice $\mathcal{L}$ of propositions of a physical system, we prefer to speak of compatibility instead of commutativity. Thus two propositions are called compatible if represented by commuting elements of $\mathcal{L}$. Two observables, say x, y, will be called compatible if any proposition in the range of one of them is compatible with every proposition in the range of the other: formally, if

$$(x(E), y(F))C \qquad \text{for any} \quad E, F \in \mathcal{B}(\mathbb{R}).$$

We proceed to mention a number of facts that will clarify the physical and the mathematical content of the notion of compatibility.

(i) Two propositions $a, b \in \mathcal{L}$ are compatible if and only if the sublattice they generate admits a separating set of dispersion-free states (Exercises 2 and 3). Recall that a state is called dispersion-free if the only values it can take are 0 and 1. Pure states of a classical system are known to be dispersion-free. On the contrary, everybody but the advocates of hidden variables (see next chapter) believes that the states of a quantum system, even the pure ones, have dispersion, that is, they take all values from 0 to 1 (this is indeed the case with Hilbert-space quantum mechanics). Thus the occurrence of dispersion for pure states appears as a branching point between classical and quantum behavior. On the other hand, the occurrence of noncompatible propositions also appears as such a branching point. The result above says that there is an intimate link between the two facts.

(ii) Two propositions $a, b \in \mathcal{L}$ are compatible if and only if they are contained in the range of a single observable, that is, if and only if they can be written as $a = x(E)$, $b = x(F)$ for some observable x (Exercise 4). Referring to the physical interpretation of propositions discussed in Section 13.6, this result says that the yes-no experiments that correspond to two compatible propositions can both be derived from a single measuring instrument relative to a physical quantity, just putting two appropriate windows on the reading scale.

(iii) The propositions in the range of any observable form a Boolean sub-σ-algebra of $\mathcal{L}$, and, conversely, every (separable) Boolean sub-σ-algebra of $\mathcal{L}$ is the range of some observable. This is an important result due to Varadarajan.[1, 2] Notice that, according to this result, the propositions in the range of an observable not only are compatible, as already said in (ii), but form a set closed under countable meets and joins, as well as under orthocomplementation.

(iv) An observable y is said to be a function of another observable x if there exists a real Borel function f on $\mathbb{R}$ such that $y(E) = x(f^{-1}(E))$ for all $E \in \mathcal{B}(\mathbb{R})$, in which case we write $y = f(x)$. Notice that f^{-1} preserves unions, intersections, and complements, so that for every real Borel function f and every observable x, $f(x)$ is a new observable. Now we have that every observable is compatible with all its functions. Indeed, if $y = f(x)$, the proposition $y(F) = x(f^{-1}(F))$ is, by item (ii), compatible with the proposition $x(E)$ for any $E, F \in \mathcal{B}(\mathbb{R})$. Notice that the expectation $\mathcal{E}(f(x), \alpha)$ can also be written as an integral over the distribution of x, namely

$$\mathcal{E}(f(x), \alpha) = \int_{\mathbb{R}} f(\lambda) \alpha(x(d\lambda)),$$

by the use of a standard property of integration theory.[3]

(v) If the range of an observable y is contained in the range of an observable x, then y is a function of x. For the proof of this fact we refer to Gudder.[4]

(vi) Given an index set $\mathcal{I}$, $\{y_i : i \in \mathcal{I}\}$ is a collection of pairwise compatible observables if and only if there exist an observable x and Borel functions f_i, $i \in \mathcal{I}$, such that $y_i = f_i(x)$ for all $i \in \mathcal{I}$. In other words, compatible observables can be thought of as functions of one and the same observable. This is a fundamental result on compatible observables and constitutes a generalization of a theorem of von Neumann on commuting self-adjoint operators in Hilbert space (see Section 3.2); it was conjectured in this form by Mackey, and then proved by Varadarajan.[1, 2] Actually, the hypotheses (Section 14.1) are slightly redundant: $\mathcal{L}$ need not be a lattice; it is enough that $\mathcal{L}$ is a poset with the property that if a, b, c are pairwise compatible elements of $\mathcal{L}$, then $(a, b \vee c)C$, a property whose role has been discussed in Section 12.2. Let us also remark that, in case the index set $\mathcal{I}$ is countable, then the separability of $\mathcal{L}$ is not required; in case $\mathcal{I}$ contains just two marks, then the foregoing property is useless (see again Section 12.2).

The notion of a function of several observables can now be investigated. Technically, one must first raise the following question: given the observables $x_1, \ldots, x_n$, is there a σ-homomorphism of $\mathcal{B}(\mathbb{R}^n)$ into $\mathcal{L}$ that, on the elements of $\mathcal{B}(\mathbb{R}^n)$ having the product form $E_1 \times \cdots \times E_n$ $[E_1, \ldots, E_n \in \mathcal{B}(\mathbb{R})]$, reduces to $E_1 \times \cdots \times E_n \mapsto x_1(E_1) \wedge \cdots \wedge x_n(E_n)$? The answer[1, 4] is that such an homomorphism exists and is unique if and only if the observables $x_1, \ldots, x_n$ are pairwise compatible. Then assume this compatibility and denote the homomorphism by*

$$\tau_{x_1, \ldots, x_n}(\Delta), \qquad \Delta \in \mathcal{B}(\mathbb{R}^n).$$

Now we say that the observable y is function of $x_1, \ldots, x_n$ if there exists a measurable real function $f(\lambda_1, \ldots, \lambda_n)$ on $\mathbb{R}^n$ such that

$$y(E) = \tau_{x_1, \ldots, x_n}\left(f^{-1}(E)\right), \qquad E \in \mathcal{B}(\mathbb{R}),$$

in which case we write $y = f(x_1, \ldots, x_n)$. Notice the strict analogy between this definition and the one that applies to operators in Hilbert space [see Section 3.2, item (i)]. Notice also that, in case $x_1, \ldots, x_n$ are functions of one and the same observable x, say $x_i = g_i(x)$, then $f(x_1, \ldots, x_n)$ reduces, as expected, to a composite function of x:

$$f(x_1, \ldots, x_n) = \left(f \circ (g_1, \ldots, g_n)\right)(x).$$

Finally remark that two observables that are different functions of the same compatible observables are compatible [compare Section 3.2, item (ii)].

The notion of a complete set of compatible observables, which plays a fundamental role in quantum theory, can now be given. We say that the compatible observables $x_1, \ldots, x_n$ form a *complete set* when every observable y that is compatible with all of them is a function of them. It would be the same to say that the compatible observables $x_1, \ldots, x_n$ form a complete set when the range of $\tau_{x_1, \ldots, x_n}$ is a maximal Boolean sub-σ-algebra of $\mathcal{L}$ (see Exercise 5). And it would also be the same to say that the range of the observable of which $x_1, \ldots, x_n$ are functions is a maximal Boolean sub-σ-algebra of $\mathcal{L}$. The reader will again notice the analogy with the characterization of complete sets of self-adjoint operators in Hilbert space (see Section 3.2).

*Notice that it is the lattice-theoretic counterpart of what in Section 3.2 we have denoted by $P_{A_1, \ldots, A_n}(\Delta)$.

For compatible observables it is meaningful to speak of joint distributions. Given the compatible observables $x_1, \ldots, x_n$ and the state α on $\mathfrak{L}$, the *joint distribution* of $x_1, \ldots, x_n$ in the state α, denoted $\alpha_{x_1, \ldots, x_n}$, is the probability distribution on $\mathfrak{B}(\mathbb{R}^n)$ defined by

$$\alpha_{x_1, \ldots, x_n}(\Delta) = \alpha\left(\tau_{x_1, \ldots, x_n}(\Delta)\right), \qquad \Delta \in \mathfrak{B}(\mathbb{R}^n).$$

14.4 Spectral Properties of Observables

It is possible to develop, in the general framework of the logic associated with a physical system, a spectral theory of observables that resembles in many respects the usual spectral theory of self-adjoint operators in Hilbert space. Here we sketch some results without entering into mathematical details.

The *point spectrum* $\sigma_p(x)$ of x is defined by

$$\sigma_p(x) = \left\{\lambda \in \mathbb{R} : \exists \alpha \in \mathcal{S} \text{ such that } \alpha\left(x(\{\lambda\})\right) \neq 0\right\},$$

while the *continuous spectrum* $\sigma_c(x)$ is all the rest:

$$\sigma_c(x) = \sigma(x) \backslash \sigma_p(x).$$

If $\lambda \in \sigma_p(x)$, then any state such that $\alpha(x(\{\lambda\})) = 1$ is called an *eigenstate* of x belonging to λ.

An observable x is called *invertible* when $x(\{0\}) = 0$; in this case we consider the real function $f(\lambda) = 1/\lambda$ [for $\lambda \neq 0$, while $f(0) =$ any real number] and define x^{-1} as $f(x)$. Notice that $x^{-1}x = I$, where the product of x^{-1} and x must be understood in the sense of a function of compatible observables, and I denotes the unique observable whose spectrum reduces to the number 1.

The spectrum of an observable x can be classified using the observable $x - \lambda$ in exactly the same way as is usually done in operator theory:

THEOREM 14.4.1.

(a) λ is outside $\sigma(x)$ if and only if $x - \lambda$ is invertible and $(x - \lambda)^{-1}$ is bounded;

(b) λ is contained in $\sigma_p(x)$ if and only if $x - \lambda$ is not invertible;

(c) λ is contained in $\sigma_c(x)$ if and only if $x - \lambda$ is invertible and $(x - \lambda)^{-1}$ is not bounded.

When an observable is a function of another one, say $y = f(x)$, there are several results that allow us to characterize the spectrum of y in terms of the spectrum of x. The main facts, discovered by Gudder,[4, 5] are known as the "spectral mapping theorem" and read:

THEOREM 14.4.2.

(a) If E is a Borel set containing $\sigma(x)$, then $\sigma(f(x))$ is contained in the closure of $f(E)=\{\lambda\in\mathbb{R}:\lambda=f(\xi),\ \xi\in E\}$, where the closure is understood in the sense of the usual topology of real numbers.
(b) If f is a continuous function, then $\sigma(f(x))$ equals the closure of $f(\sigma(x))$.
(c) If f is continuous and x is bounded, then $\sigma(f(x))=f(\sigma(x))$.

Part (c), which refers to the most familiar case, says, in physical language, that in order to measure the physical quantity represented by y we can simply measure the one represented by x and then operate on the observed results according to the given function f.

In Hilbert-space quantum mechanics, where observables correspond to self-adjoint operators, it is known[6] that the self-adjoint operators are in one-to-one correspondence with the so called "spectral families of projections"; one might notice[7] that an analogous fact holds in the more general context here considered, but we shall not make use of it.

14.5 Linearity Properties of Observables

When $\mathcal{L}$ is the lattice $\mathcal{P}(\mathcal{H})$ of projectors on a Hilbert space, the bounded observables (that is, the bounded self-adjoint operators) form a linear manifold with respect to the sum of operators. In the more general framework here considered, we have at hand a notion of the sum of bounded observables only when we deal with compatible observables: in that case the sum is defined via the notion of a function of compatible observables. But when the observables are noncompatible we must resort to some other way out. An obvious possibility is to have recourse to the notion of expectation and to say that the bounded observable z is the sum of the bounded observables x and y whenever

$$\mathcal{E}(z,\alpha)=\mathcal{E}(x,\alpha)+\mathcal{E}(y,\alpha) \qquad \text{for all} \quad \alpha\in\mathcal{S}. \qquad (14.5.1)$$

We must say that it is an open problem to determine the minimal conditions on the logic $\mathcal{L}$ and on the set $\mathcal{S}$ of its states able to secure the existence of the sum of bounded observables, defined according to (14.5.1). The orthomodularity of $\mathcal{L}$ and the strongly ordering property of $\mathcal{S}$ are certainly not sufficient, as shown by counterexamples.[4, 8] It is clear, however, that in general z is not uniquely determined by the equality (14.5.1): take for instance x and y with $\mathcal{E}(x,\alpha)=\mathcal{E}(y,\alpha)$ for all α; then $2x$ and $2y$

both satisfy (14.5.1). Thus, in order to have z uniquely defined we need the assumption that the expectations separate the set of bounded observables, that is,

$$\mathscr{E}(x_1,\alpha)=\mathscr{E}(x_2,\alpha) \quad \text{for all} \quad \alpha\in\mathbb{S} \quad \text{implies} \quad x_1=x_2. \quad (14.5.2)$$

This condition, which is known to be fulfilled in Hilbert-space quantum mechanics (see Exercise 6) as well as in classical mechanics (see Exercise 7), is however not ensured by the $(\mathcal{L},\mathbb{S})$ structure depicted in Section 14.1, though it works for particular classes of observables—for instance for a set of compatible observables, and for the class of observables whose spectrum has at most one limit point. This last fact has been proved by Gudder,[4, 8] who has also proved that the equality $\mathscr{E}(f(x_1),\alpha)=\mathscr{E}(f(x_2),\alpha)$ for all Borel functions f and all $\alpha\in\mathbb{S}$ would indeed imply $x_1=x_2$.

It is interesting to note, on the other hand, that assuming for $\mathcal{L}$ only the structure of an orthomodular poset and for $\mathbb{S}$ the property of being strongly ordering, the existence and uniqueness of the sum of bounded observables ensures that $\mathcal{L}$ is a lattice and that the states have property (iii) of Section 14.1: $\alpha(a)=\alpha(b)=1$ implies $\alpha(a\wedge b)=1$.

Let us add a few remarks.

Remark 1. In familiar cases in which a direct notion of the sum of bounded observables is available, there is perfect equivalence with the sum defined via the expectation values. This happens when two observables are compatible, in which case the sum is directly available as a function of the two observables; or when $\mathcal{L}$ is $\mathcal{P}(\mathcal{H})$, in which case the sum is directly given as a sum of operators; or when we deal with a classical system, in which case the sum is directly given as a sum of real functions. The reader is referred to Exercises 8, 9.

Remark 2. In case the sum of observables exists and is unique, then the properties

$$(x+y)+z=x+(y+z),$$
$$x+y=y+x,$$
$$\lambda(x+y)=\lambda x+\lambda y \qquad \text{for all} \quad \lambda\in\mathbb{R}$$

are immediate consequences, so that the set of bounded observables is a real vector space. Moreover,[8] it is normed with respect to the norm $\|x\|=\sup\{|\lambda|:\lambda\in\sigma(x)\}$. Notice that the assumption that the bounded observables form a normed real vector space is the starting point of the formulation of quantum mechanics proposed by Segal,[9] in which the states are identified with the normalized positive linear functionals on that normed vector space.

Remark 3. Whenever the sum of observables is uniquely defined, it becomes possible to introduce a notion of product by setting

$$x \circ y = \frac{(x+y)^2 - x^2 - y^2}{2}.$$

This product, usually called the symmetrized product, is commutative by definition but, in general, is not associative nor distributive with respect to the sum. In case $\mathcal{L} = \mathcal{P}(\mathcal{H})$ the distributivity is recovered and one gets, for the set of bounded observables, a commutative nonassociative algebra, usually called a Jordan algebra. This structure has been studied in connection with some early axiomatizations of quantum mechanics and is at the root of the so-called "algebraic approaches to quantum mechanics".[10]

14.6 Noncompatible Observables and Heisenberg Inequalities

In this section we reconsider in a more general framework the problem of Heisenberg inequalities, which has already been discussed in Section 3.4 in terms of the Hilbert-space formulation of quantum mechanics. As in Section 3.4, we particularly refer to Lahti.[11]

The occurrence of a nonvanishing variance $\mathcal{V}(x, \alpha)$ of an observable x in a pure state α is a distinguishing feature of quantum systems: we have often referred to it by saying that quantum states have dispersion. Nevertheless, for any fixed observable x, it is always possible to find at least one state α that makes $\mathcal{V}(x, \alpha)$ arbitrarily small; the proof of this fact closely follows the analogous one given in the remark at the end of Section 3.1.

Here, our main concern is to examine the product of the variances of two observables, say x and y, in the same state (of course we suppose that in this state both variances exist). We are faced with two alternatives:

(I) for every $\varepsilon > 0$ there exists a state α such that $\mathcal{V}(x, \alpha)\mathcal{V}(y, \alpha) \leqslant \varepsilon$,
(II) there exists $\varepsilon \geqslant 0$ such that for every $\alpha \in \mathcal{S}$, $\mathcal{V}(x, \alpha)\mathcal{V}(y, \alpha) > \varepsilon$.

Alternative (II) is precluded for classical systems: for them alternative (I) occurs in a stronger form, since for every pure state α we have that $\mathcal{V}(x, \alpha)\mathcal{V}(y, \alpha) = 0$. On the contrary, alternative (II) is accessible to quantum systems, though it is not a necessity. Whenever two observables give rise to the situation (II), it is customary to say that they satisfy a Heisenberg inequality. But, as stated, not every pair of observables of a quantum system gives rise to a Heisenberg inequality; indeed we have the following facts:

(i) if one of the observables, say x, has a nonempty point spectrum and at least one of its eigenstates makes the variance of y well defined, then alternative (I) occurs;

(ii) if one of the observables is bounded, then alternative (I) occurs;
(iii) if x and y are compatible, then alternative (I) occurs.

The proofs of these facts follow closely the corresponding ones given in Section 3.4, noticing that everything can be repeated without explicit reference to the Hilbert-space structure but using only the properties of observables considered in this chapter. All this specifies the cases in which Heisenberg inequalities can occur: the two observables have to be noncompatible, unbounded, with purely continuous spectrum (a circumstance not sufficiently emphasized by the teaching tradition in quantum mechanics). But when must Heisenberg inequalities occur? This is an open problem in the general descriptive scheme here considered. In the Hilbert-space formulation it was possible to give a sharper characterization of the Heisenberg inequalities, the one provided by (3.4.2). But that formula made essential use of the notion of scalar product, which has no counterpart, at this stage, in our scheme.

The formalism we are here considering is too general to take care of the features of particular quantum systems; rather, it accounts for the general features that are common to all quantum systems. It points to the fact that Heisenberg inequalities are not accessible to classical systems, thus being a typical quantum phenomenon, but it is not structured enough to make a clear cut between quantum systems in which Heisenberg inequalities do occur, and quantum systems in which they do not.

We emphasize once more, as we did in Sections 3.4 and 3.5, that the actual occurrence of Heisenberg inequalities does not imply any notion of "simultaneous measurability": it merely refers to the spreading of the distributions of two observables in the same state.

14.7 Complementarity

The notion of complementary physical quantities has been introduced in Section 3.5 within the Hilbert-space framework. As already remarked there, the heavy but wavering use of the word "complementarity" in the pioneering writings on quantum mechanics partially reflected the attitude of describing quantum phenomena by concepts and words of classical mechanics. Though it is generally necessary for the initial growth of a new theory to use the language of the theory to be supplanted, the habit of speaking about quantum systems in terms of classical concepts perhaps lasted too long, as the use of the work "complementarity" over half a century may epitomize.

In Section 3.5 we isolated a precise, though admittedly onesided, notion of complementarity; the purpose of the present section is to show that that idea of complementarity can be founded on the general $(\mathcal{L}, \mathcal{S})$ structure here adopted. For simplicity, we shall initially refer to the propositions of the system, rather than to the wider class of all physical quantities (observables).

We say that two propositions $a, b \in \mathfrak{L}$ are complementary if it is impossible to find a state that causes with certainty the yes outcome of both a and b, namely, if there is no $\alpha \in \mathfrak{S}$ such that $\alpha(a) = \alpha(b) = 1$. This translates into unambiguous terms the intuitive idea that a and b cannot be "simultaneously true". We can rephrase this definition by saying that a and b are complementary whenever $\mathfrak{S}_1(a) \cap \mathfrak{S}_1(b) = 0$, and in our hypotheses this is also equivalent to saying $a \wedge b = 0$.

Of course, in an orthocomplemented lattice there are trivial examples of complementary elements, for any two orthogonal elements are complementary. The interesting question is whether there are, in $\mathfrak{L}$, propositions that are complementary without being orthogonal.

As an example of nontrivial complementary propositions, consider the lattice of polarizations of a spin-$\frac{1}{2}$ system (see Figure 10.7): any two elements different from 0 and 1 are complementary, but they are orthogonal only when they are orthocomplements of each other. Another familiar example has been touched on in Section 4.4: in the projection lattice $\mathcal{P}(\mathcal{H})$ of a free particle without spin, if two projectors belong, respectively, to the spectral decompositions of the position Q and of the momentum P and are images of bounded Borel sets, then they are complementary without being, in general, orthogonal.[11]

There is some similarity between complementarity and noncompatibility. Indeed, by simple inspection of the definition of commutativity (see Section 12.1), we have that $a \wedge b = 0$, with a nonorthogonal to b, implies that a cannot commute with b: thus, nontrivial complementarity implies noncompatibility. But the converse is not so sharp. For instance, from Hilbert-space quantum mechanics we know that the proposition $a_x = $ "the x-component of the orbital angular momentum of a particle is zero" is noncompatible with a_y, but the meet $a_x \wedge a_y$ is not zero, for the state of zero angular momentum assigns the yes outcome with certainty to both a_x and a_y. Thus, we can have noncompatible propositions that are not complementary.

Nevertheless, it holds true that if the proposition lattice $\mathfrak{L}$ contains a pair, say a, b, of noncompatible propositions, then it certainly contains a pair (not necessarily the pair a, b itself) of nontrivial complementary propositions. This is a consequence of the following fact (Exercise 10):

THEOREM 14.7.1. *An orthomodular lattice $\mathfrak{L}$ is Boolean if and only if $a \wedge b = 0$ implies $a \perp b$.*

Thus we can say that the following are equivalent statements:

   (i) $\mathfrak{L}$ is non-Boolean,
   (ii) $\mathfrak{L}$ contains at least one pair of noncompatible propositions,
   (iii) $\mathfrak{L}$ contains at least one pair of complementary but nonorthogonal propositions.

The notion of complementarity is naturally transferred from propositions to observables (or physical quantities). We say that two observables x, y associated with $\mathcal{L}$ are complementary if every proposition different from I in the range of x is complementary to every proposition different from I in the range of y. In more explicit terms, x and y are complementary if, for any two bounded Borel sets E and F not containing respectively $\sigma(x)$ and $\sigma(y)$, we have $x(E) \wedge y(F) = 0$.

Familiar examples of complementary physical quantities are, in Hilbert-space quantum mechanics, the position Q and the momentum P of a free particle.

In analogy to what was said for propositions, when dealing with observables we also have that nontrivial complementarity excludes compatibility. This is made precise by the following result, due to Lahti[11] (Exercise 11):

THEOREM 14.7.2. *Two complementary observables x and y are compatible only in case one of them is a constant observable.*

Concluding, we can say that if a physical system exhibits nontrivial complementarity (of propositions or of observables), then the lattice $\mathcal{L}$ associated with it cannot be Boolean. Since classical systems do belong to the Boolean case, we have that nontrivial complementarity represents a typical quantum feature, a branching point between classical and quantum behavior.

14.8 Superpositions of Pure States

The crucial role played in quantum mechanics by superpositions of pure states has been already outlined in Part I. In this section we want to argue that the notion of quantum superposition of pure states does not need the full machinery of the Hilbert-space formalism, but can be dealt with at the more fundamental level of the logic associated with quantum systems.

Assume that $\mathcal{S}$ contains a nonempty subset $\mathcal{S}^P$ of pure states. In Section 11.5 we have introduced the notion of superposition relative to a generic ambient set of states: if we identify the ambient set with $\mathcal{S}^P$, we get the notion of superposition with respect to $\mathcal{S}^P$, that is, the desired notion of quantum superposition. In the rest of this section the closure $\overline{S}$ of a subset of S of $\mathcal{S}^P$ will be always understood relative to the ambient $\mathcal{S}^P$.

Hence, given two distinct pure states α, β, we say that the pure state γ is a quantum superposition of α and β when $\gamma \in \overline{\{\alpha, \beta\}}$ with $\gamma \neq \alpha, \beta$. As seen in Section 11.5, we can rephrase this definition in terms of supports of states by saying that γ is a quantum superposition of α and β when it is pure and $s(\gamma) \leqslant s(\alpha) \vee s(\beta)$. Of course, this notion of superposition is immediately extended to the case of superposition of any subset of pure states. Notice also that when $\mathcal{L} = \mathcal{P}(\mathcal{H})$, this definition of quantum superposition reduces to the usual linear combination of vectors. Indeed, we know that the pure states on $\mathcal{P}(\mathcal{H})$ are identified with the one-dimensional subspaces of $\mathcal{H}$ that are their supports and are the atoms of $\mathcal{P}(\mathcal{H})$; therefore, denoting by φ_1, φ_2 two (nonproportional) unit vectors in $\mathcal{H}$, we have that the state (represented by) φ is a quantum superposition of φ_1 and φ_2 if and only if it has the form $c_1\varphi_1 + c_2\varphi_2$ with c_1 and c_2 complex numbers $(|c_1|^2 + |c_2|^2 = 1)$.

To discuss the properties, meaning, and existence of quantum superpositions in the $(\mathcal{L}, \mathcal{S})$ structure we have to add some assumptions to the ones of Section 14.1. We need the following:

(1) $\mathcal{S}^P$ is sufficient on $\mathcal{L}$: that is, for every $a \in \mathcal{L}$, $a \neq 0$, there exists $\alpha \in \mathcal{S}^P$ such that $\alpha(a) = 1$. This assumption is well founded from the physical point of view, at least for quantum systems with a finite number of degrees of freedom. However, it might put some limitations on the class of quantum systems we can describe (e.g., statistical systems associated with type-III factors).

(2) A single pure state cannot produce quantum superpositions, that is, $\overline{\{\alpha\}} = \{\alpha\}$. This fact seems physically obvious.

Assumptions (1) and (2) above imply that:

(i) $\mathcal{S}^P$ is strongly ordering on $\mathcal{L}$ [by Theorem 11.4.4].
(ii) $\mathcal{L}$ is atomic, and the restriction to $\mathcal{S}^P$ of the support function defined on $\mathcal{S}$ determines a one-to-one correspondence between $\mathcal{S}^P$ and the set of atoms of $\mathcal{L}$ (see Section 11.6).
(iii) The lattice $\mathcal{K}^P$ of closed subsets of $\mathcal{S}^P$ is isomorphic to $\mathcal{L}$ (see Section 11.5).

The last item is often regarded as the abstract formulation, in the $(\mathcal{L}, \mathcal{S})$ structure, of the familiar "superposition principle" of quantum mechanics. We think however that the isomorphism between $\mathcal{K}^P$ and $\mathcal{S}^P$ does not catch all the flavor of the superposition principle: in particular, it does not imply that (neglecting superselection rules) any two pure states can be superposed to generate a new pure state, which would appear to be the main point of the superposition principle.

Thus we come to the problem of understanding when two pure states can generate quantum superpositions. This problem will be clarified by the following series of facts.

THEOREM 14.8.1. *Two pure states α, β admit quantum superpositions if and only if the join of the atoms $s(\alpha)$ and $s(\beta)$ contains at least a third, different atom.*

THEOREM 14.8.2. *$\mathcal{L}$ is distributive if and only if no pair of pure states admits quantum superpositions.*

(See Exercise 12.)

THEOREM 14.8.3. *If $\mathcal{L}$ is not distributive, then any two nonorthogonal pure states admit at least a quantum superposition.*

(This is a by-product of Exercise 12.)

THEOREM 14.8.4. *If any two pure states admit at least one quantum superposition, then $\mathcal{L}$ is irreducible.*

(See Exercise 13.)

THEOREM 14.8.5. *If $\mathcal{L}$ is reducible, but not Boolean, so that $\mathcal{L}$ is the direct sum of the segments $\mathcal{L}[0, z_n]$ where z_n is the set of atoms of the Boolean algebra $Z(\mathcal{L})$, then, given a pure state $\alpha^\sim$ on $\mathcal{L}[0, z_n]$, the function α on $\mathcal{L}$ defined by $\alpha(a) = \alpha^\sim(a \wedge z_n)$ is a pure state on $\mathcal{L}$. Moreover, all pure states on $\mathcal{L}$ are generated in this way; hence, denoting by $\mathcal{P}_n$ the set of pure states on $\mathcal{L}$ generated by the pure states on $\mathcal{L}[0, z_n]$, we have $\mathcal{P} = \bigcup_n \mathcal{P}_n$.*

THEOREM 14.8.6. *Two pure states belonging to different $\mathcal{P}_n$'s do not admit quantum superpositions.*

(See Exercise 14.)

THEOREM 14.8.7. *Each $\mathcal{P}_n$ is closed under superpositions, that is, if two pure states belonging to the same $\mathcal{P}_n$ admit quantum superpositions, then these superpositions still belong to that $\mathcal{P}_n$.*

(See Exercise 15.)

The facts above are not sufficient to solve entirely the problem of specifying when two pure states do admit quantum superpositions, even though Theorem 14.8.3 constitutes a remarkable step. To fill the gap we shall add one further hypothesis, namely the covering property of $\mathcal{L}$. As we are going to see, it is a sufficient condition to recover the main content of the superposition principle, though its necessity has not been proved and one might have the feeling that it is indeed unnecessary.

THEOREM 14.8.8. *If $\mathcal{L}$ also has the covering property, then any two states in the same $\mathcal{P}_n$ certainly admit at least one quantum superposition, which by Theorem 14.8.7 still belongs to $\mathcal{P}_n$.*

(See Exercise 16.)

---

*Note that $\mathcal{L} = \bigcup \mathcal{L}[0, z_n]$, and that if $\mathcal{L}$ is separable then the set of atoms of $Z(\mathcal{L})$ is countable. The properties of α and the equality $\mathcal{P} = \bigcup_n \mathcal{P}_n$ have been proved by Varadarajan.[1]

THEOREM 14.8.9. *If $\mathcal{L}$ also has the covering property, then any two pure states admit quantum superpositions if and only if $\mathcal{L}$ is irreducible.*

(See Exercise 17.)

Let us stress that in our context the covering property finds a natural physical motivation: according to what was said in Section 11.7, in particular formulation (CP), we have in fact

THEOREM 14.8.10. *$\mathcal{L}$ has the covering property whenever the quantum superposition operation has the exchange property, i.e., whenever the assertion that γ is a superposition of α and β entails the assertion that α (respectively β) is a superposition of γ and β (respectively α).*

The above theorems show that the notion of quantum superposition here used, which makes no reference to any Hilbert-space structure, adheres to the main features empirically demanded by the description of quantum physical systems. Irreducible proposition lattices correspond to the unrestricted validity of the superposition principle, and thus correspond to physical systems with unlimited quantum behavior. Reducible proposition lattices correspond to the usual pattern of superselection rules: the set of pure states is partitioned into pieces, and the superposition principle holds within each piece, but not between states belonging to different pieces. We are thus faced with the situation corresponding to physical systems with limited quantum behavior, as discussed in Chapter 5.

Exercises

1. If $\mathcal{L} = \mathcal{P}(\mathcal{H})$, if $\langle P_k \rangle$ is an orthogonal sequence of projectors, and if $\langle \lambda_k \rangle$ is a corresponding sequence of distinct real numbers, then the observable x whose projection-valued measure is $x(E) = +(P_k : \lambda_k \in E)$ corresponds to the self-adjoint operator A_x whose spectral decomposition is $A_x = \Sigma_k \lambda_k P_k$.
   [*Hint.* Use the spectral theorem for operators with pure point spectrum.]

2. Show that the pure states on a Boolean algebra $\mathcal{B}$ are dispersion-free and form a separating set.
   [*Hint.* See p. 116 of Reference 1. See also Example C of Section 11.6.]

3. If the sublattice generated by $a, b \in \mathcal{L}$ admits a separating set S of dispersion-free states with the property (11.4.1), then $(a, b)C$.
   [*Hint.* If $\alpha(a) = 0$, $\alpha \in S$, then $\alpha(a \wedge b^{\perp}) = \alpha(a \wedge b) = 0$; if $\alpha(a) = 1$, then either $\alpha(a \wedge b^{\perp}) = 1$ or $\alpha(a \wedge b) = 1$. In both cases $\alpha(a) = \alpha((a \wedge b^{\perp}) + (a \wedge b))$. Then use the sufficiency of S to conclude $(a, b)C$.]

4. $(a, b)C$ if and only if $a = x(E)$, $b = x(F)$ for some observable x.
   [*Hint.* If $(a, b)C$, write $c_1 = a \wedge b^{\perp}$, $c_2 = a \wedge b$, $c_3 = a^{\perp} \wedge b$, $c_4 = a^{\perp} \wedge b^{\perp}$, and define $x(E) = +(c_k : k \in E)$, so that $a = x(\{1, 2\})$, $b = x(\{2, 3\})$. If $a =$

$x(E)$ and $b=x(F)$, notice that $x(E)=x(E\cap F')+x(E\cap F)$ and similarly for $x(F)$, where the prime denotes set complement. Hence $(a, b)C$.]

5. Given the compatible observables $x_1,\ldots, x_n$, the range of $\tau_{x_1,\ldots, x_n}$ is a maximal Boolean sub-σ-algebra of $\mathcal{L}$ if and only if every observable y that is compatible with all the x_i's is a function of them.

[*Hint.* Suppose the range of $\tau_{x_1,\ldots, x_n}$ is maximal; remark that every proposition in the range of y is compatible with every proposition in the range of $\tau_{x_1,\ldots, x_n}$; then use maximality to conclude that the range of y is contained in the range of $\tau_{x_1,\ldots, x_n}$, so that $y=f(x_1,\ldots, x_n)$. Conversely, proceed *ad absurdum* and denote by a an element of $\mathcal{L}$ that is supposed to be compatible with every element in the range of $\tau_{x_1,\ldots, x_n}$ without belonging to it; then remark that the observable x_a defined by $x_a(\{1\})=a$, $x_a(\{0\})=a^\perp$ would be compatible with $x_1,\ldots, x_n$ without being a function of them.]

6. Show that for the bounded observables associated with $\mathcal{P}(\mathcal{H})$, the equality $\mathcal{E}(x_1,\alpha)=\mathcal{E}(x_2,\alpha)$ for all pure states implies $x_1=x_2$.

[*Hint.* $\mathcal{E}(x_1,\alpha)$ takes the form $(\psi, A_1\psi)$, A_1 being a bounded self-adjoint operator and ψ a unit vector; then $(\psi, A_1\psi)=(\psi, A_2\psi)$ for all unit vectors implies $A_1-A_2=0$ and hence $A_1=A_2$.]

7. Show that for the bounded observables associated with a Boolean σ-algebra Σ of subsets of a phase space Ω, the equality $\mathcal{E}(x_1,\alpha)=\mathcal{E}(x_2,\alpha)$ for all probability measures concentrated at points of Ω implies $x_1=x_2$.

[*Hint.* Recall Example C in Section 14.2 and conclude that if α is a probability measure concentrated at $\bar\omega$, then the equality $\mathcal{E}(x_1,\alpha)=\mathcal{E}(x_2,\alpha)$ reduces to $f_1(\bar\omega)=f_2(\bar\omega)$, where f_1 and f_2 are the function notation for the observables x_1 and x_2.]

8. Show that when two bounded observables x, y are compatible, the sum defined via the function of compatible observables coincides with the sum defined via the expectation values.

[*Hint.* Write $x=f(z)$ and $y=g(z)$; remark that if $h(\lambda_1,\lambda_2)=\lambda_1+\lambda_2$, then $h(x, y)=(h\cdot(f, g))(z)$, so that $\mathcal{E}(h(x, y),\alpha)=\int[f(\lambda)+g(\lambda)]\alpha(z(d\lambda))$.]

9. When $\mathcal{L}$ is identified with $\mathcal{P}(\mathcal{H})$ or with a σ-algebra Σ of a set Ω, the sum of two bounded observables defined via the expectation values coincides with the sum of bounded operators or, respectively, with the sum of Borel functions.

[*Hint.* By direct inspection.]

10. Prove Theorem 14.7.1.

[*Hint.* If $\mathcal{L}$ is Boolean and $a\wedge b=0$, then $a=(a\wedge b^\perp)+(a\wedge b)=a\wedge b^\perp$; hence $a\perp b$. Conversely, from the identity $[a-(a\wedge b)]\wedge[b-(a\wedge b)]=0$ deduce $[a-(a\wedge b)]\perp[b-(a\wedge b)]$; then use orthomodularity to write $a=(a$

$\wedge b)+[a-(a\wedge b)]$ and $b=(a\wedge b)+[b-(a\wedge b)]$, and conclude $(a,b)C$.]

11. Prove Theorem 14.7.2.

[*Hint.* If x, y are complementary and compatible, write $x=f(z)$, $y=g(z)$, and remark that the ranges of x, y are contained in the range of z, so that, unless x or y is constant, there would be a partition of $\mathbb{R}$, say $\mathbb{R}=\cup_i E_i$, such that there is no E_i containing $\sigma(x)$ or $\sigma(y)$; hence the contradiction
$$1=x(\cup_i E_i)\wedge y(\cup_i E_i)=[\vee_i x(E_i)]\wedge[\vee_j y(E_j)]=\vee_i\{\vee_j[x(E_i)\wedge y(E_j)]\}$$
$$=\vee_i\vee_j 0=0.]$$

12. Prove Theorem 14.8.2.

[*Hint.* Suppose $\overline{\{\alpha,\beta\}}=\{\alpha,\beta\}$, and write $p=s(\alpha)$, $q=s(\beta)$, $\varphi_{p^\perp}(q)=(p\vee q)\wedge p^\perp$. Since $p^\perp \not\perp q$, it follows, by known properties of Sasaki projections,[12] that $\varphi_{p^\perp}(q)\neq 0$, so that there is in $\varphi_{p^\perp}(q)$ an atom r. If $r\neq q$ (of course $r\neq p$ and $r\leqslant p\vee q$), then r would be the support of a quantum superposition of α, β, contrary to the hypothesis. Thus $r=q$; hence $\varphi_{p^\perp}(q)\geqslant q$; hence $p\perp q$. This proves that any two atoms of $\mathfrak{L}$ are orthogonal; hence by atomisticity $\mathfrak{L}$ is Boolean. Suppose now $\mathfrak{L}$ is Boolean; if p,q are atoms and $r\leqslant p\vee q$, then $r=r\wedge(p\vee q)=(r\wedge p)\vee(r\wedge q)=0$, so that there are no quantum superpositions.]

13. Prove Theorem 14.8.4.

[*Hint.* Suppose, if possible, that $z\in Z(\mathfrak{L})$, $z\neq 0,1$. Choose an atom p in z and an atom q in $z^\perp$, let r be a third atom in $p\vee q$, and use orthomodularity to write $p\vee q=r+[(p\vee q)\wedge r^\perp]$. Distinguish the two cases $r\leqslant z$ and $r\leqslant z^\perp$. If $r\leqslant z$, write $p\vee q=r+[(p\vee q)\wedge(r^\perp\vee z^\perp)]$, and use the Foulis-Holland theorem (see Section 12.3) twice to get $p\vee q=r\vee q$, whence $z\wedge(p\vee q)=z\wedge(r\vee q)$ and, using the Foulis-Holland theorem once more, $(z\wedge p)\vee(z\wedge q)=(z\wedge r)\vee(z\wedge q)$, so that $r=p$. If $r\leqslant z^\perp$, similarly conclude $r=q$.]

14. Prove Theorem 14.8.6.

[*Hint.* If $\alpha\in\mathcal{P}_n$ and $\beta\in\mathcal{P}_m$ ($n\neq m$), we have $s(\alpha)\leqslant z_n$ and $s(\beta)\leqslant z_m\leqslant z_n^\perp$; then apply the arguments of the preceding exercise.]

15. Prove Theorem 14.8.7.

[*Hint.* If γ is a quantum superposition of two states in $\mathcal{P}_n$, use Theorem 14.8.5 to write $\gamma(a)=\gamma^\sim(a\wedge z_m)$ for some m and some pure state $\gamma^\sim$ on $\mathfrak{L}[0,z_m]$. Then remark that $\gamma(z_n)=1$, hence $\gamma(z_m\wedge z_n)=1$, and hence $m=n$.]

16. Prove Theorem 14.8.8.

[*Hint.* See Section 12.6 and notice that the equivalence relation $\approx$ partitions the set of atoms of $\mathfrak{L}$ into equivalence classes that correspond precisely to the $\mathcal{P}_n$'s.]

17. Prove Theorem 14.8.9.

[*Hint.* See the preceding exercise.]

References

1. V. S. Varadarajan, *Geometry of Quantum Theory*, Vol. I, Van Nostrand, Princeton, N.J., 1968.
2. V. S. Varadarajan, *Comm. Pure and Appl. Math.* 15 (1962) 189.
3. H. L. Royden, *Real Analysis*, 2nd ed., Macmillan, New York, 1963.
4. S. P. Gudder, in *Probabilistic Methods in Applied Mathematics*, Vol. 2, A. T. Bharucha-Reid, ed., Academic Press, New York, 1970.
5. S. P. Gudder, *Trans. Amer. Math. Soc.* 119 (1965) 428.
6. E. Prugovecki, *Quantum Mechanics in Hilbert Space*, Academic Press, New York and London, 1971.
7. D. E. Catlin, *Int. J. Theor. Phys.* 1 (1968) 285.
8. S. P. Gudder, *Pacific J. Math.* 19 (1966) 81; correction, *Pacific J. Math.* 19 (1966) 588.
9. I. E. Segal, *Ann. Math.* 48 (1947) 930.
10. P. Jordan, J. von Neumann, and E. P. Wigner, *Ann. Math.* 35 (1934) 29.
11. P. Lahti, "Uncertainty and Complementarity in Axiomatic Quantum Mechanics", Thesis, Univ. of Turku, Series N.D2, 1979.
12. F. Maeda and S. Maeda, *Theory of Symmetric Lattices*, Springer, Berlin, Heidelberg, New York, 1970.

Additional Bibliography for Chapters 10–14

1. V. Berzi and A. Zecca, *Commun. Math. Phys.* 35 (1974) 93.
2. K. Bugajska and S. Bugajski, *Bull. Acad. Pol. Sci.* 20 (1972) 231.
3. K. Bugajska and S. Bugajski, *Bull. Acad. Pol. Sci.* 21 (1973) 873.
4. K. Bugajska and S. Bugajski, *Rep. Math. Phys.* 4 (1973) 1.
5. K. Bugajska and S. Bugajski, *Ann. Inst. Henri Poincaré* 19A (1973) 333.
6. F. Gallone and A. Manià, *Ann. Inst. Henri Poincaré* 15A (1971) 37.
7. F. Gallone and A. Zecca, *Int. J. Theor. Phys.* 8 (1973) 51.
8. S. P. Gudder, *Trans. Amer. Math. Soc.* 119 (1965) 420.
9. S. P. Gudder, *J. Math. Mech.* 18 (1968) 296.
10. S. P. Gudder, *J. Math. Phys.* 11 (1970) 1037.
11. S. P. Gudder, *Can. J. Math.* 23 (1971) 659.
12. S. P. Gudder, *Int. J. Theor. Phys.* 6 (1972) 369.
13. J. Gunson, *Ann. Inst. Henri Poincaré* 11A (1972) 295.
14. W. Guz, *Rep. Math. Phys.* 6 (1974) 445.
15. W. Guz, *Rep. Math. Phys.* 7 (1975) 313.
16. W. Guz, *Int. J. Theor. Phys.* 16 (1977) 299.
17. W. Guz, *Ann. Inst. Henri Poincaré* 28A (1978) 1.
18. F. Jenč, *J. Math. Phys.* 13 (1972) 1675.
19. F. Jenč, *Rep. Math. Phys.* 6 (1974) 253.
20. M. J. Maczynski, *Int. J. Theor. Phys.* 8 (1973) 353.
21. M. J. Maczynski, *Int. J. Theor. Phys.* 11 (1974) 149.
22. W. Ochs, *Z. Naturforsch.* 27 (1972) 893.
23. R. Plymen, *Helv. Phys. Acta* 41 (1968) 132.
24. S. Pulmannova, *Acta Phys. Slov.* 25 (1975) 234.
25. S. Pulmannova, *Commun Math. Phys.* 49 (1976) 47.

CHAPTER 15

On the Hidden-Variables Issue

15.1 The Question of the Completeness of Quantum Mechanics

When a classical system is in a pure state, that is, when its preparation represents maximal information, the value of every physical quantity is uniquely predicted. In other words, classical pure states are dispersion-free, and classical mechanics is deterministic. Only in case the state is nonpure, thus representing nonmaximal information about the system, does probability enter into the theory.

Things are different with quantum systems. Even in case the state of the system is pure, it is impossible to predict a definite value of every physical quantity: only the probability distributions of the physical quantities are given. Pure quantum states have dispersion (though, of course, they can be dispersion-free for particular physical quantities), and probability enters into quantum theory at a much more fundamental level.

The basic idea of so-called hidden-variable theories is to reject the fundamental role of probability in quantum theory and to argue that probability arises because quantum states, even the pure ones, do not represent the ultimate information about the system. Accordingly, it is conjectured that there are certain hidden variables, not yet subject to experimental detection, which would complete the information carried by the quantum states, thus giving rise to so precise a knowledge of the system that probabilities would disappear and precise values of all physical quantities would be uniquely determined. We shall call the states so conjectured *completed states*; they embody the information coded in the usual quantum states and also the additional information coded by the hidden variables. Completed states are thus, by the very aim of hidden-variable theories,

ENCYCLOPEDIA OF MATHEMATICS and Its Applications, Gian-Carlo Rota (ed.). Vol. 15: E. G. Beltrametti and G. Cassinelli, The Logic of Quantum Mechanics

ISBN 0-201-13514-0

dispersion-free. In this way there would be, beyond quantum mechanics, an underlying classical deterministic theory such that the predictions of quantum mechanics would correspond to averaging over the hidden variables. Of course, this idea is reminiscent of the relationship between thermodynamics and the kinetic theory of gases.

The question of hidden variables has a long tradition, which goes back to the earliest days of quantum mechanics and even engaged some of the founding fathers of the theory. But it must be remarked that it is hard to find in the literature a unique or even a dominant prescription of what a hidden-variable theory ought to be. In this respect, a remarkable clarification of the subject took place in the sixties, as a result of the collision between believers in hidden-variable theories and the inventors of so-called "no-go theorems". The literature on the subject is quite rich: for extensive reviews, also in historical perspective, we refer in particular to the books of Belinfante[1] and of Jammer.[2]

Without claim of exhaustiveness, and leaving aside the more philosophical and epistemological aspects, we shall touch upon the problem of hidden variables partly in this chapter and partly in Chapter 25 where reference will be made to the Hilbert-space structure. Here the more general $(\mathcal{L}, \mathcal{S})$ structure, as summarized in Section 14.1, will be enough.

15.2 Contextual and Noncontextual Hidden-Variable Theories

A natural way of formalizing the starting point of the simplest (the noncontextual) version of hidden-variable theories goes as follows. To the pair $(\mathcal{L}, \mathcal{S})$ associated with the quantum system under discussion, a space Ω is added whose elements are interpreted as the hidden variables; since averagings over the hidden variables are needed, the minimal structure required for Ω is that of a classical probability space. This means that a family Σ (technically, a Boolean σ-algebra) of measurable subsets of Ω is given together with a probability measure μ on Σ. With usual notation, we write $\langle \Omega, \Sigma, \mu \rangle$ for that probability space. A completion of a quantum state, for simplicity a pure state $\alpha \in \mathcal{S}^P$, is then defined as a pair (α, ω) with $\omega \in \Omega$, and it is understood to be a dispersion-free state on $\mathcal{L}$, that is, a function that maps $\mathcal{L}$ into $\{0, 1\}$. In order to have consistency with the predictions of quantum mechanics, it is necessary that the probability distributions assigned to propositions (hence to physical quantities) by the quantum state α be recovered as averages over the hidden variables of the values assigned by the completed states (α, ω). Thus, for every $a \in \mathcal{L}$ and every $\alpha \in \mathcal{S}^P$, the function $\omega \mapsto (\alpha, \omega)(a)$ has to be measurable and

$$\alpha(a) = \int_{\Omega} (\alpha, \omega)(a) \mu(d\omega) \tag{15.2.1}$$

[notice that $(\alpha, \omega)(a)$ is either 0 or 1].

If the pattern above occurs, we say that the pair $(\mathcal{L}, \mathcal{S})$ admits an underlying noncontextual hidden-variable theory. The term noncontextual is motivated by the fact that a single space Ω of hidden variables has been introduced to provide dispersion-free states defined on the whole $\mathcal{L}$, so that each of these states assigns zero variance to all observables. The noncontextuality is indeed a very stringent link for hidden-variable theories, as the next section will outline. But not all advocates of hidden-variable theories are pursuing this kind of theory: some of them have, and have had all along, a different idea of what a hidden-variable theory should be. Their idea is that the choice of the hidden-variable space can, and actually must, depend upon the physical quantity to be dealt with. Thus, the choice of the hidden-variable space becomes contextual to the choice of a physical quantity, i.e. to the choice of the Boolean sub-σ-algebra of $\mathcal{L}$ that is associated with that physical quantity.[3] For sake of simplicity it is usual to consider only maximal Boolean sub-σ-algebras, corresponding to complete sets of compatible physical quantities. Then we say that the pair $(\mathcal{L}, \mathcal{S})$ admits a contextual hidden-variable theory if there is a family $\langle \Omega_{\mathcal{B}}, \Sigma_{\mathcal{B}}, \mu_{\mathcal{B}} \rangle$ of probability spaces labeled by the maximal Boolean sub-σ-algebras of $\mathcal{L}$ such that, for every $\mathcal{B}$ and every quantum state α, the pair $(\alpha, \omega_{\mathcal{B}})$, with $\omega_{\mathcal{B}} \in \Omega_{\mathcal{B}}$, defines a dispersion-free probability measure not on the whole of $\mathcal{L}$, but only on the maximal Boolean sub-σ-algebra $\mathcal{B}$ of $\mathcal{L}$, and the function $\omega_{\mathcal{B}} \mapsto (\alpha, \omega_{\mathcal{B}})(a)$, which maps $\Omega_{\mathcal{B}}$ into $\{0, 1\}$, is measurable and satisfies, for all $a \in \mathcal{B}$,

$$\alpha(a) = \int_{\Omega_{\mathcal{B}}} (\alpha, \omega_{\mathcal{B}})(a) \mu_{\mathcal{B}}(d\omega_{\mathcal{B}}) \qquad (15.2.2)$$

Notice that this equality represents for contextual theories what (15.2.1) represented for noncontextual ones, namely the requirement that the probability distributions assigned to propositions by quantum states are recovered as averages over the hidden variables of the values (0 or 1) assigned by completed states.

As will be seen in Section 15.4, the generalization that accompanies the transition from noncontextual to contextual theories is so significant as to make all the no-go theorems that have been invented for noncontextual theories irrelevant for contextual theories. However, despite the absence of mathematical obstacles against contextual hidden-variable theories, it must be stressed that their calling for completed states that are probability measures not on the whole proposition lattice $\mathcal{L}$ but only on a subset of $\mathcal{L}$ is rather far from intuitive physical ideas of what a state of a physical system should be. Thus, contextual hidden-variable theorists, in their search for the restoration of some classical deterministic aspects, have to pay, on other sides, in quite radical departures from properties of classical states.

One might argue, as a way out from these difficulties, that a completed state on the whole $\mathcal{L}$ could be constructed by gluing together the functions $(\alpha, \omega_{\mathcal{B}})$ with $\mathcal{B}$ running over the maximal Boolean sub-σ-algebras of $\mathcal{L}$. But this is in general impossible, for a single proposition a can be an element of two different maximal Boolean sub-σ-algebras, say $\mathcal{B}_1$ and $\mathcal{B}_2$, and it is perfectly conceivable that $(\alpha, \omega_{\mathcal{B}_1})(a) \neq (\alpha, \omega_{\mathcal{B}_2})(a)$.

15.3 No-Go Theorems for Noncontextual Hidden-Variable Theories

In dealing with noncontextual hidden-variable theories a conservative attitude would require that the completed states preserve the properties shared by the quantum states. In particular, this would mean requiring that the (dispersion-free) completed states form a sufficient set of probability measures on $\mathcal{L}$ [that is, for every $a \in \mathcal{L}$ there exists a completed state (α, ω) such that $(\alpha, \omega)(a) = 1$] with the property [analogous to (11.4.1)] that

$$(\alpha, \omega)(a) = 1 \text{ and } (\alpha, \omega)(b) = 1 \quad \text{implies} \quad (\alpha, \omega)(a \wedge b) = 1. \quad (15.3.1)$$

However, the possibility of a hidden-variable theory of this kind is ruled out by a no-go theorem first proved by Jauch and Piron[4] and then, under slightly weaker hypotheses, by Gudder [5]. This theorem, which generalizes a much older fact proved by von Neumann in the Hilbert-space framework (see Section 25.1), says that the mentioned requirements for the completed states unavoidably lead to the conclusion that all elements of $\mathcal{L}$ are mutually compatible. Since on the other hand the presence in $\mathcal{L}$ of noncompatible propositions is a necessary condition for the description of quantum systems, the possibility of hidden-variable theories of that kind is ruled out.

As a matter of fact, one might argue that it is too conservative and physically questionable to require of the completed states, which involve the hidden variables, the same properties fulfilled by the quantum states, the only ones under actual empirical control. In particular, one might consider it questionable to require the property (15.3.1), and one might ask what happens when it is dropped. From the mathematical point of view, this is a legitimate question, for it is known that there exist orthomodular lattices endowed with a strongly ordering set of dispersion-free probability measures that do no satisfy the condition (15.3.1): such lattices are particular examples of the so-called σ-classes studied by Gudder[6] and Neubrunn,[7] and their use as models of noncontextual hidden-variable theories has been advanced by Ochs[8] and Gudder.[9] However, it is an open problem to decide whether the noncontextual hidden-variable theories that become available by the abandonment of the condition (15.3.1) have any physical interest.

A different reinterpretation of noncontextual hidden-variable theories has been given by Zierler and Schlessinger.[10] Essentially, their idea is as follows: try to embed the proposition lattice $\mathcal{L}$ in some Boolean algebra $\mathcal{B}$, and then construct the completed dispersion-free states on $\mathcal{L}$ via the pure states on $\mathcal{B}$, which are known to be dispersion-free. The crucial step is the first one, the embedding of $\mathcal{L}$ in $\mathcal{B}$. Let us make this point more precise. We say that $\mathcal{L}$ is *faithfully embedded* into $\mathcal{B}$ when there exists a mapping f from $\mathcal{L}$ into $\mathcal{B}$ such that for $a, b \in \mathcal{L}$,

(i) $a \leqslant b$ if and only if $f(a) \leqslant f(b)$,
(ii) $f(a^\perp) = f(a)^\perp$,
(iii) $f(a \vee b) = f(a) \vee f(b)$.*

Now Zierler and Schlessinger have proved that such a faithful embedding never exists unless $\mathcal{L}$ is Boolean. This result constitutes a no-go theorem for that particular interpretation of noncontextual hidden-variable theories.

Again, the advocate of hidden-variable theories might repropose the question under relaxed hypotheses. Zierler and Schlessinger themselves have shown, in this respect, that if condition (iii) above is demanded not for any pair a, b but only for elements in the center of $\mathcal{L}$, then a (nonfaithful) embedding of $\mathcal{L}$ into a Boolean algebra $\mathcal{B}$ always exists. It is however doubtful whether the existence of such an embedding represents a physically meaningful way out for hidden-variable theories.

15.4 On Contextual Hidden-Variable Theories

The no-go theorems for noncontextual hidden-variable theories are irrelevant for the much more flexible case of contextual theories. The first construction of a contextual hidden-variable theory is probably the one of Wiener and Siegel[11,12] (see also Ochs[13] for comments on it). The possibility of inventing contextual theories has been further illuminated by a sharp theorem of Gudder[14] according to which the $(\mathcal{L}, \mathcal{S})$ pair, even under hypotheses slightly weaker than the ones we have here assumed, always admits a contextual hidden-variable theory. Needless to say, this result does not secure for these contextual theories the status of physical theories. Gudder's theorem opens a mathematical possibility but does not solve the physical aspect of the hidden-variables issue.

Notice that it is not ensured that a hidden-variable theory (if it exists) agrees in all respects with every physical requirement. As we shall see in Section 25.3, certain contextual hidden-variable theories fail to satisfy a criterion of causality unless one gives up a condition like (15.2.2), in which case discrepancies with the predictions of quantum mechanics may occur. Under these circumstances there are aspects of the hidden-variables issue that admit experimental tests. Anticipating some conclusions of Chapter 25,

---

*In the terminology of Section 12.5, f would be an injective morphism that maps the unit of $\mathcal{L}$ into the unit of $\mathcal{B}$.

we can however say that, presently, no clear empirical evidence at all has been found in favor of a hidden-variable theory.

References

1. F. J. Belinfante, *A Survey of Hidden Variables Theories*, Pergamon Press, Oxford, 1973.
2. M. Jammer, *The Philosophy of Quantum Mechanics*, Wiley, New York, 1974.
3. A. Shimony, in *Foundations of Quantum Mechanics* (International School of Physics "E. Fermi", 49 Course), B. d'Espagnat, ed., Academic Press, New York, 1971.
4. J. M. Jauch and C. Piron, *Helv. Phys. Acta* 36 (1963) 827.
5. S. P. Gudder, *Proc. Amer. Math. Soc.* 19 (1968) 319.
6. S. P. Gudder, *Proc. Amer. Math. Soc.* 21 (1969) 296; *Found. Phys.* 3 (1973) 399; in *Foundations of Probability Theory, Statistical Inference and Statistical Theories of Science*, Vol. III, W. Harper and C. A. Hooker, eds., Reidel, Dordrecht, 1976.
7. T. Neubrunn, *Proc. Amer. Math. Soc.* 25 (1970) 672.
8. W. Ochs, *Nuovo Cimento*, 10B (1972) 172.
9. S. P. Gudder, *Nuovo Cimento* 10B (1972) 518.
10. N. Zierler and M. Schlessinger, *Duke Math. J.* 32 (1965) 251.
11. N. Wiener and A. Siegel, *Nuovo Cimento Suppl.* 2 (1955) 982.
12. N. Wiener, A. Siegel, B. Rankin, and W. T. Martin, *Differential Space, Quantum Systems and Prediction*, M.I.T. Press, Cambridge, Mass., 1966.
13. W. Ochs, *Über die Wiener-Siegelsche Formulierung der Quantentheorie*, Thesis, University of Frankfurt-am-Mainz, 1964.
14. S. P. Gudder, *J. Math. Phys.*, 11 (1970) 431.

CHAPTER 16

Proposition-State Structure and Idealized Measurements

16.1 A Lattice-Semigroup Connection

The proposition-state structure $(\mathcal{L}, \mathcal{S})$ considered in preceding chapters generates in a natural way a family of mappings of $\mathcal{S}$ into itself labeled by the elements of $\mathcal{L}$. Given $a \in \mathcal{L}$, the corresponding mapping has all the properties that the usual quantum theory of measurement prescribes for the transformations suffered by the state of the physical system as result of an idealized measurement procedure for a. These facts will be examined in the next two sections. Also a reverse argument holds true: if one takes from the outset that to each proposition there corresponds a mapping of $\mathcal{S}$ into $\mathcal{S}$ with the properties of an idealized measurement, then very little of the usual structure of $\mathcal{L}$ needs to be assumed; the rest can be proved. This aspect will be examined in Sections 16.4 and 16.5. Mappings can be composed, and the natural mathematical structure they generate is a semigroup; it is thus plausible to ask whether behind the facts we have mentioned there is some lattice-semigroup connection. The present section is devoted to pointing out such a connection. We anticipate that it will be explicitly needed in Sections 16.4 and 16.5, and will appear as background in Sections 16.2 and 16.3.

The lattice-semigroup connection under discussion refers to orthomodular lattices and a special class of involutive semigroups, the so-called Baer *-semigroups.[†] The connection was discovered by Foulis[1] and is twofold:

[†] The term "Baer *" is perhaps more familiar as applied to rings rather than to semigroups. Baer *-rings have their roots in operator algebra and appear as a natural tool in axiomatizing certain parts of the theory of von Neumann algebras. The reader who has in mind Baer *-rings can find an example of Baer *-semigroups in the multiplicative semigroups of those rings.

ENCYCLOPEDIA OF MATHEMATICS and Its Applications, Gian-Carlo Rota (ed.). Vol. 15: E. G. Beltrametti and G. Cassinelli, The Logic of Quantum Mechanics

ISBN 0-201-13514-0

every Baer *-semigroup contains a subset that is an orthomodular lattice; conversely, every orthomodular lattice generates, in a canonical way, a Baer *-semigroup.

A semigroup T, whose binary operation we denote by $\cdot$, is called involutive if it is equipped with a mapping $*: T \to T$ (called the involution), such that for $x, y \in T$, $(x \cdot y)^* = y^* \cdot x^*$ and $(x^*)^* = x$. The structure of involutive semigroup is itself sufficient to carry some ordered structure. In fact, if an element $e \in T$ with the property that $e = e^2 = e^*$ is called a projection, the set $P(T)$ of all projections of T is partially ordered by

$$e \leqslant f \quad \text{whenever} \quad e \cdot f = e \text{ (equivalently } f \cdot e = e), \qquad (16.1.1)$$

for this relation is reflexive, transitive, and antisymmetric (see also Exercise 1). Notice that if T has a two-sided zero element 0, then it is the least element of $P(T)$, and if it has a two-sided unit 1, then it is the greatest element of $P(T)$.

We come now to Baer *-semigroups. A *-semigroup (short for involutive semigroup) T with a two-sided zero 0 is called a Baer *-semigroup if for each element $x \in T$ there exists a projection e such that

$$\{y \in T: x \cdot y = 0\} = e \cdot T, \qquad (16.1.2)$$

in other words, if the right annihilator of x equals the right ideal generated by e. The projection e is uniquely determined by x (see Exercise 1), and following Foulis we denote it by x'. A Baer *-semigroup always has a two-sided unit (Exercise 2) and $0' = 1$, $1' = 0$. The mapping $x \mapsto x'$ of T into $P(T)$, when restricted to $P(T)$, has some similarity to an orthocomplementation; in fact we have[1,2] (i) $e \wedge e' = 0$, (ii) $e \leqslant f$ implies $f' \leqslant e'$, (iii) $e \leqslant e''$ but $e' = e'''$.*

Now consider the class of projections of a Baer *-semigroup that have the property

$$e = e'',$$

which we call the *closed* projections. The set $P'(T)$ they form is ordered by the restriction of the ordering (16.1.1) holding in $P(T)$, and it is much more than a poset: it is an orthomodular lattice. Specifically:[1,2]

THEOREM 16.1.1. $P'(T)$ *is orthocomplemented by the mapping* $e \mapsto e'$.

THEOREM 16.1.2. *The meet in* $P'(T)$ *has the form* $e \wedge f = e \cdot (f' \cdot e)'$, *while* $e \cdot f = f \cdot e$ *if and only if* $e \wedge f = e \cdot f$.

---

*These properties are reminiscent of the so-called pseudo-Boolean or Heyting algebras, which are the algebraic models of intuitionistic logics.[3] In Chapter 18 we shall encounter a similar structure in the context of so-called transition probability spaces.

Another remarkable fact is that the commutativity in the sense of the semigroup operation is fully equivalent to the commutativity in the sense of orthomodular lattices. Symbolically, for $e, f \in P'(T)$,

$$e \cdot f = f \cdot e \quad \text{if and only if} \quad e = (e \wedge f) \vee (e \wedge f'), \qquad (16.1.3)$$

or, in our notation, $e \cdot f = f \cdot e$ if and only if $(e, f)C$.

Up to now, in talking about the orthomodular lattice $P'(T)$, we have used the word "lattice" in a somewhat loose sense, without specifying whether this lattice is complete or not. Actually, the following is true:[2] the lattice $P'(T)$ is complete if and only if the Baer *-semigroup T is complete in the sense that for any nonempty subset S of T there exists a projection e of T such that

$$\{x \in T: y \cdot x = 0 \quad \text{for every} \quad y \in S\} = e \cdot T.$$

Having seen that every Baer *-semigroup determines an orthomodular lattice, we can ask whether every orthomodular lattice arises in this way. The answer is yes, as Foulis has shown.[1] Let us briefly list the main steps of this result.

(1) Given the orthomodular lattice $\mathcal{L}$, consider the family of monotone mappings of $\mathcal{L}$ into $\mathcal{L}$ [a mapping h is *monotone* if $a, b \in \mathcal{L}$ with $a \leqslant b$ implies $h(a) \leqslant h(b)$]. Two monotone mappings h, g are said to be *mutually adjoint* if $h(g(a^\perp)^\perp) \leqslant a$ and $g(h(a^\perp)^\perp) \leqslant a$ for every $a \in \mathcal{L}$; if there exists an adjoint of h, then there exists only one, say h^*. Then consider the set $T(\mathcal{L})$ of all monotone mappings that possess an adjoint; we call them the *residuated* mappings, according to Derderian's terminology,[4] and we note the following property, to be used in the sequel:

$$h(a) = 0 \quad \text{if and only if} \quad a \perp h^*(1). \qquad (16.1.4)$$

(2) Recognize that $T(\mathcal{L})$ is a semigroup under mapping composition and that the *-operation is an involution. $T(\mathcal{L})$ has a two-sided zero element given by the mapping that transforms every $a \in \mathcal{L}$ into the least element of $\mathcal{L}$.

(3) To each $a \in \mathcal{L}$ associate the mapping φ_a (the Sasaki projection) defined by

$$\varphi_a(b) = (b \vee a^\perp) \wedge a, \qquad b \in \mathcal{L} \qquad (16.1.5)$$

and recognize that $T(\mathcal{L})$ is a Baer *-semigroup under the prime operation $h \mapsto h' = \varphi_{h(1)^\perp}$ and that the closed projections of $T(\mathcal{L})$ are precisely the mappings of the form φ_a with a running over $\mathcal{L}$.

(4) Recognize that the correspondence $a \mapsto \varphi_a$ determines an isomorphism (in the sense of Section 12.5) between $\mathcal{L}$ and the orthomodular lattice $P'(T(\mathcal{L}))$ of the closed projections of $T(\mathcal{L})$.

Summing up, we have seen that starting from an arbitrary orthomodular lattice $\mathcal{L}$, we can go to its canonical Baer *-semigroup $T(\mathcal{L})$ and then go to the closed projections, getting an isomorphic replica of the original lattice $\mathcal{L}$. This motivates for $T(\mathcal{L})$ the name of coordinate Baer *-semigroup of $\mathcal{L}$. Similarly, we can ask whether starting from a Baer *-semigroup, going to the orthomodular lattice of its closed projections, and then constructing the associated coordinate Baer *-semigroup, we get an isomorphic replica of the original semigroup. This is not the case, as shown by Foulis[1] (there is only an involution-preserving semigroup homomorphism). Thus, in the correspondence between orthomodular lattices and Baer *-semigroups the semigroup is not uniquely determined by the lattice; additional conditions on the semigroup to ensure uniqueness have recently been attained.[5]

16.2 Ideal First-Kind Measurement

We assume for the proposition-state structure $(\mathcal{L}, \mathcal{S})$ the properties discussed in Chapter 14: specifically, we adopt the hypotheses put forth in Section 14.1 as well as the ones of Section 14.8. Thus, in particular, $\mathcal{S}$ contains a subset $\mathcal{S}^P$ of pure states sufficient on $\mathcal{L}$, and $\mathcal{L}$ is an atomic complete lattice. Here we shall deal with properties relative to pure states, so that the ambient set of states is not the whole of $\mathcal{S}$ but only $\mathcal{S}^P$. Recall that in this context the support function s determines a bijection between $\mathcal{S}^P$ and the set $\mathcal{C}(\mathcal{L})$ of the atoms of $\mathcal{L}$, and, with slight abuse of notation, we still write s for the restriction of the support to the pure states. Since s is now invertible, we shall write $s^{-1}(p)$ for the unique pure state that is the support of the atom p.

The Sasaki projections, defined in (16.1.5) and equivalently expressed by

$$\varphi_a(b) = (b \vee a^\perp) - a^\perp = a - (a \wedge b^\perp),$$

will be heavily used in this section, and we list here a number of their properties:[2]

THEOREM 16.2.1. φ_a is monotone and $\varphi_a(b) \leqslant a$.

THEOREM 16.2.2. $\varphi_a(b) = 0$ if and only if $a \perp b$.

THEOREM 16.2.3. $b \leqslant a$ if and only if $\varphi_a(b) = b$.

THEOREM 16.2.4. $(a, b)C$ if and only if $\varphi_a(b) = a \wedge b$.

THEOREM 16.2.5. $(a, b)C$ if and only if $\varphi_a \cdot \varphi_b = \varphi_b \cdot \varphi_a = \varphi_{a \wedge b}$.

THEOREM 16.2.6. *An atomic orthomodular lattice $\mathcal{L}$ has the covering property if and only if, for every atom p such that $p \not\perp a$, $\varphi_a(p)$ is an atom of $\mathcal{L}$.*

Having established these premises, we shall now introduce a particular class of mappings of $\mathcal{S}^P$ into itself that admit a suggestive physical interpretation. To do so, we assume that $\mathcal{L}$ has also the covering property. In this case, given any nonzero element a of $\mathcal{L}$, if $\alpha \in \mathcal{S}^P$ with $s(\alpha) \not\perp a$, then $\varphi_a(s(\alpha))$ is an atom by Theorem 16.2.6. Then we can define the mapping G_a of $\mathcal{S}^P$ into itself by setting

$$G_a\alpha = s^{-1}(\varphi_a(s(\alpha))) \qquad \text{for all} \quad \alpha \in \mathcal{S}^P \quad \text{such that} \quad s(\alpha) \not\perp a.$$

$$(16.2.1)$$

Equivalently, we can define G_a as the only mapping whose domain $\mathcal{D}(G_a)$ is

$$\mathcal{D}(G_a) = \{\alpha \in \mathcal{S}^P : s(\alpha) \not\perp a\} = \{\alpha \in \mathcal{S}^P : \varphi_a(s(\alpha)) \neq 0\}$$

$$= \{\alpha \in \mathcal{S}^P : \alpha(a) \neq 0\} \qquad (16.2.2)$$

and that makes commutative the diagram

A first set of properties of the G_a's, which translate properties of Sasaki projections (see Exercise 3), reads as follows: for $\alpha \in \mathcal{D}(G_a)$,

$$\alpha(a) = 1 \quad \text{implies} \quad G_a\alpha = \alpha, \qquad (16.2.3)$$

$$(G_a\alpha)(a) = 1, \qquad (16.2.4)$$

$$(a,b)C \text{ and } \alpha(b) = 1 \quad \text{imply} \quad (G_a\alpha)(b) = 1. \qquad (16.2.5)$$

These facts suggest for G_a a natural physical interpretation: if α represents the initial state of a physical system, then $G_a\alpha$ is the state of the system after a measurement with yes outcome of the proposition a. With this interpretation the property (16.2.3) says that whenever the yes outcome is certain, then the state is left unchanged by the measurement procedure, while (16.2.4) says that after a measurement of a with yes outcome, the emerging state assigns probability 1 to the yes outcome of a. These two properties imply that the repetition of the measurement procedure corresponding to G_a does not further modify the state of the system; in other words G_a is idempotent, that is, $G_a^2 = G_a$. This corresponds to the notion of first-kind measurement given by Pauli.[6] The property (16.2.5) says that if a state assigns probability 1 to (the yes outcome of) b and if a is compatible with b, then the state emerging from a measurement of a still assigns probability 1 to b; this fact is just the commonly accepted definition of ideal measurement. Summing up, the mapping G_a appears to be the abstract counterpart of an ideal first-kind measurement, with yes outcome, of the proposition a. Recall that we have already introduced, in the framework of Hilbert-space quantum mechanics, the notion of first-kind measurement, relating it to Lüders's version of the projection postulate: the state transformation prescribed by Lüders's postulate [see (8.1.1)] is indeed an example of the mappings G_a.

We turn now to another set of properties of the G_a's, which have to do with the composition of these mappings. The composition of G_a with G_b, denoted $G_a \cdot G_b$, is defined in the usual way:

$$\mathcal{D}(G_a \cdot G_b) = \{\alpha \in \mathcal{D}(G_b), \, G_b \alpha \in \mathcal{D}(G_a)\}$$

and

$$G_a \cdot G_b(\alpha) = G_a(G_b \alpha) \qquad \text{for all} \quad \alpha \in \mathcal{D}(G_a \cdot G_b).$$

Of course, attention to the domains is necessary, for they do not coincide with the whole of $\mathcal{S}^P$; notice that the mappings of the form G_a generate under composition a larger family of mappings, and the equality of two mappings means that they have equal domains and that they coincide on this common domain (something similar to what happens with unbounded operators in Hilbert space). We have the following properties (see Exercises 4–6):

THEOREM 16.2.7. $(a, b)C$ implies $G_a \cdot G_b = G_b \cdot G_a = G_{a \wedge b}$.

THEOREM 16.2.8. If there are in $\mathcal{L}$ two sequences $\langle a_1, \ldots, a_n \rangle$ and $\langle b_1, \ldots, b_m \rangle$ such that $G_{a_1} \cdot \cdots \cdot G_{a_n} = G_{b_1} \cdot \cdots \cdot G_{b_m}$, then $G_{a_n} \cdot \cdots \cdot G_{a_1} = G_{b_m} \cdot \cdots \cdot G_{b_1}$.

THEOREM 16.2.9. If $G_{a_1} \cdot \cdots \cdot G_{a_n}$ has nonempty domain, then there exists a unique nonzero element $a \in \mathcal{L}$ such that $\mathcal{D}(G_{a_1} \cdot \cdots \cdot G_{a_n}) = \mathcal{D}(G_a)$.

In view of the interpretation of the G_a in terms of ideal first-kind measurements, the composed mapping $G_a \cdot G_b$ has to be interpreted as the state transformation caused by two ideal first-kind measurements performed in succession: first the physical system undergoes the transformation represented by G_b, and then the emerging state is further transformed by G_a. With this interpretation, Theorem 16.2.7 says that whenever the propositions are compatible, the order in which the two measurements are performed is immaterial, a fact that matches the operational notion of compatibility; moreover we have that when $G_a \cdot G_b = G_b \cdot G_a$, the composed mapping is still of the form G_c for some $c \in \mathcal{L}$ and this c is precisely the meet $a \wedge b$. Thus we see that whenever a and b are compatible we have an explicit recipe for constructing an ideal first-kind measurement of $a \wedge b$ out of the ideal first-kind measurements of a and b: it is simply the composition in series, in either order. In case $\mathcal{L}$ is Boolean, the set of the G_a's is then closed under composition. This fact shows that Theorems 16.2.8 and 16.2.9 are trivial in the Boolean case; what is not trivial is that they survive when the structure of $\mathcal{L}$ is relaxed to an orthomodular lattice. Notice also that in case $\mathcal{L}$ is not Boolean the set of the G_a's is definitely not closed under composition: in fact, should we have $G_a \cdot G_b = G_c$ for some c, we would deduce, by Theorem 16.2.8, that $G_b \cdot G_a = G_c = G_a \cdot G_b$, and hence $(a, b)C$ and $c = a \wedge b$. Taking this fact into account, we have that $G_{a_1} \cdot \cdots \cdot G_{a_n}$ is not in general of the form G_a for some a; nevertheless Theorem 16.2.9 ensures that there still exists a correspondence between compositions and elements of $\mathcal{L}$, but in a weaker sense.

16.3 Covering Property and Ideal First-Kind Measurements

In the last section we have deduced, from a sufficiently structured $(\mathcal{L}, \mathcal{S})$ pair, the existence and a number of properties of the mappings that we denoted G_a, to which a physical interpretation was given in terms of ideal first-kind measurements. Notice that the covering property of $\mathcal{L}$ was included in the hypotheses and it was essential for the very definition of the G_a. Here we partially reverse the point of view and show that the covering property can be deduced at the cost of assuming the existence of a family of mappings of $\mathcal{S}^P$ into $\mathcal{S}^P$ endowed with certain characteristic properties that allow for them the same physical meaning of the G_a. This fact will throw light on the physical meaning of the covering property of $\mathcal{L}$, a property which at first sight looks rather technical.*

We assume for $\mathcal{L}$ and $\mathcal{S}$ the structure recalled in the last section, with the sole exception of the covering property of $\mathcal{L}$. We further assume that for all $a \in \mathcal{L}$, $a \neq 0$, there exists a mapping G_a of $\mathcal{S}^P$ into itself defined by the properties (16.2.2)–(16.2.4) and

$$(a, b)C \text{ implies } G_a \cdot G_b = G_b \cdot G_a = G_{a \wedge b}. \tag{16.3.1}$$

We are using for these mappings the same symbol used in last section for mappings defined in a different way; this abuse of notation is however justified by the following result, which ensures that we are actually talking about the same thing. In fact, our present hypotheses allow us to prove (see Exercises 7–10) that $\mathcal{L}$ has the covering property and that G_a has the explicit

---

*Recall however that in Section 14.8 another way of giving physical naturalness to the covering property was discussed.

form

$$G_a\alpha = s^{-1}(\varphi_a(s(\alpha))) \qquad \text{for all} \quad \alpha \in \mathcal{D}(G_a),$$

which was taken as definition in the preceding section.

This result shows that the covering property of $\mathcal{L}$ is deeply connected with the possibility of associating with the elements of $\mathcal{L}$ a transformation of pure states into pure states reminiscent of the commonly accepted idealizations of the measurement process in quantum mechanics. This connection has been exploited in a series of papers by Pool[7], Ochs,[8] Cassinelli and Beltrametti,[9] and Guz[10] that form the background of the present and preceding sections.

16.4 Ideal First-Kind Measurements and the Structure of $\mathcal{L}$

In this section we continue the study of connections between the $(\mathcal{L},\mathcal{S})$ structure and the existence of mappings of states that admit an interpretation in terms of idealized measurements; we shall push forward along the line of the last section, showing that not only the covering property but also other properties of $\mathcal{L}$ and $\mathcal{S}$ come from the above mappings.

Propositions and states are taken as primitive notions, but we do not assume any ordered structure for $\mathcal{L}$, and accordingly we cannot think of the states as probability measures on $\mathcal{L}$: the elements of $\mathcal{S}$ are simply functions from $\mathcal{L}$ into $[0,1]$, and $\alpha(a)$ is interpreted as the probability of the yes outcome of the proposition a when the initial state of the system is α. $\mathcal{S}$ has the natural structure of a σ-convex set, and it is meaningful to suppose that $\mathcal{S}$ has a nonempty set $\mathcal{S}^P$ of extremal points that we can interpret as the physical pure states of the system. For simplicity, and to be closer to the context of the last sections, we restrict our attention to the set $\mathcal{S}^P$ of pure states (however, what will be said in this section is not affected by this restriction). The only structure we assume for $\mathcal{L}$ and $\mathcal{S}^P$ is:

AXIOM 16.4.1.   There are in $\mathcal{L}$ two trivial elements 0 and 1 such that for every $\alpha \in \mathcal{S}^P$, $\alpha(0)=0$ and $\alpha(1)=1$.

AXIOM 16.4.2.   For each $a \in \mathcal{L}$ there exists in $\mathcal{L}$ a unique element, denoted $a^\perp$, such that $\mathcal{S}_1^P(a)=\mathcal{S}_0^P(a^\perp)$ and $\mathcal{S}_1^P(a^\perp)=\mathcal{S}_0^P(a)$.

We have used our common notation $\mathcal{S}_1^P(a)=\{\alpha \in \mathcal{S}^P: \alpha(a)=1\}$, $\mathcal{S}_0^P(a)=\{\alpha \in \mathcal{S}^P: \alpha(a)=0\}$, and we have also used the notation $a^\perp$, which recalls the orthocomplementation; notice however that at this stage it would be meaningless to talk of orthocomplementations, for $\mathcal{L}$ has no ordered structure; our notation is in fact an anticipation of a result to be found in this section.

To each proposition a we associate a transformation G_a of $\mathbb{S}^P$ into $\mathbb{S}^P$, and we interpret G_a as the transformation of the state of the system caused by an ideal first-kind measurement, with yes outcome, of the two-valued physical quantity represented by a. With reference to the discussion of Section 16.2, we formalize this point of view by the following series of assumptions:

AXIOM 16.4.3. For every $a \in £$ there exists exactly one mapping G_a of $\mathbb{S}^P$ into itself with the properties listed below; conversely, every mapping with these properties can be uniquely associated to an element of $£$ (G_0 is the map with empty domain, and G_1 acts as the identity on the whole $\mathbb{S}^P$):

(a) $\mathcal{D}(G_a) = \{\alpha \in \mathbb{S}^P : \alpha(a) \neq 0\}$;
(b) $\alpha(a) = 1$ implies $G_a \alpha = \alpha$;
(c) $G_a \alpha(a) = 1$ for all $\alpha \in \mathcal{D}(G_a)$;
(d) if $a_1, \ldots, a_n, b_1, \ldots, b_m$ are such that $G_{a_1} \cdot \cdots \cdot G_{a_n} = G_{b_1} \cdot \cdots \cdot G_{b_m}$, then we have also $G_{a_n} \cdot \cdots \cdot G_{a_1} = G_{b_m} \cdot \cdots \cdot G_{b_1}$.
(e) for every sequence $a_1, \ldots, a_n$ there exists a unique element a of $£$ such that $\mathcal{D}(G_{a_1} \cdot \cdots \cdot G_{a_n}) = \mathcal{D}(G_a)$.

The reader will note the abuse of notation in this section: we are denoting with the same symbols used in previous sections objects that have the same physical meaning but have different mathematical structure: in particular, the mappings defined by Axiom 16.4.3 are still denoted by G_a, though from the mathematical point of view they are not the same as the ones considered in Sections 16.2 and 16.3.

These hypotheses on the G_a's are rich in consequences for the structure of $£$ and $\mathbb{S}$, which will come from the lattice-semigroup connection examined in Section 16.1.

Starting from the collection of the G_a's, we define

$$T = \left\{ G_{a_1} \cdot \cdots \cdot G_{a_n} : a_1, \ldots, a_n \in £, \forall n \right\}.$$

T has the natural structure of a semigroup with respect to composition of mappings; clearly, G_0 is a two-sided zero and G_1 is a two-sided unit of T. The semigroup T is naturally equipped with an involution:

$$\left(G_{a_1} \cdot \cdots \cdot G_{a_n} \right)^* = G_{a_n} \cdot \cdots \cdot G_{a_1},$$

and thanks to Axiom 16.4.3(d) the function $x \mapsto x^*$, $x \in T$, is well defined. As seen in Section 16.1, the set $P(T)$ of projections of the involutive semigroup T has a poset structure under the ordering (16.1.1); since the G_a's are obviously projections, the order relation in $P(T)$ induces an ordering in the set $\{G_a\}$ and hence in $\mathcal{L}$ itself:

$$a \leqslant b \quad \text{whenever} \quad G_a \cdot G_b = G_a \ (\text{equivalently}, \ G_b \cdot G_a = G_a). \quad (16.4.1)$$

It is worth remarking that this order relation, which makes explicit reference to the semigroup operation, can also be expressed in the following equivalent way (see Exercise 11):

$$G_a \cdot G_b = G_a \quad \text{whenever} \quad \mathcal{S}_1^P(a) \subseteq \mathcal{S}_1^P(b); \quad\quad (16.4.2)$$

This fact recovers for $\mathcal{L}$ the familiar notion of order encountered in previous chapters, and allows us to state that $\mathcal{S}_1^P(a) = \mathcal{S}_1^P(b)$ implies $a = b$, so that the elements of $\mathcal{L}$ are uniquely determined by their certainly-yes domain. By the last remark we have also that the function $a \mapsto a^\perp$ introduced in Axiom 16.4.2 has the property

$$a^{\perp\perp} = a; \quad\quad (16.4.3)$$

in fact we have $\mathcal{S}_1^P(a^{\perp\perp}) = \mathcal{S}_1^P(a)$.

We can now make T a Baer *-semigroup (see Exercise 12) with respect to the function of T into $P(T)$ defined by

$$G_{a_1} \cdot \ \cdots \ \cdot G_{a_n} \mapsto (G_{a_1} \cdot \ \cdots \ \cdot G_{a_n})' = G_{a^\perp}, \quad\quad (16.4.4)$$

where a is the unique element whose existence is stated by Axiom 16.4.3(e). In particular $G_a' = G_{a^\perp}$, and hence by (16.4.3) $G_a'' = G_a$, so that the G_a's are recognized as closed projections of T. Conversely, it is known from the theory of Baer *-semigroups[2] that every closed projection takes the form x' for some $x \in T$; hence, by (16.4.4), every closed projection of T has the form G_a for some $a \in \mathcal{L}$. Thus we come to the important conclusion that

$$P'(T) = \{G_a : a \in \mathcal{L}\}.$$

By this fact and by the one-to-one correspondence between $\mathcal{L}$ and $\{G_a\}$, all the properties of closed projections reviewed in Section 16.1 can be restated in terms of $\mathcal{L}$. Thus:

(i) $\mathcal{L}$ is a complete lattice under the ordering (16.4.1), (16.4.2),
(ii) $a \mapsto a^\perp$ is an orthocomplementation in $\mathcal{L}$,
(iii) $\mathcal{L}$ is orthomodular,
(iv) the meet in $\mathcal{L}$ is characterized by $G_{a \wedge b} = G_a \cdot (G_{b^\perp} \cdot G_a)'$,
(v) a and b commute in $\mathcal{L}$ if and only if $G_a \cdot G_b = G_b \cdot G_a$, in which case $G_{a \wedge b} = G_a \cdot G_b$.

Roughly speaking, these results close the loop between the approach of Sections 16.2 and 16.3 and the approach of this section: most of the usual structure of the $(\mathcal{L}, \mathcal{S})$ pair has been in fact recovered. The simplified context here adopted leaves aside atomicity and covering property of $\mathcal{L}$; they could be encompassed by enlarging the starting assumptions, but this would complicate matters. We have also another nontrivial gap: the properties of the G_a's adopted in this section are not sufficient to prove that the elements of $\mathcal{S}^P$, and hence of $\mathcal{S}$, are probability measures on $\mathcal{L}$ (recall that in the present section the elements of $\mathcal{S}^P$ were merely functions from $\mathcal{L}$ into $[0, 1]$): the missing point is the additivity on orthogonal sets of propositions. However, this fact does not prevent $\mathcal{S}^P$ from having further regularity properties, such as (see Exercise 13)

$$\mathcal{S}_1^P(a \wedge b) = \mathcal{S}_1^P(a) \cap \mathcal{S}_1^P(b). \tag{16.4.5}$$

This property has been already met in Section 11.4 [see (11.4.1)], and its importance has been commented on there; it is essential for the existence of supports of states and for the characterization of the notion of superposition.*

The basic idea of the approach considered in this section, resting on the properties of the G_a's and on the connection between Baer *-semigroups and orthomodular lattices, was proposed by Pool[11, 7] and further developed by the present authors.[12, 13]

16.5 Active and Passive Pictures of Propositions

The one-to-one correspondence between the elements of $\mathcal{L}$ and the maps G_a, discussed from different points of view in the preceding sections, provides two physical interpretations of the notion of proposition. We have a passive picture, in which propositions are viewed as two-valued physical quantities, and states enter only to assign the probability of a yes outcome: this is the picture developed mainly in Section 13.6. We have also an active picture, in which propositions are viewed as state transformations induced by an idealized measurement procedure.

This fact allows us to overcome some of the difficulties met with in Chapter 13, where we tried to give physical content to the notion of proposition. The idea was to interpret a proposition as a yes-no experiment, but we realized that not every yes-no experiment could represent a proposition (see for instance Section 13.3). Now we have a recipe for picking out the good yes-no experiments: if the yes outcome occurs, the experimental device must transform the initial state of the physical system according to the properties of the G_a's. Only in this way we can get a consistent correspondence between the passive and the active picture. If, for instance, we refer to a spin-$\frac{1}{2}$ system and to the yes-no experiments of Figure 13.2, we

*Should it be possible to extend (16.4.5) to $\mathcal{S}_1^P(\bigwedge_i a_i) = \bigcap_i \mathcal{S}_1^P(a_i)$ for every index set, the existence of supports would be ensured, and very much of what was said in Chapters 11 and 14 could be worked out.

easily see that only e_2 satisfies the properties of the G_a's, thus being the good candidate to represent the proposition "the z-component of the spin is $+\frac{1}{2}$", a fact well known to every experimentalist.

The connection between the passive and the active picture can also clarify an old issue on the physical interpretation of the lattice operations in $\mathfrak{L}$. Essentially, the problem is to give an operational meaning to the meet, that is, to express the yes-no experiment for $a \wedge b$ in terms of the yes-no experiments for a and b. We have a natural physical notion of composition for the yes-no experiments: put them in series and perform them in succession. Now the representation of $\mathfrak{L}$ in terms of the lattice of closed projections of the Baer *-semigroup generated by the G_a's ensures that when a and b are compatible, the operational interpretation of their meet is immediate: since in this case $G_{a \wedge b} = G_a \cdot G_b$, the yes-no experiment associated to $a \wedge b$ is just the composition in series of the yes-no experiments for a and b (irrespective of the order). When a and b are not compatible the situation is not so simple: we have $G_{a \wedge b} = G_a \cdot G_c$ with $G_c = (G_{b^\perp} \cdot G_a)'$, so that the yes-no experiment for $a \wedge b$ is still a series composition of two yes-no experiments, but one of them, G_c, though uniquely determined by a and b, is not expressible in terms of G_a and G_b by the use of series composition alone.*

Exercises

1. Show that the order (16.1.1) can be equivalently expressed by $e \leqslant f$ whenever $e \cdot T \subseteq f \cdot T$.

[*Hint.* If $e = f \cdot e$ then $e \cdot T \subseteq f \cdot T$. Conversely, if $e \cdot T \subseteq f \cdot T$, then since $e = e^2 \in e \cdot T \subseteq f \cdot T$, there exists x such that $e = f \cdot x$; hence $f \cdot e = f \cdot f \cdot x = f \cdot x = e$.]

2. In a Baer *-semigroup $0'$ is a two-sided unit.

[*Hint.* $0' \in P(T)$ and $0' \cdot T = \{x \in T : 0 \cdot x = 0\} = T$, so that every $x \in T$ has the form $x = 0' \cdot y$; hence $0' \cdot x = 0' \cdot 0' \cdot y = 0' \cdot y = x$. Moreover $x \cdot 0' = (0' \cdot x^*)^* = x^{**} = x$).

3. Prove (16.2.3)–(16.2.5).
[*Hint.* Use Theorems 16.2.1–16.2.5.]

4. Prove Theorem 16.2.7.
[*Hint.* Note that $G_a \cdot G_b = s^{-1} \cdot \varphi_a \cdot s \cdot s^{-1} \cdot \varphi_b \cdot s = s^{-1}(\varphi_a \cdot \varphi_b)s$; then use Theorem 16.2.5.]

5. Prove Theorem 16.2.8.
[*Hint.* Write $G_{a_1} \cdot \, \cdots \, \cdot G_{a_n} = s^{-1} \cdot (\varphi_{a_1} \cdot \, \cdots \, \cdot \varphi_{a_n}) \cdot s = G_{b_1} \cdot \, \cdots \, \cdot G_{b_m} =$

---

*On the ground of a formal analogy with the explicit expression for the meet of two (noncommuting) projectors in $\mathcal{P}(\mathcal{H})$, Jauch[14] and Watanabe[15] have advanced an operational interpretation of the meet of a and b in terms of an infinite alternating sequence of the yes-no experiments for a and b.

$s^{-1} \cdot (\varphi_{b_1} \cdot \ \cdots \ \cdot \varphi_{b_m}) \cdot s$; note that φ_{a_i} and φ_{b_j} are projections in the Baer *-semigroup $T(\mathcal{L})$ of residuated mappings of $\mathcal{L}$, so that $(\varphi_{a_1} \cdot \ \cdots \ \cdot \varphi_{a_n})^* = \varphi_{a_n} \cdot \ \cdots \ \cdot \varphi_{a_1}$ and $(\varphi_{b_1} \cdot \ \cdots \ \cdot \varphi_{b_m})^* = \varphi_{b_m} \cdot \ \cdots \ \cdot \varphi_{b_1}$.

6. Prove Theorem 16.2.9.
   [*Hint.* $\alpha \in \mathcal{D}(G_{a_1} \cdot \ \cdots \ \cdot G_{a_n})$ if and only if $(\varphi_{a_1} \cdot \ \cdots \ \cdot \varphi_{a_n})(s(\alpha)) \neq 0$. Note that $\varphi_{a_1} \cdot \ \cdots \ \cdot \varphi_{a_n} \in T(\mathcal{L})$, and use (16.1.4) to deduce $\alpha \in \mathcal{D}(G_{a_1} \cdot \ \cdots \ \cdot G_{a_n})$ if and only if $\alpha((\varphi_{a_1} \cdot \ \cdots \ \cdot \varphi_{a_n})^*(I)) \neq 0$. Then take $a = (\varphi_{a_1} \cdot \ \cdots \ \cdot \varphi_{a_n})^*(I)$.]

7. From (16.2.2)–(16.2.4) and (16.3.1) deduce that if $a, b \in \mathcal{L}$, $a \neq 0$, $a \leqslant b$, then $G_b \cdot G_a = G_a$.
   [*Hint.* If $\alpha \in \mathcal{D}(G_b \cdot G_a)$ then $\alpha \in \mathcal{D}(G_a)$; conversely if $\alpha \in \mathcal{D}(G_a)$, use (16.2.4) to get $s(G_a \alpha) \leqslant a \leqslant b$ and hence $G_a \alpha \in \mathcal{D}(G_b)$ and $\alpha \in \mathcal{D}(G_b \cdot G_a)$. From $G_a \alpha(b) = 1$ deduce $G_b \cdot G_a(\alpha) = G_a \alpha$.]

8. From (16.2.2)–(16.2.4), (16.3.1) and $(a, b)C$ deduce $G_a \cdot G_b = G_b \cdot G_a = G_{a \wedge b}$.
   [*Hint.* The case $a \wedge b = 0$ (hence $a \perp b$) is trivial; thus suppose $a \wedge b \neq 0$. Let $\alpha \in \mathcal{D}(G_a \cdot G_b)$, $\gamma = G_a \cdot G_b \alpha$, and note that $\gamma(a) = 1$; write $a = a_1 + c$, $b = b_1 + c$, $c = a \wedge b$, consider $G_{a_1} \cdot G_a \cdot G_b$, and use (16.3.1) and Exercise 7 to get $G_{a_1} \cdot G_a = G_{a_1}$, so that $\mathcal{D}(G_{a_1} \cdot G_a \cdot G_b) = \mathcal{D}(G_{a_1} \cdot G_b) = \varnothing$, since $a_1 \perp b$. This entails $\gamma(a_1) = 0$, hence $\gamma(c) = 1$, hence $\gamma \in \mathcal{D}(G_c)$, and hence $\mathcal{D}(G_a \cdot G_b) \subseteq \mathcal{D}(G_c \cdot G_a \cdot G_b) = \mathcal{D}(G_c)$, where the equality follows from (16.3.1) and Exercise 7. Conversely let $\alpha \in \mathcal{D}(G_c)$; then since $c \leqslant b$, we have $\alpha \in \mathcal{D}(G_b)$ and $G_c \cdot G_b = G_c$; therefore $G_b \alpha(c) \neq 0$, whence $G_b \alpha(a) \neq 0$, that is, $\alpha \in \mathcal{D}(G_a \cdot G_b)$. This proves that $\mathcal{D}(G_a \cdot G_b) = \mathcal{D}(G_c)$. Let $\alpha \in \mathcal{D}(G_a \cdot G_b)$, and recall that $\gamma(c) = 1$, so that $G_c \gamma = \gamma$; this shows that $G_c \alpha = G_c \cdot G_a \cdot G_b \alpha = G_a \cdot G_b \alpha$, where the first equality follows from (16.3.1) and Exercise 7. This proves $G_a \cdot G_b = G_{a \wedge b}$; in the same way one proves $G_b \cdot G_a = G_{a \wedge b}$.]

9. On the hypotheses of Section 16.3 for $\mathcal{L}$ and $\mathcal{S}$, if for every $a \in \mathcal{L}$, $a \neq 0$, there exists a mapping $G_a : \mathcal{S}^P \to \mathcal{S}^P$ defined by (16.2.2)–(16.2.4) and (16.3.1), then $\mathcal{L}$ has the covering property.
   [*Hint.* By Theorem 16.2.6 one has to show that if $p \in \mathcal{Q}(\mathcal{L})$, $p \not\perp a$, then $\varphi_a(p) \in \mathcal{Q}(\mathcal{L})$. Put $\alpha = s^{-1}(p)$; since $p \not\perp a$, we have $\alpha \in \mathcal{D}(G_a)$; put $\beta = G_a \alpha$. From $(a^\perp \vee p, a)C$ it follows that $G_{a^\perp \vee p} \cdot G_a = G_a \cdot G_{a^\perp \vee p}$. Since $\alpha(a^\perp \vee p) = 1$, we have $G_{a^\perp \vee p} \alpha = \alpha$ and hence $\alpha \in \mathcal{D}(G_a \cdot G_{a^\perp \vee p})$; therefore $G_{a^\perp \vee p} \cdot G_a \alpha = G_{a^\perp \vee p} \beta = G_a \cdot G_{a^\perp \vee p} \alpha = G_a \alpha = \beta$, whence $\beta(a^\perp \vee p) = 1$. This gives $s(\beta) \leqslant a^\perp \vee p$, and hence $s(\beta) + a^\perp \leqslant a^\perp \vee p$, having noticed that $\beta(a) = 1$, that is, $s(\beta) \perp a^\perp$. On the other hand, since $(s(\beta)^\perp, a)C$, we have $G_{s(\beta)^\perp \wedge a} = G_{s(\beta)^\perp} \cdot G_a$ by Exercise 8; now observe that $\alpha \notin \mathcal{D}(G_{s(\beta)^\perp} \cdot G_a)$ because $\alpha(s(\beta)^\perp) = 0$, and conclude $\alpha(s(\beta)^\perp \wedge a) = 0$, so that $p = s(\alpha) \leqslant (s(\beta)^\perp \wedge a)^\perp = s(\beta) + a^\perp$ and also $a^\perp \vee p \leqslant s(\beta) + a^\perp$. This and the preceding result prove that $a^\perp \vee p = s(\beta) + a^\perp$; hence by orthomodularity $s(\beta) = (a^\perp \vee p) - a^\perp = \varphi_a(p)$, and $\varphi_a(p)$ is an atom, for $s(\beta)$ is an atom.]

10. Under the conditions of Exercise 9 show that $G_a\alpha = s^{-1}(\varphi_a(s(\alpha)))$ for all $\alpha \in \mathcal{D}(G_a)$.

[*Hint.* If $\alpha \in \mathcal{D}(G_a)$ one has, by Exercise 9, $s(G_a\alpha) = \varphi_a(s(\alpha))$.]

11. Prove (16.4.2).

[*Hint.* Suppose $G_b \cdot G_a = G_a$, and let $\alpha \in \mathcal{S}_1^P(a)$; then $\alpha \in \mathcal{S}_1^P(b)$ by Axiom 16.4.3(b),(c). Suppose $\mathcal{S}_1^P(a) \subseteq \mathcal{S}_1^P(b)$. Clearly $\mathcal{D}(G_b \cdot G_a) \subseteq \mathcal{D}(G_a)$; moreover, if $\alpha \in \mathcal{D}(G_a)$ then $G_a\alpha(a) = 1$, hence $G_a\alpha(b) = 1$, hence $\mathcal{D}(G_a) \subseteq \mathcal{D}(G_b \cdot G_a)$, and also $G_b \cdot G_a\alpha = G_a\alpha$).]

12. Prove that T is a Baer *-semigroup under the prime operation (16.4.4).

[*Hint.* Suppose $x \cdot y = G_0$ and write $x' = G_{a^\perp}$; then notice that $\mathcal{R}(y) \subseteq \mathcal{CD}(x) = \mathcal{CD}(G_a) = \mathcal{R}(G_{a^\perp})$, where $\mathcal{R}$ denotes the range, so that $y = G_{a^\perp} y$. Conversely, suppose $y = G_{a^\perp} z$ for some $z \in T$; then $\mathcal{R}(y) \subseteq \mathcal{R}(G_{a^\perp}) = \mathcal{CD}(G_a) = \mathcal{CD}(x)$, and hence $x \cdot y = G_0$.]

13. Prove (16.4.5).

[*Hint.* Remark that $\mathcal{S}_1^P(a)$ is the range of G_a; then proceed by simple check.]

References

1. D. J. Foulis, *Proc. Amer. Math. Soc.*, 11 (1960) 648.
2. F. Maeda and S. Maeda, *Theory of Symmetric Lattices*, Springer, Berlin, Heidelberg, New York, 1970.
3. M. C. Fitting, *Intuitionistic Logic, Model Theory and Forcing*, North-Holland, Amsterdam, 1969.
4. J. C. Derderian, *Pacific J. Math.* 20 (1967) 35.
5. G. Kalmbach, *Omolattices*, Academic Press, New York, in press (Section 18).
6. W. Pauli, *Die Allgemeinen Prinzipien der Wellenmechanik*, in *Handbuch der Physik*, Vol. V, Springer, Berlin, Göttingen, Heidelberg, 1958 (Part I, pp. 1–168).
7. J. C. T. Pool, *Commun. Math. Phys.* 9 (1968) 212.
8. W. Ochs, *Commun. Math. Phys.* 25 (1972) 245.
9. G. Cassinelli and E. G. Beltrametti, *Commun. Math. Phys.* 40 (1975) 7.
10. W. Guz, *Rep. Math. Phys.* 16 (1979) 125.
11. J. C. T. Pool, *Commun. Math. Phys.* 9 (1968) 118.
12. E. G. Beltrametti and G. Cassinelli, *Rivista Nuovo Cimento* 6 (1976) 321.
13. E. G. Beltrametti and G. Cassinelli, in *Italian Studies in the Philosophy of Science*, M. L. Dalla Chiara, ed., Reidel, Dordrecht, 1981.
14. J. M. Jauch, *Foundations of Quantum Mechanics*, Addison-Wesley, Advanced Book Program, Reading, Mass., 1968 (Section 5.3).
15. S. Watanabe, *Knowing and Guessing*, Wiley, New York, 1969 (Chapter 9).

Additional Bibliography

1. P. Deliyannis, *J. Math. Phys.* 19 (1978) 2341.

CHAPTER 17

Superpositions of States and Closure Spaces

17.1 A General Characterization of State Superpositions

In the usual Hilbert-space formulation of quantum mechanics the notion and the existence of quantum superpositions of states reflect the linear-vector-space structure underlying the theory. As seen in Section 11.5, a notion of superposition can be introduced also in the more general $(\mathcal{L}, \mathcal{S})$ structure, viewing the elements of $\mathcal{S}$ as probability measures on $\mathcal{L}$, and in this context the existence of quantum superpositions of states finds its roots in the combined effects of atomicity, the covering property, and the irreducibility of the proposition lattice $\mathcal{L}$: we refer to the discussion of Sections 11.6, 11.7, and 14.8.

In the present section we introduce from the very beginning a notion of superposition, in a scheme that uses states alone as explicit primitive, undefined ingredients.* By necessity, this notion of superposition will be very general and will embody only some general features of what is commonly meant by superposition. We know that the occurrence of quantum superpositions of states marks a crucial novelty and peculiarity of quantum systems; hence it should not be surprising that several structures of quantum theory rest solely on an axiomatic notion of superposition. The framework that best emphasizes the role of quantum superpositions, and that avoids any overlapping with the notion of mixtures, is the one in which

---

*We are not claiming that a self-consistent approach to quantum mechanics can be developed by adopting states alone as primitive notions: we are merely anticipating that in this and the following sections we shall need to refer only to states.

ENCYCLOPEDIA OF MATHEMATICS and Its Applications, Gian-Carlo Rota (ed.). Vol. 15: E. G. Beltrametti and G. Cassinelli, The Logic of Quantum Mechanics

ISBN 0-201-13514-0

one restricts oneself from the outset to pure states. This is why we do not consider, in this section, the whole set $\mathcal{S}$ of states, but restrict ourselves to the subset $\mathcal{S}^P$ of pure states, which we assume rich enough to be representative of the physical system.* Of course, to talk of pure states, the convex structure of $\mathcal{S}$ has to be presupposed: here the elements of the set $\mathcal{S}$ are primitive objects, and we do not need to refer to any interpretation of them as probability measures, so that the convex structure is not derived from something else.

Let us now consider a list of facts that are characteristic of the intuitive idea of superposition and for which there is rather immediate empirical evidence:

   (i) superposing pure states, we get pure states;
   (ii) superposing states from some given subset S of $\mathcal{S}^P$, we can get states not contained in S;
   (iii) if $\alpha \in \mathcal{S}^P$ is a superposition of β_1 and β_2, and if β_1 is a superposition of γ_1 and γ_2, then α is a superposition of γ_1, γ_2, and β_2 (the same is true also in iterated form);
   (iv) superposing a pure state with itself, one gets necessarily the same state;
   (v) if α is a superposition of β and γ, then β is a superposition of γ and α (and γ is a superposition of α and β).

We do not claim that these properties exhaust all significant properties of the physical notion of superposition, and hence we do not claim that these properties are able to define unambiguously the mathematical operation of superposing states; however, as we are going to see, as soon as $\mathcal{S}^P$ is endowed with a superposition operation that satisfies (i)–(v), then a relevant geometrical structure for $\mathcal{S}^P$ is achieved. We formalize the physical requirements (i)–(v) by saying that there is a function $S \mapsto \bar{S}$ defined for all subsets of $\mathcal{S}^P$, and physically interpreted by viewing $\bar{S}$ as the set formed by adding to S all states that are superpositions of two or more states of S, such that:

$$\bar{S} \subseteq \mathcal{S}^P, \qquad\qquad (17.1.1)$$

$$S \subseteq \bar{S}, \qquad\qquad (17.1.2)$$

$$S_1 \subseteq \bar{S}_2 \quad \text{implies} \quad \bar{S}_1 \subseteq \bar{S}_2, \qquad\qquad (17.1.3)$$

$$\varnothing = \bar{\varnothing} \quad (\varnothing \text{ the empty subset of } \mathcal{S}^P), \qquad\qquad (17.1.4)$$

$$\text{for every} \quad \alpha \in \mathcal{S}^P, \quad \{\alpha\} = \overline{\{\alpha\}} \qquad\qquad (17.1.5)$$

---

*This would be the case if every mixture could be written as convex combination of pure states, as occurs in the ordinary formulation of quantum mechanics.

($\{\alpha\}$ being the subset of $\mathbb{S}^P$ formed by α alone),

$$\text{if } \alpha \in \overline{A \cup \{\beta\}} \text{ and } \alpha \notin A, \text{ then } \beta \in \overline{A \cup \{\alpha\}}. \qquad (17.1.6)$$

Notice that (17.1.4) has no explicit counterpart in (i)–(v), but is obviously presupposed on physical grounds (from nothing we get nothing); notice also that (17.1.6) is not the strict transcription of (v), but the natural generalization to the case in which γ is in turn a superposition.

17.2 The Closure Space of Pure States and the Associated Lattice Structure

Now we proceed to examine the impact of a function $S \mapsto \overline{S}$ with the properties (17.1.1)–(17.1.6) on the geometrical structure of $\mathbb{S}^P$. We make use of the classical papers by H. H. Crapo and G. C. Rota.[1, 2]

The conditions (17.1.1), (17.1.2), and (17.1.3) can be summarized by saying that the function $S \mapsto \overline{S}$ is a *closure relation*, $\overline{S}$ being called the *closure* of S. A set endowed with a closure relation is a *closure space*; thus $\mathbb{S}^P$ becomes a closure space. It easily follows that the closure relation $S \mapsto \overline{S}$ is order-preserving:

$$S_1 \subseteq S_2 \quad \text{implies} \quad \overline{S}_1 \subseteq \overline{S}_2;$$

satisfies

$$\overline{S_1 \cap S_2} = \overline{S}_1 \cap \overline{S}_2; \qquad (17.2.1)$$

and is idempotent:

$$\overline{S} = \overline{\overline{S}}; \qquad (17.2.2)$$

We shall be particularly concerned with closed subsets of $\mathbb{S}^P$, the subset S being called *closed* if and only if $S = \overline{S}$. The first relevant fact is that the closed subsets of $\mathbb{S}^P$ form a lattice, to be denoted $\mathcal{K}^P$.* We write $M, N, \ldots$ for its elements, and we notice that $\varnothing \in \mathcal{K}^P$ due to (17.1.4). The order relation in $\mathcal{K}^P$ is set-theoretic inclusion, and $\varnothing$ and the whole $\mathbb{S}^P$ are the least and greatest element of $\mathcal{K}^P$, so they will also be denoted by *0* and *1*.

---

*The reader will notice that we are continuing an abuse of notation, in that objects having equivalent physical interpretations are denoted by the same symbol even if they correspond to different mathematical definitions. So, for instance, the symbol $\mathcal{K}^P$ was used in Section 14.8 to designate the lattice of sets of probability measures closed under the explicit definition of superposition given in Section 11.5.

The meet and join are as follows:

$$M_1 \wedge M_2 = M_1 \cap M_2, \qquad M_1, M_2 \in \mathcal{K}^P, \tag{17.2.3}$$

$$M_1 \vee M_2 = \overline{M_1 \cup M_2}, \qquad M_1, M_2 \in \mathcal{K}^P. \tag{17.2.4}$$

To motivate (17.2.3) notice that, due to (17.2.1), $M_1 \cap M_2$ is closed, and it is obviously the greatest subset contained in both M_1 and M_2. To motivate (17.2.4) notice that, due to (17.2.2), $\overline{M_1 \cup M_2}$ is closed, and for every $N \in \mathcal{K}^P$ such that $N \supseteq M_1$ and $N \supseteq M_2$, we get $N \supseteq M_1 \cup M_2$ and hence, by (17.1.3), $N \supseteq \overline{M_1 \cup M_2}$. Notice that in a closure space the intersection of any collection of closed subsets is also closed; therefore we can generalize (17.2.3) to the meet of any collection of elements of $\mathcal{K}^P$. Similarly we can generalize (17.2.4), and $\mathcal{K}^P$ is thus a complete lattice.

These remarks show how intimately a closure space [satisfying (17.1.4)] is associated with an order structure. And recall that all this comes from the empirical requirements (i) to (iii) of the previous section.

The effect of (17.1.5), which formalizes requirement (iv), is to make $\mathcal{K}^P$ an atomic lattice. Indeed, (17.1.5) ensures that the subsets of $\mathcal{S}^P$ formed by only one state belong to $\mathcal{K}^P$, and hence are precisely the atoms of $\mathcal{K}^P$; obviously, every nonzero element of $\mathcal{K}^P$ majorizes one atom at least. As a matter of fact, every $M \in \mathcal{K}^P$ coincides with the join of all the atoms it contains; thus $\mathcal{K}^P$ is not only atomic but also atomistic, according to the terminology of Section 10.1. Let us argue that in this context the atoms of $\mathcal{K}^P$ come out to be, by their very construction, in one-to-one correspondence with the pure states of the system.

The condition (17.1.6) is the classical Steinitz-MacLane exchange property; it is the geometrical counterpart of Birkhoff's order-theoretic covering property. Indeed, in terms of $\mathcal{K}^P$ the condition (17.1.6) is expressed as follows: if $M \in \mathcal{K}^P$ and p, q are atoms of $\mathcal{K}^P$ with $M \wedge p = 0$, then $p \leqslant M \vee q$ implies $q \leqslant M \vee p$. This is just a way of expressing the covering property (see Section 10.1). Therefore, the effect of the condition (17.1.6), which formalizes the natural requirement (v) of the previous section, is to endow the lattice $\mathcal{K}^P$ with the covering property.

Let us finally remark that the pattern of classical mechanics is recovered as a particular case by assuming that every subset of $\mathcal{S}^P$ (the phase space of the system) is closed, that is, by taking as closure relation the identity. This would make $\mathcal{K}^P$ the Boolean algebra of all subsets of $\mathcal{S}^P$

17.3 Rank and Basis of a Set of States

The structure formed by $\mathcal{S}^P$ with a closure relation satisfying (17.1.1)–(17.1.6) constitutes a generalization of a combinatorial geometry.[1, 2] It would become precisely a combinatorial geometry if $\mathcal{S}^P$ were a finite set or if, at least, we were to add a "finite basis" hypothesis, namely, that for every $S \subseteq \mathcal{S}^P$ there exists a *finite* subset S_f of S such that $\bar{S}_f = \bar{S}$. A hypothesis of this kind will not be adopted here, for it would be too stringent for most quantum systems.

Though our structure constitutes a relaxation of combinatorial geometries, several features of the latter still hold true and correspond to significant facts of quantum mechanics. In particular, as we are going to show, we have at hand the notion of the rank, or dimension function, of subsets of $\mathcal{S}^P$, closed or not, and the related notion of the basis of a subset of states. According to a more geometrical terminology, we shall often use the term "flat" in place of "closed subset"; thus $\mathcal{K}^P$ is the lattice of the flats of $\mathcal{S}^P$.

A *chain* in a lattice is a completely ordered subset of the lattice. If $M_1, M_2 \in \mathcal{K}^P$ and $M_1 < M_2$, a maximal chain connecting M_1 with M_2 has the form $M_1 < N_1 < \cdots < N_i < \cdots < M_2$, where each term covers its antecedent. We say that M is a finite element if every maximal chain connecting it with 0 (the empty set $\varnothing$) has finite length. Now we have an important fact, whose proof, due to Crapo and Rota,[2] is suggested in Exercise 1:

THEOREM 17.3.1. *If M_1 and M_2 are finite elements of $\mathcal{K}^P$ and $M_1 < M_2$, then all maximal chains from M_1 to M_2 have the same length.*

According to this result, we define the dimension, or rank, of a finite element M of $\mathcal{K}^P$ as the common length of all maximal chains from 0 to M, and denote it by $\mu(M)$; if M is not a finite element, we define $\mu(M) = \infty$. Thus the dimension function is defined on $\mathcal{K}^P$; we have in particular

$$\mu(M) = 0 \quad \text{if and only if} \quad M = 0, \tag{17.3.1}$$

$$M_1 \leqslant M_2 \quad \text{implies} \quad \mu(M_1) \leqslant \mu(M_2). \tag{17.3.2}$$

It becomes justified to adopt a more geometric language, calling the rank-1 flats "points", the rank-2 flats "lines", the rank-3 flats "planes", etc. (see Exercise 2). The points of our geometrical structure are precisely the elements of $\mathcal{S}^P$, the pure states of the physical system. Each maximal chain from 0 to M determines a family of points contained in M and having M as join: indeed, if N_{i-1} and N_i are adjacent elements of the chain, then N_i is the join of N_{i-1} and a point not contained in N_{i-1}, so that each link in the chain is produced by the "addition" of one point. Of course, the family of points associated with a maximal chain from 0 to M contains exactly $\mu(M)$ elements. We call such a family a *basis* of M: each maximal chain from 0 to M determines a basis for M. Clearly, the bases of M can also be viewed as the minimal families of points contained in M and having M as join.

The dimension function on $\mathcal{K}^P$ satisfies a characteristic inequality: if M_1, M_2 are finite elements of $\mathcal{K}^P$, then[2] (see Exercise 3)

$$\mu(M_1 \vee M_2) + \mu(M_1 \wedge M_2) \leqslant \mu(M_1) + \mu(M_2). \tag{17.3.3}$$

Notice that this inequality is a relaxation of the condition holding in modular lattices, where it is indeed replaced the equality.*

The notion of rank, which up to now has been referred to the flats of $\mathbb{S}^P$, can now be generalzied to every subset of $\mathbb{S}^P$. Indeed, if S is any subset of $\mathbb{S}^P$, we define the rank $r(S)$ of S as the dimension of the smallest flat that contains S, that is, as the dimension of the closure of S. Symbolically,

$$r(S) = \mu(\bar{S}), \qquad S \subseteq \mathbb{S}^P.$$

The use of a symbol different from μ stresses the fact that the μ-function is defined on the flats, and hence on the lattice $\mathcal{K}^P$, while the r-function is defined on every subset of states, and hence on the Boolean algebra of all parts of $\mathbb{S}^P$.

It is easily seen that the r-function satisfies a condition analogous to (17.3.3); in fact, if S_1 and S_2 are subsets of $\mathbb{S}^P$ with finite rank, we have (see Exercise 4)

$$r(S_1 \cup S_2) + r(S_1 \cap S_2) \leqslant r(S_1) + r(S_2). \tag{17.3.4}$$

Notice that, parallel to the fact that (17.3.3) becomes an equality for a modular $\mathcal{K}^P$, we have now that (17.3.4) becomes an equality for a projective geometry. Our geometrical structure is weaker than a projective geometry; as already remarked, it is even weaker than a combinatorial geometry.

17.4 Propositions and Flats

The arguments of previous sections have shown that upon endowing the set of pure states with a notion of superposition, a remarkable geometric construction comes out and, parallel to it, also a remarkable order construction, the lattice $\mathcal{K}^P$. $\mathcal{K}^P$ has several of the properties that the set of propositions of a physical system should have: it is a complete lattice (with a dimension function), it is atomic and also atomistic, and it has the covering property. These facts constitute a strong hint to think of the elements of $\mathcal{K}^P$ as representatives of the propositions of the system. The reader will recall that in Sections 11.5 and 14.8, where a primitive notion of proposition was adopted, where states were probability measures on propositions, and where an explicit definition of superposition of probability measures was used, we found indeed that the lattice structure of propositions was mirrored by the lattice structure of the closed (under superposition) subsets of states, and this correspondence between propositions and closed subsets of states was seen to have physical naturalness and content.

There are also properties that the lattice of propositions should have but that we do not find, in any simple way, in $\mathcal{K}^P$. First of all, we need

*Recall that in Hilbert-space quantum mechanics, where $\mathcal{K}^P$ becomes the lattice $\mathcal{P}(\mathcal{K})$ of closed subspaces of $\mathcal{K}$, the modularity is ruled out when $\mathcal{K}$ is infinite-dimensional.

orthocomplementation* and the related orthomodularity. Given a closed set M, it is natural to say that the closed set M' is a complement of M if it has no points in common with M and is such that $\overline{M \cup M'} = \mathbb{S}^P$. In lattice notation, $M \wedge M' = 0$, $M \vee M' = 1$. Now, the existence of complements can be proved[2] if $\mathbb{S}^P$ has a finite basis, that is, if $\mu(1)$ is finite, but the problem is open in the infinite-dimensional case. To approach the notion of orthogonality one might look for minimal complements, but it is clear that orthocomplementation is not caught by the bare idea of superposition of states that we have used in this chapter. Of course one might consider strengthening the point of departure so as to reach all the usual $(\mathcal{L}, \mathbb{S})$ structure, but this goes beyond our interest. We content ourselves with having indicated which parts of this structure can be traced back to a general notion of state superposition.

Exercises

1. Prove Theorem 17.3.1

[*Hint.* (taken from Reference 2, Proposition 2.5). Let $M_1 = N_0 < N_1 < \cdots < N_n = M_2$, $M_1 = N_0' < N_1' < \cdots < N_m' = M_2$ be two maximal chains from M_1 to M_2. If $n = 0$ or $n = 1$, the assertion is obvious; so proceed by induction, assuming the truth of the statement for all pairs M_1', M_2' between which there exists a maximal chain of length less than n. By the covering property, $N_1 \vee N_1'$ covers both N_1 and N_1'. Select a maximal chain $N_1 \vee N_1' = N_2'' < N_3'' < \cdots < N_k'' = M_2$ from $N_1 \vee N_1'$ to M_2. Comparing the two paths from N_1 to M_2, we have $k = n$ by the induction hypothesis. Thus there is a maximal chain from N_1' to M_2 having length $n-1$, and $m = k = n$ by the induction hypothesis.]

2. Show that the lattices D_{16} and G_{12} of Figures 10.4 and 10.6, respectively, come from the following geometrical patterns of points and lines shown in Figure 17.1.

[*Hint.* By direct inspection.]

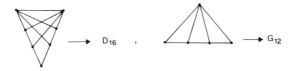

Figure 17.1. Geometries associated with D_{16} and G_{12}.

3. Prove (17.3.3).

[*Hint.* (taken from Reference 2, Proposition 2.6). Choose a maximal

---

*The question whether the lattice of propositions of a physical system must necessarily be orthocomplemented as a consequence of firm empirical evidence has been critically raised by Mielnik; we shall return to this issue in Chapter 19.

chain from $M_1 \wedge M_2$ to M_1, say $M_1 \wedge M_2 = N_0 < N_1 < \cdots < N_n = M_1$. Put $N_i' = N_i \vee M_2$, and observe that N_i' covers or equals N_{i-1}'. Thus, except for possible repetition of some elements, $N_0' < N_1' < \cdots < N_n'$ is a maximal chain from M_2 to $M_1 \vee M_2$ and $\mu(M_1) - \mu(M_1 \wedge M_2) \geqslant \mu(M_1 \vee M_2) - \mu(M_2)$.]

4. Prove (17.3.4).

[*Hint.* Use (17.3.2) to write $r(S_1 \cup S_2) = \mu(\overline{S_1 \cup S_2}) \leqslant \mu(\bar{S}_1 \vee \bar{S}_2)$ and $r(S_1 \cap S_2) = \mu(\overline{S_1 \cap S_2}) \leqslant \mu(\bar{S}_1 \wedge \bar{S}_2)$; then use (17.3.3).]

References

1. H. H. Crapo and G. C. Rota, in *Trends in Lattice Theory*, J. C. Abbott, ed., Van Nostrand–Reinhold, New York, 1970.
2. H. H. Crapo and G. C. Rota, *Studies in Appl. Math.* 49 (1970) 109.

CHAPTER 18

Transition-Probability Spaces and Quantum Systems

18.1 Transition Probabilities between Pure States

In physicists' nomenclature the expression "transition probability" generally refers to some dynamical instability, more specifically to a nonzero
probability for the system to make a transition from an initial to a final
state. Our use of the term is not directly related to dynamical instabilities;
rather we follow von Neumann's terminology,* and the transition probability between two states is meant to represent, intuitively, a measure of their
overlapping. To visualize this notion in an explicit example consider the
states of linear polarization of a photon beam, let α be the polarization state
filtered by a Nicol prism N_1, and let β be the state filtered by a Nicol prism
N_2 rotated by an angle ϑ with respect to N_1; then the transition probability
between α and β is the probability for a photon in state α to pass N_2, which
is empirically known to be $\cos^2 \vartheta$ (Malus law), and equals the probability
for a photon in state β to pass N_1.

In the Hilbert-space formulation of quantum mechanics the idea of
transition probability we have in mind is simply the modulus squared of the
scalar product (see Sections 2.5 and 9.3). The purpose of this chapter is to
formalize, without any Hilbert space in view, a notion of transition probability with the aim of pointing out which parts of the $(\mathcal{L}, \mathcal{S})$ structure
associated with quantum systems can be based on that notion. As in the
previous section, we take the states as primitive entities, without assuming

*J. von Neumann, *Continuous Geometries with a Transition Probability*, unpublished
manuscript, 1937.

ENCYCLOPEDIA OF MATHEMATICS and Its Applications, Gian-Carlo Rota (ed.).
Vol. 15: E. G. Beltrametti and G. Cassinelli, The Logic of Quantum Mechanics

ISBN 0-201-13514-0

them to be probability measures on propositions. To get a simpler situation we restrict ourselves again to pure states, denoting by $\mathbb{S}^P$ the (nonempty) set they form. Consider then a mapping $\langle \cdot | \cdot \rangle$ of $\mathbb{S}^P \times \mathbb{S}^P$ into $[0,1]$ satisfying the following properties:

(i) $\langle \cdot | \cdot \rangle$ is *separating* on $\mathbb{S}^P$, that is, $\langle \alpha | \beta \rangle = 1$ if and only if $\alpha = \beta$;
(ii) $\langle \cdot | \cdot \rangle$ is *symmetric*, that is, $\langle \alpha | \beta \rangle = \langle \beta | \alpha \rangle$ for every $\alpha, \beta \in \mathbb{S}^P$;
(iii) calling α, β *orthogonal* when $\langle \alpha | \beta \rangle = 0$, every maximal pairwise orthogonal subset R of $\mathbb{S}^P$ is a basis for $\mathbb{S}^P$, that is, $\Sigma_{\beta \in R} \langle \alpha | \beta \rangle = 1$ for every $\alpha \in \mathbb{S}^P$.

The function $\langle \cdot | \cdot \rangle$ is called the *transition probability*, and we call the set $\mathbb{S}^P$ equipped with the function $\langle \cdot | \cdot \rangle$ the *transition-probability space*.

We get a trivial example of a transition-probability space by taking $\langle \alpha | \beta \rangle = \delta_{\alpha, \beta}$, so that two distinct states are always orthogonal; this corresponds to the classical case, with $\mathbb{S}^P$ standing for the phase space of the classical system. As another example, take a separable Hilbert space $\mathcal{H}$, with $\mathbb{S}^P$ denoting the set of one-dimensional subspaces, and put $\langle \alpha | \beta \rangle = |(\varphi, \psi)|^2$, where φ, ψ are unit vectors of α and β respectively. Explicit examples of finite transition-probability spaces have been provided by Belinfante:[1] the reader is also referred to Exercises 1, 2.

Notice that the symmetry requirement would be physically untenable if the states were not restricted to pure ones. Think for instance of a photon beam, and let α be the unpolarized state and β a linearly polarized state; then the probability for a photon in state α to pass the Nicol prism that filters β is less than 1, while the probability for a photon in state β to pass the filter of α (the trivial filter transparent to all polarizations) is equal to 1.

It is a distinctive mark of quantum systems that the transition probabilities take values on the whole real segment $[0,1]$, not only 0 and 1 as classical systems do. Given two pure states α, β, the number $\langle \alpha | \beta \rangle$ establishes an absolute selectivity limit for quantum systems in the sense that every yes-no experiment whose "yes" outcome is produced with certainty by the system in state α must assign a probability not less than $\langle \alpha | \beta \rangle$ to the "yes" outcome when the system is in state β. If $\langle \alpha | \beta \rangle$ is different from zero, it is impossible to find a yes-no experiment such that the "yes" outcome is produced with certainty by the state α and the "no" outcome is produced with certainty by the state β. In other words, every filtering device that is transparent to a beam of identical physical systems in state α must accept at least a fraction $\langle \alpha | \beta \rangle$ of systems in state β. Only in case $\langle \alpha | \beta \rangle = 0$ does there exist a filter transparent to α and blind to β.

18.2 Properties of Transition-Probability Spaces

The concept of transition-probability space as a tool in the axiomatic study of quantum mechanics was introduced by Mielnik.[2] The effectiveness of this tool, outlined by Zabey, [3] rests on the properties we are going to review. A first relevant fact[1-3] is that in a transition-probability space all bases (that is, all maximal pairwise orthogonal subsets) have the same cardinality, or, loosely speaking, have the same number of elements (see Exercise 3). This common cardinality will be called the dimension, or rank, of the transition probability space.

Let us now proceed to define the orthogonal complement of a subset S of $\mathcal{S}^P$ as the set of all states orthogonal to every state of S. We write S' for this orthogonal complement: symbolically,

$$S' = \{\beta \in \mathcal{S}^P : \langle \alpha | \beta \rangle = 0 \text{ for every } \alpha \in S\}.$$

The following assertions are obvious ($S, S_1, S_2 \subseteq \mathcal{S}^P$):

$$S \cap S' = \varnothing, \quad (S \cup S')' = \varnothing, \tag{18.2.1}$$

$$\text{if } S_1 \subseteq S_2 \text{ then } S_2' \subseteq S_1', \tag{18.2.2}$$

$$S \subseteq S'', \tag{18.2.3}$$

and, combining (18.2.2) with (18.2.3),

$$S' = S'''. \tag{18.2.4}$$

By the way, recall that a structure like this (which is the algebraic model of intuitionistic logic) has been already encountered in Section 16.1 for the projections of a Baer *-semigroup.

Consider now the mapping $S \mapsto S''$, and observe that it is a closure relation in $\mathcal{S}^P$. Indeed, besides the properties $S'' \subseteq \mathcal{S}^P$ and $S \subseteq S''$ already mentioned, we have from (18.2.2) and (18.2.4) that

$$S_1 \subseteq S_2'' \text{ implies } S_1'' \subseteq S_2'',$$

so that all conditions that define a closure relation are met [see (17.1.1)–(17.1.3)].

The empty set and the whole of $\mathcal{S}^P$ are obviously invariant under this closure relation: $\varnothing = \varnothing''$ and $\mathcal{S}^P = \mathcal{S}^{P''}$. The singletons too are invariant; in fact, after remarking (see Exercise 4) that each state α belongs to at least one basis, say $\{\alpha, \beta_1, \beta_2, \ldots\}$, we have, for every $\gamma \in \mathcal{S}^P$, $\langle \gamma | \alpha \rangle + \Sigma_i \langle \gamma | \beta_i \rangle = 1$, and if $\gamma \in \{\alpha\}''$, so that $\langle \gamma | \beta_i \rangle = 0$, we get $\langle \gamma | \alpha \rangle = 1$ and hence $\gamma = \alpha$, showing that $\{\alpha\} = \{\alpha\}''$.

Writing $\bar{S}$ for S'', we have in conclusion that our closure relation $S \mapsto \bar{S}$ satisfies all properties (17.1.1)–(17.1.5) that we met in Section 17.1, with the sole exception of the Steinitz-MacLane exchange property (17.1.6), which indeed is found to fail in explicit counterexamples (see Exercise 5).

We come now to another relevant notion, that of subspace. A subset S of a transition-probability space is said to be a *subspace* if one obtains a transition probability on S by restricting $\langle \cdot | \cdot \rangle$ to $S \times S$. Since $\langle \cdot | \cdot \rangle$ is obviously separating and symmetric when restricted to $S \times S$, the only nontrivial requirement contained in the definition is:

DEFINITION 18.2.1. *S is a subspace if and only if every maximal pairwise orthogonal subset R of S is a basis for S, that is,*

$$\sum_{\beta \in R} \langle \alpha | \beta \rangle = 1 \qquad \text{for every} \quad \alpha \in S.$$

The empty set and the singletons are obvious examples of subspaces. Of course, if S is a subspace, then it is itself a transition-probability space.

We have already mentioned that all bases of a transition-probability space have the same cardinality (Exercise 3); thus, each subspace can be labeled by the common cardinality of its bases, which we call the dimension, or rank, of the subspace.

Let us also note the following[1] (see Exercises 6, 7):

THEOREM 18.2.1. *If R is a basis for the subspace S, then $R' = S'$,*

THEOREM 18.2.2. *If S is a subspace, then also S' is a subspace.*

18.3 The Order Structure Arising from Transition-Probability Spaces

As seen in the preceding section, once a transition-probability space is given, a parallel structure of closure space is naturally generated. The closure operation $S \mapsto \bar{S} = S''$ we are referring to will be called *orthoclosure*, to recall its origin; accordingly the subsets with the property $S = \bar{S}$ will be called *orthoclosed*. Having seen that the empty set and the singletons are orthoclosed, we can transplant into the present situation all the conclusions that in Section 17.2 we derived from (17.1.1)–(17.1.5). Thus we immediately conclude that the orthoclosed subsets of $\mathcal{S}^P$ form a complete atomic lattice, still denoted $\mathcal{K}^P$, with order induced by set-theoretic containment, with least and greatest elements given by the empty set and by the whole of $\mathcal{S}^P$, with meet and join given by

$$S_1 \wedge S_2 = S_1 \cap S_2, \quad S_1 \vee S_2 = \overline{S_1 \cup S_2}, \qquad S_1, S_2 \in \mathcal{K}^P, \quad (18.3.1)$$

with atoms given by the singletons, and with every element obtained as the join of the atoms it contains, so that $\mathcal{K}^P$ is not only atomic but also atomistic. We miss the covering property, for, as remarked in the last section, the exchange property (17.1.6) is not secured for our orthoclosure.

On the other hand, the orthocomplementation of $\mathfrak{K}^P$ is now secured; in fact, for the elements of $\mathfrak{K}^P$ the function $S \mapsto S'$ has all the properties of an orthocomplement [see (18.2.1), (18.2.2), and (18.3.1)].

The lattice $\mathfrak{K}^P$ has a nice substructure. In fact, if we consider the elements of $\mathfrak{K}^P$ that are also subspaces,* we get a poset, say $\tilde{\mathfrak{K}}^P$, ordered by restriction of the order of $\mathfrak{K}^P$, which contains the least and greatest elements of $\mathfrak{K}^P$ as well as all singletons, which is thus atomic, which admits a dimension function, which is still orthocomplemented by the mapping $S \mapsto S'$ (this comes from Theorem 18.2.2), and which has the further property of being orthomodular (see Exercise 9). This last fact has been proved by Belinfante;[1] a connection between orthomodularity and the transition-probability structure was also remarked on by Bugajska.[4] Now it becomes meaningful to speak of probability measures on $\tilde{\mathfrak{K}}^P$ and, more precisely, to ask whether the states behave as probability measures on $\tilde{\mathfrak{K}}^P$. Given $S \in \tilde{\mathfrak{K}}^P$, let R be a basis for S, and consider, for any $\alpha \in \mathbb{S}^P$, the number

$$\sum_{\beta \in R} \langle \alpha | \beta \rangle .$$

This number depends only on S and α, not on the particular choice of the basis R: in fact, if R_1 is another basis for S, and R_2 is a basis for S', we have that both $R \cup R_2$ and $R_1 \cup R_2$ are bases for $\mathbb{S}^P$, so that

$$\sum_{\beta \in R} \langle \alpha | \beta \rangle = 1 - \sum_{\beta \in R_2} \langle \alpha | \beta \rangle = \sum_{\beta \in R_1} \langle \alpha | \beta \rangle .$$

In view of this fact, to each $\alpha \in \mathbb{S}^P$ we associate a mapping of $\tilde{\mathfrak{K}}^P$ into $[0, 1]$, still denoted by α, defined by

$$\begin{aligned} \alpha(S) &= \sum_{\beta \in R} \langle \alpha | \beta \rangle \quad \text{if} \quad S \neq 0 \quad (R \text{ a basis for } S), \\ \alpha(0) &= 0, \end{aligned} \tag{18.3.2}$$

and observe that it is a probability measure on $\tilde{\mathfrak{K}}^P$, for besides the properties $\alpha(0) = 0$ and $\alpha(1) = 1$, the function α is additive on orthogonal elements (see Exercise 10). It is clear from (18.3.2) that $\alpha(S) = 1$ if and only if α belongs to S, in other words, each element of $\tilde{\mathfrak{K}}^P$ coincides with the set of states that, when thought of as probability measures, assign to it probability 1. Symbolically,

$$S = \{ \alpha \in \mathbb{S}^P : \alpha(S) = 1 \},$$

*If one had in mind the transition-probability space formed by the rays of a Hilbert space under the squared modulus of the scalar product, one might have the impression that orthoclosed subsets are automatically subspaces and vice versa. That this is not generally true is shown in Exercise 8.

or, with our standard notation, $S = S_1^P(S)$. The set S^P of states can thus be viewed as a strongly ordering (hence also ordering) set of probability measures on $\mathcal{K}^P$.

18.4 Transition Probabilities and the Hilbert Model

The order structure discussed in last section has many of the features one would like to have for the set of propositions of a quantum system. One could then conceive of an approach to quantum mechanics in which propositions are defined as, or identified with, the elements of $\mathcal{K}^P$. To go further in this direction and approach the Hilbert-space model, one might for instance argue a connection between the notion of orthoclosure discussed in this chapter and the notion of closure under superposition examined in the last chapter. Or one might argue a connection between the transition-probability structure and the active picture of propositions commented on in Chapter 16.[5] But we do not insist on this point, because the construction of a full-fledged approach to quantum mechanics based on an abstract notion of transition probability goes beyond our present purposes.*
We shall only point out some facts that demonstrate the distance between a general transition-probability space and the Hilbert-space structure.

A first fact is that when S^P is the set of rays of a Hilbert space $\mathcal{K}$ and $\langle \cdot | \cdot \rangle$ is the modulus squared of the scalar product, we get a transition-probability space in which orthoclosed subsets are subspaces, subspaces are orthoclosed, and $\mathcal{K}^P = \tilde{\mathcal{K}}^P$. Moreover, orthoclosure is equivalent to the closure in the topology of Hilbert space, which, in turn, is the same as closure in the sense of superposition (completion of a subset of states by their superpositions). And finally, $\mathcal{K}^P$ coincides with the orthomodular lattice $\mathcal{P}(\mathcal{K})$ of all projectors of $\mathcal{K}$.

A pictorial way of stressing that not every physically conceivable transition-probability structure is compatible with the Hilbert-space structure has been proposed by Mielnik,[2] who imagines the following hypothetical situation:

> ...Someone looked at a small spherical glass bubble: inside there was a drop of liquid. The drop occupied exactly a half of the bubble in the shape of a hemi-sphere. He was able to introduce inside a thin, flat partition dividing the interior of the bubble into two equal volumes. He tried to do this so that the drop would become split. However, the drop exhibited a quantum behaviour: instead of being

---

*The reader will note that starting from the standard $(\mathcal{L}, S)$ structure, say for definiteness the one of Section 14.1 with S restricted to pure states, one can construct a transition probability by defining $\langle \alpha | \beta \rangle$ as $\alpha(s(\beta))$ and assuming as an extra hypothesis the symmetry $\alpha(s(\beta)) = \beta(s(\alpha))$, which is not secured by the standard $(\mathcal{L}, S)$ structure.

divided into two parts the drop jumped and occupied the space on only one side of the partition. ...He began to observe this phenomenon and discovered that each time the partition is introduced the drop chooses a certain side with a definite probability. This probability depends upon the angle between the partition and the initial surface of the drop. If the drop occupied a hemi-sphere s and the partition forces it to choose between two hemi-spheres r and r' the probabilities of transition into r and r' are proportional to the volumes of $s \cap r$ and $s \cap r'$. ...He wanted to formulate the quantum theory of this phenomenon, but he realized that he could not use Hilbert spaces: the space of states of the drop was not Hilbertian....

(See Exercise 11.)

Exercises

1. Show that the set $\{\alpha_1, \ldots, \alpha_6\}$ endowed with the function $\langle \cdot | \cdot \rangle$ defined by

$\langle \cdot \| \cdot \rangle$	α_1	α_2	α_3	α_4	α_5	α_6
α_1	1	0	0	$\frac{1}{3}$	$\frac{1}{3}$	$\frac{1}{3}$
α_2	0	1	0	$\frac{1}{3}$	$\frac{1}{3}$	$\frac{1}{3}$
α_3	0	0	1	$\frac{1}{3}$	$\frac{1}{3}$	$\frac{1}{3}$
α_4	$\frac{1}{3}$	$\frac{1}{3}$	$\frac{1}{3}$	1	0	0
α_5	$\frac{1}{3}$	$\frac{1}{3}$	$\frac{1}{3}$	0	1	0
α_6	$\frac{1}{3}$	$\frac{1}{3}$	$\frac{1}{3}$	0	0	1

is a transition-probability space (this example is contained in Reference 1). [*Hint.* By simple check.]

2. Show that the set $\{\alpha_1, \ldots, \alpha_8\}$ endowed with the function $\langle \cdot | \cdot \rangle$ defined by

$\langle \cdot \| \cdot \rangle$	α_1	α_2	α_3	α_4	α_5	α_6	α_7	α_8
α_1	1	0	0	$\frac{1}{2}$	$\frac{1}{2}$	$\frac{1}{2}$	0	0
α_2	0	1	0	$\frac{1}{2}$	$\frac{1}{2}$	0	$\frac{1}{2}$	$\frac{1}{2}$
α_3	0	0	1	0	0	$\frac{1}{2}$	$\frac{1}{2}$	$\frac{1}{2}$
α_4	$\frac{1}{2}$	$\frac{1}{2}$	0	1	0	$\frac{1}{2}$	$\frac{1}{2}$	0
α_5	$\frac{1}{2}$	$\frac{1}{2}$	0	0	1	0	0	$\frac{1}{2}$
α_6	$\frac{1}{2}$	0	$\frac{1}{2}$	$\frac{1}{2}$	0	1	0	$\frac{1}{2}$
α_7	0	$\frac{1}{2}$	$\frac{1}{2}$	$\frac{1}{2}$	0	0	1	0
α_8	0	$\frac{1}{2}$	$\frac{1}{2}$	0	$\frac{1}{2}$	$\frac{1}{2}$	0	1

is a transition-probability space (this example is contained in Reference 1). [*Hint.* By simple check.]

3. All bases of a transition-probability space have the same cardinality.

[*Hint.* Let R_1, R_2 be two bases. In the finite case the proof is just that (cardinality of $R_1) = \Sigma_{\alpha \in R_1} \Sigma_{\beta \in R_2} \langle \alpha | \beta \rangle = \Sigma_{\beta \in R_2} \Sigma_{\alpha \in R_1} \langle \alpha | \beta \rangle =$ (cardinality of R_2). For the infinite case see Reference 2, Theorem 1.]

4. Show that each state α belongs to at least one basis.

[*Hint.* Choose β_1 in $\{\alpha\}'$, β_2 in $\{\alpha, \beta_1\}'$, β_3 in $\{\alpha, \beta_1, \beta_2\}'$, and so on. Use Zorn's lemma to conclude that the sequence $\alpha, \beta_1, \beta_2, \ldots$ is well defined, and observe that it is a maximal orthogonal subset.)

5. Show that in the transition-probability space of Exercise 1 the closure relation $S \mapsto \bar{S} = S''$ does not satisfy the exchange property.

[*Hint.* Verify that $\alpha_1 \in \{\alpha_3, \alpha_4\}''$ while $\alpha_4 \notin \{\alpha_1, \alpha_3\}''$.)

6. Prove Theorem 18.2.1.

[*Hint.* (derived from Reference 1, Section 3). Clearly $R' \supseteq S'$. If $\beta \in R'$, then $R \cup \{\beta\}$ is pairwise orthogonal, and hence $\langle \alpha | \beta \rangle + \Sigma_{\gamma \in R} \langle \alpha | \gamma \rangle \leq 1$ for any α; taking α in S, we have $\Sigma_{\gamma \in R} \langle \alpha | \gamma \rangle = 1$; hence $\langle \alpha | \beta \rangle = 0$ and $\beta \in S'$. Thus $R' \subseteq S'$.]

7. Prove Theorem 18.2.2.

[*Hint.* (derived from Reference 1, Section 3). If R_1 is a basis for S, and R_2 is a maximal pairwise orthogonal subset in S', then $R_1 \cup R_2$ is a basis for the whole of $\mathcal{S}^P$. For arbitrary $\alpha \in S'$ we have $\Sigma_{\beta \in R_1 \cup R_2} \langle \alpha | \beta \rangle = \Sigma_{\beta \in R_2} \langle \alpha | \beta \rangle = 1$, so that R_2 is a basis for S'.]

8. Show that in a transition-probability space the orthoclosed subsets do not need to be subspaces and subspaces do not need to be orthoclosed.

[*Hint.* Refer to Exercise 2 and observe that $\{\alpha_3, \alpha_7\}$ is orthoclosed without being a subspace, while $\{\alpha_1, \alpha_2\}$ is a subspace without being orthoclosed.]

9. Show that $\tilde{\mathcal{K}}^P$ is orthomodular.

[*Hint.* If $S_1, S_2 \in \mathcal{K}^P$ and $S_1 \subseteq S_2$, take a basis R_2 for S_2 and denote by R_1 the part of R_2 that belongs to S_1. $R_2 - R_1$ is a basis for $S_2 \cap S_1'$, and the latter is a subspace. Then show that $S_1 \vee (S_2 \cap S_1')$ exists in $\mathcal{K}^P$, and R_2 is a basis for it.]

10. Prove that the function α defined by (18.3.2) is additive on orthogonal elements of $\tilde{\mathcal{K}}^P$.

[*Hint.* Let $\{S_i\}$ be a family of orthogonal elements of $\tilde{\mathcal{K}}^P$, let R_i be a basis for S_i, and let $R = \cup_i R_i$. Notice that R is a basis for R'' and that $R'' = (\cap_i R_i')' = (\cap_i S_i')' = (\cup_i S_i)'' = \vee_i S_i$ (see Theorem 18.2.1); then conclude

$$\alpha\left(\bigvee_i S_i\right) = \sum_{\beta \in R} \langle \alpha | \beta \rangle = \sum_i \sum_{\beta \in R_i} \langle \alpha | \beta \rangle = \sum_i \alpha(S_i).]$$

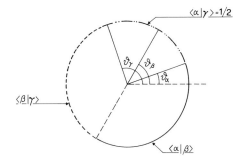

Figure 18.1. Non-Hilbertian transition probabilities.

11. Let S^P be a two-dimensional transition-probability space, let each element $\alpha \in S^P$ be labeled by an angle ϑ_α such that $0 \leqslant \vartheta_\alpha \leqslant 2\pi$, and let $\langle \alpha | \beta \rangle = |1 - |\vartheta_\beta - \vartheta_\alpha|/\pi|$. Show that this transition probability space is non-Hilbertian.

[*Hint.* Consider three states α, β, γ as in Figure 18.1, so that $\langle \alpha | \beta \rangle + \langle \beta | \gamma \rangle = 2 - \langle \alpha | \gamma \rangle$, and observe that such a linear relation cannot be reproduced by squared moduli of scalar products in a (2-dimensional) Hilbert space.]

References

1. J. G. F. Belinfante, *J. Math. Phys.* 17 (1976) 285.
2. B. Mielnik, *Commun. Math. Phys.* 9 (1968) 55.
3. P. C. Zabey, *Found. Phys.* 5(1975) 323.
4. K. Bugajska, *Int. J. Theor. Phys.* 9 (1974) 93.
5. P. C. Deliyannis, *J. Math. Phys.* 17 (1976) 653.

Additional Bibliography

1. V. Cantoni, *Commun. Math. Phys.* 44 (1975) 125.
2. V. Cantoni, *Commun. Math. Phys.* 56 (1977) 189.
3. S. P. Gudder, *Commun. Math. Phys.* 63 (1978) 265.

CHAPTER 19

On the Convex-Set Approach

19.1 The Convex Structure of States and the Collection of Yes-No Experiments

The approach to quantum theory alluded to in the title has been developed in the last ten or fifteen years and can be traced back, at least partly, to the theory of Ludwig and his school.[1,2] There has been impressive and promising research work on this approach,* but an exhaustive review of it certainly goes beyond the scope of this volume. The purpose of this chapter is only to provide an introduction to the convex-set approach and catch its main ideas, novelties, and problems. In the present section we mainly refer to the work of Mielnik.[6–8]

Here the primitive ingredients are the states of the physical system, and the operation postulated in the set S they form is that of forming mixtures. The convex structure of S is the point of departure. When states are viewed as probability measures on propositions, the notion of convex combination is naturally accounted for, but this is not the case with the convex-set approach, where propositions are viewed as derived elements. A possibility† of accommodating the notion of convexity for S would then be to represent S as being embedded in a vector space V, so that the convex combinations in S are deduced from the notion of linear combination in V, and the

---

*The expression "operational approach" is sometimes used in connection with, or as a substitute for, "convex-set approach". The use of "operational" here is not in the sense of the strictly operational scheme of Foulis, Randall,[3,4] and their school (for on overall bibliography see Reference 5).

†As a matter of fact, this problem was present also in the last two chapters, but there we skipped it because the convex structure of S played a less central role.

ENCYCLOPEDIA OF MATHEMATICS and Its Applications, Gian-Carlo Rota (ed.). Vol. 15: E. G. Beltrametti and G. Cassinelli, The Logic of Quantum Mechanics

ISBN 0-201-13514-0

assertion that $\mathcal{S}$ is convex appears as follows: if $\alpha, \beta \in \mathcal{S}$, then for any two weights w_1, w_2 ($w_1, w_2 \geq 0$, $w_1 + w_2 = 1$), the vector $w_1\alpha + w_2\beta$ belongs to $\mathcal{S}$. While the points of $\mathcal{S}$ represent pure and mixed states of the system, the points of V outside $\mathcal{S}$ have no physical interpretation.

Regularity requirements suggest assuming that V is not an arbitrary vector space but a topological vector space, and that $\mathcal{S}$ is closed in this topology.

The ambitious program of the convex-set approach is a complete geometrization of quantum mechanics, in the sense that the physical information on the system should be read from the geometry of $\mathcal{S}$. Though the structure of $\mathcal{S}$ translates in explicit form only the information about which states are mixtures of which other states, the shape of $\mathcal{S}$, in particular the shape of its boundary, implicitly reflects a richer phenomenology of the physical system. As we are going to sketch, the shape of $\mathcal{S}$ determines the structure of the set of yes-no experiments on the system.

We need first the concept of normal functional on $\mathcal{S}$. Consider a continuous linear function e that maps the topological vector space V into the real line $\mathbb{R}$: continuity is, of course, referred to the topology of V, and linearity means that $e(t_1 x_1 + t_2 x_2) = t_1 e(x_1) + t_2 e(x_2)$ for every $x_1, x_2 \in V$, $t_1, t_2 \in \mathbb{R}$. Then we say that e is a *normal functional* on $\mathcal{S}$ if and only if it maps $\mathcal{S}$ onto the segment $[0, 1]$:

$$0 \leq e(\alpha) \leq 1 \qquad \text{for every} \quad \alpha \in \mathcal{S}.$$

Therefore a normal functional on $\mathcal{S}$ is such that $\mathcal{S}$ is inside the region of V limited by the parallel hyperplanes $e = 0$ and $e = 1$ (Figure 19.1). The collection of all normal functionals on $\mathcal{S}$ is thus represented by a collection of pairs of parallel hyperplanes enclosing $\mathcal{S}$ and tangent to it. And it is clear that this collection of pairs of hyperplanes is determined by the shape of $\mathcal{S}$, or better, by the shape of the boundary, or surface, of $\mathcal{S}$.

In order to find a physical interpretation for the normal functionals on $\mathcal{S}$, recall that a yes-no experiment determines, as discussed in Section 13.1, a

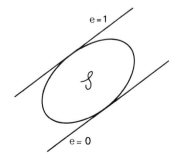

Figure 19.1. A normal functional on the set of states.

probability function on $\mathbb{S}$ (the probability of getting the yes outcome when the incoming state of the system runs over $\mathbb{S}$). So one has a hint to interpret the normal functionals on $\mathbb{S}$, which are indeed probability functions on $\mathbb{S}$, as the yes-no experiments. If e is a normal functional, the points of $\mathbb{S}$ that belong to the hyperplane $e=1$ are the states that cause with certainty the "yes" outcome of the associated yes-no experiment, while the points of $\mathbb{S}$ belonging to $e=0$ are the states that cause with certainty the "no" outcome. Thus, the structure of the collection of all yes-no experiments is not independently given, but is determined by the shape of $\mathbb{S}$, and hence by the more primitive information concerning which states of the system are mixtures of which other states.*

19.2 Faces

Another concept that plays a crucial role is that of a face of the set $\mathbb{S}$ of states. $S \subseteq \mathbb{S}$ is called a *face* when for any two weights w_1, w_2,

$$\alpha, \beta \in S \qquad \text{implies} \quad w_1\alpha + w_2\beta \in S, \qquad (19.2.1)$$

$$w_1\alpha + w_2\beta \in S \quad \text{implies} \quad \alpha, \beta \in S. \qquad (19.2.2)$$

The former property simply says that faces are convex subsets of $\mathbb{S}$. The latter says that faces are convex in a strong way, for whenever a face contains an internal point of a "segment", then it must contain the whole "segment". The term face is motivated by the fact that in case $\mathbb{S}$ became a polyhedron, its faces, in the ordinary sense, would have precisely the properties under discussion. Alternative expressions are "wall",[6] "completely convex subset",[1] and "extremal subset".[9]

Any convex set $\mathbb{S}$ has two improper faces: $\mathbb{S}$ itself and the empty set $\varnothing$. If $\mathbb{S}$ has extreme points, as here assumed, then each extreme point (that is, each pure state) is a face (in a polyhedron these are the vertices).

The set $\mathcal{F}$ of all the faces of $\mathbb{S}$ is partially ordered by set-theoretical inclusion. $\varnothing$ is the least element, denoted 0, while the whole of $\mathbb{S}$ is the greatest element, denoted 1. The faces formed by just one (pure) state are atoms of $\mathcal{F}$, and $\mathcal{F}$ is atomic if every mixture can be expressed as a mixture of pure states. But we have more: $\mathcal{F}$ is a lattice. Indeed, the intersection of any family of faces is still a face, so that for $S_1, S_2 \in \mathcal{F}$, we get $S_1 \wedge S_2 = S_1 \cap S_2$. Similarly, the join $S_1 \vee S_2$ is obtained by taking the intersection of all faces containing (i.e. majorizing) both S_1 and S_2.

The notion of face is reminiscent of the notion of the certainly-yes (and certainly-no) domain of a yes-no experiment. Indeed, if we think of S as the

---

*The reader will recall that, as outlined in Chapter 13, the notion of yes-no experiment is not to be confused with what we have called a proposition of the physical system: only yes-no experiments of a rather specialized kind correspond to propositions.

certainly-yes domain of a yes-no experiment, then the property (19.2.1) asserts the physically obvious fact that when two states cause with certainty the "yes" outcome of a yes-no experiment, their mixtures also do, while (19.2.2) asserts the similarly obvious fact that when a mixture causes with certainty the "yes" outcome, so must the states of which it is a mixture: to cause with certainty the "yes" outcome (or the "no" outcome) is strictly hereditary under convex combinations. Thus one is led to consider a particular class of faces: a face S is called *detectable* if there exists a normal functional e on $\mathbb{S}$ such that S is the geometrical contact of $\mathbb{S}$ with the hyperplane $e = 1$ — formally, if $S = e^{-1}(1)$. According to what was said in last section, we can also say that a face is detectable when it is the certainly-yes domain of a yes-no experiment. Thus the detectable faces represent operationally verifiable properties of the physical system. Conversely, a yes-no experiment e can be regarded as a measuring instrument of some property of the system provided the hyperplane $e = 1$ is tangent to $\mathbb{S}$ at a face. Again we see that the shape of $\mathbb{S}$ with its collection of faces, notably the detectable ones, contains rich information on the physical behavior of the system.

Though detectable faces resemble in many respects what we have throughout called propositions, we must however realize that in fact they provide a generalization of the idea of proposition. This will become clear in the next section, where it will be seen that, in general, detectable faces cannot be endowed with such a crucial property of propositions as ortho-complementation (which, in logical terms, corresponds to negation), and in order to catch this property one has to restrict oneself to a very specialized family of detectable faces. The convex-set approach is thus seen to provide a format that is more general and more flexible than the usual propositions-states structure.

19.3 Stable Faces and Propositions

In this section the convex set $\mathbb{S}$ of states will be subject to simplifying hypotheses: it will be a polytope. This means, essentially, that the vector space where $\mathbb{S}$ is embedded is finite-dimensional. Although a very special case of a convex set, the study of the structure of faces associated with a polytope gives a beginning insight and poses an abundance of problems in its own right. We shall refer throughout to papers and results by G. Rüttimann.[10-12]*

Technically, a polytope is defined as a convex compact subset of an n-dimensional Euclidean space that has finitely many extreme points (notice that the cardinality of the set of extreme points of the polytope is not less than $n+1$). When $\mathbb{S}$ is a polytope there is no longer a distinction between

*We are indebted to him for many valuable discussions and for his kind assistance.

detectable and nondetectable faces: every face is detectable, and thus represents an operationally verifiable property of the system. As in the last section, we write $\mathcal{F}$ for the complete lattice (relative to set inclusion) formed by the faces of $\mathcal{S}$. Viewing faces as properties of the physical system, we shall also use the following nomenclature: when $\alpha \in S \in \mathcal{F}$, we say that the system in state α *has* property S; when $S_1 \leqslant S_2$, we say that property S_2 *generalizes* property S_1.

We come now to a notion of orthogonality in $\mathcal{F}$. Two faces S_1, S_2 are called orthogonal, and we write $S_1 \perp S_2$, whenever there is a normal functional e on $\mathcal{S}$ such that the hyperplane $e=1$ contains S_1 and the hyperplane $e=0$ contains S_2. In other words, S_1 and S_2 are orthogonal whenever there exists a yes-no experiment having S_1 in its certainly-yes domain and S_2 in its certainly-no domain, so that this experiment is able to distinguish S_1 from S_2 without probabilistic uncertainty. Immediately we have: (i) $S_1 \perp S_2$ implies $S_2 \perp S_1$; (ii) $0 \perp S$ for all $S \in \mathcal{F}$; (iii) $S_1 \perp S_2$ implies $S_1 \wedge S_2 = 0$; (iv) $S \perp S$ implies $S=0$; (v) $S_1 \leqslant S_2$ and $S_2 \perp S_3$ imply $S_1 \perp S_3$. Viewing faces as properties of the physical system, if $S_1 \perp S_2$, we shall say that property S_1 *excludes* property S_2.

Given $S \in \mathcal{F}$, we denote by $\{S\}'$ the collection of all faces that are orthogonal to S, and we remark that $\{S\}'$ is an ideal in the lattice $\mathcal{F}$ (every element smaller than an element of $\{S\}'$ is itself an element of $\{S\}'$). When $\{S\}'$ is a principal ideal, that is, when it admits a (unique) maximal element (to be denoted $S^{\perp}$), we shall say that S is *orthoclosed*. The reason to look for orthoclosed faces is that $S^{\perp}$, as the notation anticipates, approaches the intuitive idea of an orthocomplement of S. Note that $0, 1$ are orthoclosed. To visualize the condition of being an orthoclosed face, suppose that $\mathcal{S}$ is the polygon of Figure 19.2, embedded in the real plane: its nontrivial faces are the four vertices $\alpha, \beta, \gamma, \delta$ and the four edges. Clearly the faces orthogonal to α are $\beta, \gamma, \delta, \beta\gamma, \gamma\delta$ (as seen by the drawn pairs of parallel lines), and they do not admit a greatest element: thus α is not orthoclosed. On the contrary, the face $\beta\gamma$ is orthoclosed, for the only (nonempty) face orthogonal to it is the vertex α.

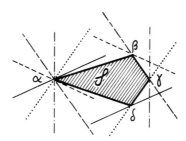

Figure 19.2. Orthoclosed and non-orthoclosed faces of a polygon.

Let us list some further pertinent definitions. A key role is played by the notion of stable face. A face S_1 is said to be *stable* provided there exists an orthoclosed face S_2 such that $S_1 = S_2^\perp$. $\mathfrak{L}(\mathcal{S})$ will denote the set of stable faces of $\mathcal{S}$. Note that $0, 1$ are stable. A pair $S_1, S_2 \in \mathfrak{L}(\mathcal{S})$ is called *mutually stabilizing*, in symbols $S_1 \perp\!\perp S_2$, if $S_1 = S_2^\perp$ and $S_2 = S_1^\perp$.*

An interesting class of polytopes, from the point of view of relating the convex approach to the usual propositions-states structures, is the following one. A polytope is said to be *strong* provided

(i) every stable face is orthoclosed,
(ii) if two stable faces S_1, S_2 are orthogonal but not mutually stabilizing, then $\{S_1, S_2\}'$ is a principal ideal in $\mathcal{F} \backslash 1$.[†]

As we are going to see, when $\mathcal{S}$ is a strong polytope, the stable faces are endowed with much of the standard structure of propositions; this motivates us to denote them by $a, b, c, \ldots$, the notation that we have throughout reserved for the objects that match the idea of propositions of the physical system. Now, if $\mathcal{S}$ is a strong polytope we have the following facts:

THEOREM 19.3.1. $\mathfrak{L}(\mathcal{S})$ *is closed under the mapping* $a \mapsto a^\perp$.

THEOREM 19.3.2. $\mathfrak{L}(\mathcal{S})$ *is an orthomodular poset with respect to the order given by set-theoretical inclusion, and the orthocomplementation is* $a \mapsto a^\perp$.

THEOREM 19.3.3. $\mathfrak{L}(\mathcal{S})$ *contains the join of orthogonal elements* (*and the join coincides with the join in* $\mathcal{F}$).

THEOREM 19.3.4. $\mathfrak{L}(\mathcal{S})$ *is an orthomodular lattice if and only if* $\{a, b\}'$ *is a principal ideal in* $\mathcal{F}$ *for all* $a, b \in \mathfrak{L}(\mathcal{S})$.

An important family of strong polytopes consists of the simplexes. A simplex is a polytope for which the cardinality of the extreme points equals $n + 1$, n being the dimension of the Euclidean space where the polytope is embedded. A triangle in the plane and a tetrahedron in 3-space are examples of simplexes. In a simplex every face is stable and $\mathfrak{L}(\mathcal{S})$ ($= \mathcal{F}$) is a Boolean algebra. We have also the converse: if $\mathcal{S}$ is a strong polytope and $\mathfrak{L}(\mathcal{S})$ is Boolean, then $\mathcal{S}$ is a simplex.

Further examples are given below.

---

*One easily verifies: (i) $S_1 \perp\!\perp S_2$ implies $S_2 \perp\!\perp S_1$; (ii) $S_1 \perp\!\perp S_2$ implies $S_1 \perp S_2$; (iii) $0 \perp\!\perp S$ if and only if $S = 1$; (iv) $1 \perp\!\perp S$ if and only if $S = 0$; (v) $S_1 \perp\!\perp S_2$, $S_3 \perp S_4$, $S_1 \leqslant S_3$, $S_2 \leqslant S_4$ imply $S_1 = S_3$, $S_2 = S_4$. From concrete examples one realizes that, loosely speaking, if one holds a polytope by two mutually stabilizing faces then it is stable against twisting.

[†]The prime of a set of faces is the collection of the faces that are orthogonal to all the elements of the set. The notation $\mathcal{F} \backslash 1$ denotes the set $\mathcal{F}$ with the exclusion of the trivial element 1.

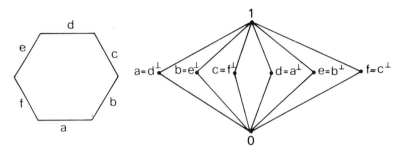

Figure 19.3.   The hexagon and the lattice of its stable faces.

(1) A hexagon is a strong polytope whose nontrivial stable faces are the six edges. The associated lattice is shown in Figure 19.3.

(2) Consider the strong polytope consisting of a pyramid with a square base. Its nontrivial stable faces are labeled in Figure 19.4, and the associated $\mathfrak{L}(\mathcal{S})$ is the lattice G_{12} that we met with in Section 10.2 (see Figure 10.6 and refer also to Figure 11.1).

(3) The triangular prism is a strong polytope with nontrivial stable faces as shown in Figure 19.5, and the associated $\mathfrak{L}(\mathcal{S})$ is the lattice that we have already mentioned in Section 10.2 (Figure 10.3).

We proceed to mention another series of nice facts. If $\mathcal{S}$ is a strong polytope, then for any stable face a there exists a unique normal functional e_a such that a lies in the hyperplane $e_a = 1$ and $a^\perp$ in the hyperplane $e_a = 0$. In other words, there is a yes-no experiment such that the states belonging to a cause with certainty the "yes" outcome while the states belonging to $a^\perp$ cause with certainty the "no" outcome. Thus, if α is an element of $\mathcal{S}$ we can define

$$m_\alpha(a) = e_a(\alpha).$$

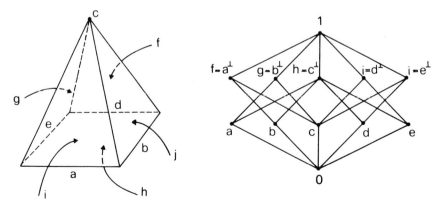

Figure 19.4.   The square pyramid and the lattice of its stable faces.

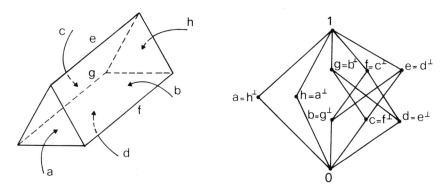

Figure 19.5.   The triangular prism and the lattice of its stable faces.

Let $\tilde{\mathcal{S}}$ be the set of functions on $\mathcal{L}(\mathcal{S})$ induced by the elements of $\mathcal{S}$, in symbols $\tilde{\mathcal{S}} = \{m_\alpha : \alpha \in \mathcal{S}\}$. We have the following results ($\mathcal{S}$ is always a strong polytope):

THEOREM 19.3.5.   *For every $\alpha \in \mathcal{S}$ the function $a \mapsto m_\alpha(a)$ is a probability measure on the orthomodular poset $\mathcal{L}(\mathcal{S})$; moreover the mapping $\alpha \mapsto m_\alpha$ is one-to-one.*

THEOREM 19.3.6. $\tilde{\mathcal{S}}$ *is strongly ordering on $\mathcal{L}(\mathcal{S})$.*

THEOREM 19.3.7. m_α *is a pure probability measure if and only if α is an extreme point of $\mathcal{S}$.*

THEOREM 19.3.8. m_α *is a superposition, in the sense of Section 11 (see Definition 11.5.1), of the family $\{m_{\alpha_i} : i \in \mathcal{I}\}$ of probability measures, if and only if α is contained in the smallest face containing $\{\alpha_i : i \in \mathcal{I}\}$.*

The series of results quoted in this section shows that the convex-set approach can be related to the usual proposition-state structure. But at the same time we see that the recovery of this structure is accomplished at the cost of a severe restriction of the shape of the convex set $\mathcal{S}$ (restriction to a strong polytope) and of a similar restriction of the class of faces to be taken into account (restriction to stable faces). If unrestricted, the convex-set scheme still accommodates the notions of yes-no experiments on the system (normal functionals on $\mathcal{S}$) and of physical properties of the system (detectable faces), but it is a relaxation of the structure required by quantum mechanics, for in general it does not allow the introduction of a notion of orthocomplement (see for instance the convex set of Figure 19.1).*

---

*It has been argued[8] that the abandonment of orthocomplementation might be physically tolerable and that it might open the way to interesting generalizations of quantum mechanics, notably in the direction of nonlinear theories.

In a slightly different point of view we have the following result of Shultz:[14] every compact convex subset of a locally convex topological vector space is affinely homeomorphic to the set of all states of an orthomodular lattice. Whether or not there is a corresponding result for quantum logics is still an open question.

Let us add a remark. We have already noticed that when $\mathfrak{S}$ is a simplex, then $\mathfrak{L}(\mathfrak{S})$ is Boolean, and vice versa. Let us here note that in a simplex a nonextreme element has a unique convex decomposition into extreme elements: in other words, a nonpure state has a unique decomposition into pure states. This feature is typical of classical systems, which are indeed associated with Boolean structures, but is not found in quantum systems (see Section 2.3), which are instead associated with orthomodular, non-Boolean structures. While a simplex is thus the format for the states of a classical system,[†] the simplex structure is off limits for quantum systems. As an example of a typical quantum shape, consider the spin-$\frac{1}{2}$ system, whose states form a sphere (see Section 4.2).

References

1. G. Ludwig, *Z. Naturforsch.* 22a (1967) 1303; *ibid.* 22a (1967) 1324; *Commun. Math. Phys.* 4 (1967) 331; *ibid.* 9 (1968) 1; *ibid.* 26 (1972) 78.
2. G. Ludwig, *Deutung des Begriffs "Physikalische Theorie" und axiomatische Grundlegung der Hilbertraumstruktur der Quantenmechanik durch Hauptsätze des Messen*, Lecture Notes in Physics, Springer-Verlag, Berlin, Heidelberg, New York, 1970.
3. D. Foulis and C. Randall, *J. Math. Phys.* 13 (1972) 1167.
4. C. Randall and D. Foulis, *J. Math. Phys.* 14 (1972) 1472.
5. D. Foulis and C. Randall, in *Current Issues in Quantum Logic*, E. G. Beltrametti and B. van Fraassen, eds., Plenum, New York, 1981.
6. B. Mielnik, *Commun. Math. Phys.* 15 (1969) 1.
7. B. Mielnik, *Commun. Math. Phys.* 37 (1974) 221.
8. B. Mielnik, in *Quantum Mechanics, Determinism, Causality and Particles*, M. Flato et al., ed., Reidel, Dordrecht, 1976.
9. G. Dähn, *Commun. Math. Phys.* 9 (1968) 192.
10. G. T. Rüttimann, *Bull. Amer. Math. Soc.* 82 (1976) 314.
11. G. T. Rüttimann, in *Interpretations and Foundations of Quantum Theories*, G. Ludwig and H. Neumann, eds., Bibliographisches Institut, München, 1980.
12. G. T. Rüttimann, *Non-Commutative Measure Theory*, Habilitationsschrift, Universität Bern, 1979.
13. E. M. Alfsen, *Compact Convex Sets and Boundary Integrals*, Springer-Verlag, Berlin and New York, 1971.
14. F. W. Shultz, *J. Comb. Theory* (A) 17 (1974) 317.

Additional Bibliography

1. E. B. Davies and J. Lewis, *Commun. Math. Phys.* 17 (1970) 239.
2. C. M. Edwards, *Commun. Math. Phys.* 16 (1970) 207.
3. C. M. Edwards, *Commun. Math. Phys.* 20 (1971) 26.
4. S. P. Gudder, *Commun. Math. Phys.* 29 (1973) 249.
5. S. P. Gudder and W. Cornette, *J. Math. Phys.* 15 (1974) 842.

[†] The generalization of the notion of simplex to the infinite-dimensional case is provided by Choquet simplexes.[13]

CHAPTER 20

Introduction to a Quantum Logic

20.1 Sentences Associated with a Quantum System and Their Truth Values

In a sense this chapter is marginal to the content of this volume; the reader can skip it without prejudice to the comprehension of chapters to follow. Quantum logic is a discipline that branched off from the 1936 paper of Birkhoff and von Neumann[1] and has the orthomodular-poset structures encountered in previous chapters as basic mathematical carriers: that is why we think it worthwhile to devote some pages to the subject. But the main interests quantum logic calls into play are in the territory of logicians and philosophers: that is why we (who are neither logicians nor philosophers) shall confine ourselves to a very simplified introduction to the subject.

Roughly, the starting question is whether the propositions of a quantum system can be associated with, or can be interpreted as, sentences of a language (or propositional calculus) and which rules this language inherits from the ordered structure of propositions. In raising this question one has in mind the fact that when the physical system is classical its propositions form a Boolean algebra, and Boolean algebras are the algebraic models of the calculus of classical logic. Thus the question above can also be phrased as follows: when a Boolean algebra is relaxed into an orthomodular nondistributive lattice, which logic is it the model of? "Quantum logic" is the name that designates the answer, but there are several views about the content of this name. Here we sketch one approach to a quantum logic.

The typical structure of the set $\mathcal{L}$ of propositions that will be used in the sequel is that of an orthomodular lattice. From one side this is something more than the poset structures emerging from some of the approaches

ENCYCLOPEDIA OF MATHEMATICS and Its Applications, Gian-Carlo Rota (ed.). Vol. 15: E. G. Beltrametti and G. Cassinelli, The Logic of Quantum Mechanics

ISBN 0-201-13514-0

considered in preceding chapters; from the other side it is something less, for atomicity, the covering property, and related facts will remain rather untreated.

To the pair formed by the state α and the proposition a we can associate the sentence

$$\langle \alpha, a \rangle = \frac{\text{"the physical system in state } \alpha \text{ causes}}{\text{the yes outcome of the proposition } a \text{ "},} \qquad (20.1.1)$$

so that propositions generate the predicates of the object language, while states play the role of individuals to which predicates apply, the physical system being fixed.

For the sake of simplicity let us restrict ourselves to pure states, writing S^P for the set they form, and recall that propositions correspond to probability measures on S^P, the value of a at $\alpha \in S^P$ being the probability of a yes outcome of a when the system is in state α. Of course, a basic departure between the classical and the quantum case is already present at this stage. In the classical case every $a \in \mathcal{L}$ maps S^P onto $\{0,1\}$, thus splitting S^P into two parts: the set of states that are mapped into 0 and the set of those that are mapped into 1. Accordingly, the semantical values of the sentence $\langle \alpha, a \rangle$ are unambiguously two: $\langle \alpha, a \rangle$ is either true or false. In the quantum case every $a \in \mathcal{L}$ maps S^P onto the whole segment $[0,1]$,* thus injecting some ambiguity into the semantical values of $\langle \alpha, a \rangle$. The knowledge that a measurement of a with the system in state α has given the yes outcome is not sufficient to say freely that $\langle \alpha, a \rangle$ is true, for in a replica of the same measurement with the system in the same state, the occurrence of the yes outcome is no longer guaranteed.

Thus one is led to define the truth of $\langle \alpha, a \rangle$ as the situation in which the yes outcome of a is certain. With this definition $\langle \alpha, a \rangle$ is true provided α belongs to the certainly-yes domain $S_1^P(a)$, the set associated with a that we have encountered throughout previous chapters. Notice that it still holds true, as in the classical case, that a predicate determines a subset of S^P (its truth set), but it is no longer true that every subset of S^P is the truth set of some predicate. Now, with the above convention about the truth of $\langle \alpha, a \rangle$, what is its falsehood?

One might take the attitude of saying that $\langle \alpha, a \rangle$ is false whenever it is not true, namely that $\langle \alpha, a \rangle$ is false whenever the state of the system does not give with certainty the yes outcome of a. This way out is the one we shall adopt in the sequel; it has the advantage of reducing the semantical values to true or false, but it must be noticed that it puts truth and falsehood in an unsymmetrical relation. Indeed, having asserted that $\langle \alpha, a \rangle$

*The pair formed by S^P and a function from S^P onto $[0,1]$ defines a fuzzy subset of S^P. Propositions of a quantum system might be associated with fuzzy subsets of the set of states.

is false, we are not guaranteed that the "no" outcome of a is certain: we only known that the "yes" outcome is not certain but nevertheless possible.

Another attitude might be the one of saying that $\langle \alpha, a \rangle$ is false whenever the "no" outcome of a is certain. Truth and falsehood are now in a symmetrical relation, but nonclassical features reemerge in other aspects. We cannot say that the sentence $\langle \alpha, a \rangle$ is either true or false, for a quantum system admits states such that $\alpha(a)$ is neither 0 nor 1: truth and falsehood do not exhaust all possibilities. Thus one is faced with a many-valued logic: to each numerical value of the probability of getting the yes outcome of a one associates a semantical entry, which becomes the "true" entry if the probability is 1 and the "false" entry if the probability is 0. In principle, this is an infinite-valued logic. Should one amalgamate all entries that are neither "true" nor "false" into a unique entry, say "indeterminate", one would get a three-valued logic, as in Reichenbach's well-known suggestion.[2] The idea of a three-valued logic has not had great influence in recent research on the logic of quantum mechanics. It has been superseded by the idea of using modal logic as a format for the interpretation of the quantum situation.[3-6] An account of these issues goes beyond our purposes; we refer to recent overviews by van Fraassen[7] and Mittelstaedt[8] and, for an historical perspective, to Jammer.[9]

20.2 Conjunction, Disjunction, Negation

We return, hereinafter, to the convention of using the word "false" as a synonym of "not true"; explicitly, we say that $\langle \alpha, a \rangle$ is false if $\alpha(a) \neq 1$. Thus we recover a two-valued semantics. Within this scheme we briefly examine, in this and next section, the elementary logical connectives, with the main purpose of pointing out the nonclassical features they might exhibit.

We deal first with the conjunction: given the sentences $\langle \alpha, a \rangle$ and $\langle \alpha, b \rangle$, the aim is of constructing the new sentence "$\langle \alpha, a \rangle$ and $\langle \alpha, b \rangle$". Recall that in previous chapters we have found many arguments in favor of characterizing the meet operation in $\mathcal{L}$ by*

$$\mathcal{S}_1^P(a \wedge b) = \mathcal{S}_1^P(a) \cap \mathcal{S}_1^P(b). \qquad (20.2.1)$$

If we generate, as usual, the "and" connective from the meet operation, thus defining

$$\langle \alpha, a \rangle \text{ and } \langle \alpha, b \rangle = \langle \alpha, a \wedge b \rangle, \qquad (20.2.2)$$

*In Section 11.4 this was the regularity condition (11.4.1); in Sections 11.5 and 14.8 it was a by-product of the isomorphism between $\mathcal{L}$ and $\mathcal{K}^P$; in Section 16.4 it came from properties of ideal first-kind measurements; and in Sections 17.2, 18.3, and 19.3 it was included in the very notion of proposition there derived.

then the property (20.2.1) means that the sentence (20.2.2) is true if and only if the sentences $\langle \alpha, a \rangle$, $\langle \alpha, b \rangle$ are both true. This is precisely the classical rule for the truth of conjunction.

Of course the operational interpretation of the conjunction inherits the problematic aspects that one finds for the corresponding meet operation when a and b do not commute (see Section 16.5). It is probably due to these problems that the conjuction connective has been for a while a debated issue in the literature on the logic of quantum mechanics.[†] But also, some misunderstanding of the so-called Heisenberg uncertainty principle and some confusion between noncompatibility and complementarity has aggravated an ill-posed issue on the quantum conjunction: we are referring, in particular, to the wrong claim that in case a, b do not commute, the sentences $\langle \alpha, a \rangle, \langle \alpha, b \rangle$ can never (that is, for no state) be both true (see the discussion in Sections 3.4, 3.5, 14.6, 14.7).

Next we consider the "or" connective, or disjunction, which translates the join operation of the lattice structure. Given the sentences $\langle \alpha, a \rangle, \langle \alpha, b \rangle$, we define their disjunction by

$$\langle \alpha, a \rangle \text{ or } \langle \alpha, b \rangle = \langle \alpha, a \vee b \rangle. \qquad (20.2.3)$$

In previous chapters we have motivated, from different standpoints, the relation

$$\mathcal{S}_1^P(a \vee b) \supseteq \mathcal{S}_1^P(a) \cup \mathcal{S}_1^P(b), \qquad (20.2.4)$$

which generalizes the equality holding in the classical case. Recall in particular that the existence of states belonging to the certainly-yes domain of $a \vee b$ without belonging to the certainly-yes domain of either a or b represents the very existence of quantum superpositions of a state in $\mathcal{S}_1^P(a)$ and a state in $\mathcal{S}_1^P(b)$ (see Sections 11.5 and 14.8). From (20.2.4) we have that in order to have the disjunction (20.2.3) true it is sufficient, but not necessary, that at least one of $\langle \alpha, a \rangle, \langle \alpha, b \rangle$ be true; for a quantum system we can have (20.2.3) true but $\langle \alpha, a \rangle, \langle \alpha, b \rangle$ both false. Quantum disjunction, though syntactically defined as in the classical case, has nonclassical features on the semantical level.

As an example, consider a spin-$\frac{1}{2}$ system and let $a=$"the spin is up along the z-axis", $b=$"the spin is up along the x-axis". The join $a \vee b$ is the trivial proposition 1, so that $\langle \alpha, a \vee b \rangle$ is a tautology: it is always true, even when $\langle \alpha, a \rangle$ as well as $\langle \alpha, b \rangle$ is false.

We come now to the quantum negation, which is the counterpart of the orthocomplementation of $\mathcal{L}$. After recalling that $\mathcal{S}_1^P(a^\perp)$, the certainly-yes domain of $a^\perp$, coincides with the certainly-no domain of a, the natural

---

[†] The reader is referred to Jammer's overview.[9]

candidate for the negation of $\langle \alpha, a \rangle$, which as usual we denote $\neg \langle \alpha, a \rangle$, is

$$\neg \langle \alpha, a \rangle = \langle \alpha, a^\perp \rangle. \qquad (20.2.5)$$

Syntactically this is again like the classical negation, but we are faced with nonclassical features at the semantical level. This is directly related to the meaning we have given to "true" and "false", according to which the truth of $\langle \alpha, a \rangle$ implies the falsehood of $\neg \langle \alpha, a \rangle$, but the falsehood of $\langle \alpha, a \rangle$ does not imply the truth of $\neg \langle \alpha, a \rangle$. Indeed, $\mathcal{S}_1^P(a^\perp)$ is not the set complement (relative to $\mathcal{S}^P$) of $\mathcal{S}_1^P(a)$, but in general is strictly contained in it; in other words, there are states that make false $\langle \alpha, a \rangle$ as well as $\neg \langle \alpha, a \rangle$.

Clearly we have the double-negation identity

$$\neg\neg \langle \alpha, a \rangle = \langle \alpha, a \rangle, \qquad (20.2.6)$$

which translates the identity $a^{\perp\perp} = a$ of lattice orthocomplementation. From the identity $a \vee a^\perp = 1$, we get, on account of (20.2.3) and (20.2.5), that the disjunction

$$(\langle \alpha, a \rangle \text{ or } \neg \langle \alpha, a \rangle) \text{ is true for every } \alpha \in \mathcal{S}^P, \qquad (20.2.7)$$

that is, it is tautologically true. This looks like the "excluded middle" law of classical logic. But now (20.2.7) cannot be interpreted as asserting that either $\langle \alpha, a \rangle$ is true or $\neg \langle \alpha, a \rangle$ is true, for there are states that make both false. In a sense, the nonclassical features of the quantum negation and of the quantum disjunction compensate each other, giving rise, formally, to the excluded-middle law (20.2.7).

In any orthocomplemented lattice, join and meet are related through orthocomplementation:

$$(a \vee b)^\perp = a^\perp \wedge b^\perp, \qquad (a \wedge b)^\perp = a^\perp \vee b^\perp.$$

In terms of the logical connectives (20.2.2), (20.2.3), (20.2.5) these identities read

$$\neg(\langle \alpha, a \rangle \text{ or } \langle \alpha, b \rangle) = \neg \langle \alpha, a \rangle \text{ and } \neg \langle \alpha, b \rangle,$$
$$\neg(\langle \alpha, a \rangle \text{ and } \langle \alpha, b \rangle) = \neg \langle \alpha, a \rangle \text{ or } \neg \langle \alpha, b \rangle,$$

and reproduce the familiar De Morgan's laws of ordinary logic. (One must however pay attention to the quantum pathologies implicit in the semantics of negations and disjunctions.)

20.3 Quantum Conditional

The *conditional*, or *implication*, is a binary connective that, operating on two sentences $\langle \alpha, a \rangle, \langle \alpha, b \rangle$, with a, b arbitrary elements of $\mathcal{L}$, forms the new sentence "if $\langle \alpha, a \rangle$ then $\langle \alpha, b \rangle$", for which the alternative notation $\langle \alpha, a \rangle \rightarrow \langle \alpha, b \rangle$ is often preferred. Since the sentences of our object language are generated by the elements of the proposition lattice $\mathcal{L}$, there must be an element of $\mathcal{L}$, depending on a, b and denoted $i(a, b)$, such that

$$(\langle \alpha, a \rangle \rightarrow \langle \alpha, b \rangle) = \langle \alpha, i(a, b) \rangle. \qquad (20.3.1)$$

The function i from $\mathcal{L} \times \mathcal{L}$ into $\mathcal{L}$ is precisely the lattice-theoretic counterpart of the conditional connective.

Let us recall, as background, that in the classical case, where $\mathcal{L}$ is a Boolean algebra, the conditional takes the Whitehead-Russell form

$$(\langle \alpha, a \rangle \rightarrow \langle \alpha, b \rangle) = (\neg \langle \alpha, a \rangle \text{ or } \langle \alpha, b \rangle), \qquad (20.3.2)$$

which corresponds to the choice

$$i(a, b) = a^{\perp} \vee b. \qquad (20.3.2')$$

It is easily seen that in the quantum case, where $\mathcal{L}$ is orthomodular but no longer distributive, this choice for the conditional becomes untenable, for it misses the most basic requirements a conditional must fulfill. First of all, it misses the natural link between the conditional and the order relations of the underlying lattice structure:

$$(\langle \alpha, a \rangle \rightarrow \langle \alpha, b \rangle) \text{ is true for every } \alpha \in \mathbb{S}^{P}$$

$$\text{if and only if} \quad a \leqslant b, \qquad (20.3.3)$$

or, in terms of the proposition lattice $\mathcal{L}$,

$$i(a, b) = 1 \quad \text{if and only if} \quad a \leqslant b. \qquad (20.3.3')$$

It is only in case $\mathcal{L}$ is Boolean that the choice (20.3.2) and (20.3.2') satisfies (20.3.3) and (20.3.3'). For an orthomodular, but nondistributive lattice, $a^{\perp} \vee b = 1$ does not imply $a \leqslant b$ (though, of course, $a \leqslant b$ still implies $a^{\perp} \vee b = 1$). To realize how decisive this fact is, we remark that, according to what was said in the last section about quantum disjunction, we can have, for a quantum system, that $\langle \alpha, a \rangle$ is true, $\langle \alpha, b \rangle$ is false, and $(\neg \langle \alpha, a \rangle$ or $\langle \alpha, b \rangle)$ is true, a fact that makes $(\neg \langle \alpha, a \rangle$ or $\langle \alpha, b \rangle)$ an inadmissible candidate for the conditional.

The choice of a conditional determines, and is determined by, a number of inferential schemes. Dismissing the classical choice (20.3.2), (20.3.2'), not every inferential scheme of classical logic can be retained; the search for a quantum substitute for the classical conditional rests on which ones one wants to preserve. The so-called *modus ponendo ponens* law is usually considered unavoidable in building up a significant logical calculus: it states that

$$[(\langle \alpha, a \rangle \text{ and } (\langle \alpha, a \rangle \to \langle \alpha, b \rangle)) \to \langle \alpha, b \rangle] \text{ is true}$$

$$\text{for every } \alpha \in \mathbb{S}^P, \quad (20.3.4)$$

or, in terms of the *i*-function,

$$a \wedge i(a, b) \leqslant b. \quad (20.3.4')$$

The joint requirement of the conditional-ordering connection (20.3.3), or (20.3.3'), and of the *modus ponendo ponens* (20.3.4), or (20.3.4'), restricts the choice of the conditional to the following five possibilities: [10-12]

$$i_1(a, b) = a^{\perp} \vee (a \wedge b),$$

$$i_2(a, b) = b \vee (a^{\perp} \wedge b^{\perp}),$$

$$i_3(a, b) = (a \wedge b) \vee (a^{\perp} \wedge b) \vee (a^{\perp} \wedge b^{\perp}),$$

$$i_4(a, b) = (a \wedge b) \vee (a^{\perp} \wedge b) \vee [(a^{\perp} \vee b) \wedge b^{\perp}],$$

$$i_5(a, b) = (a^{\perp} \wedge b) \vee (a^{\perp} \wedge b^{\perp}) \vee [a \wedge (a^{\perp} \vee b)].$$

Notice that each of these choices fulfills the consistency criterion of collapsing into the classical function (20.3.2') when the lattice $\mathcal{L}$ is assumed to be not only orthomodular but also distributive. It has to be emphasized that the orthomodularity of $\mathcal{L}$ is essential to prove that the functions i_1-i_5 satisfy (20.3.3') and (20.3.4').*

As an example of a classical law that loses its validity in the quantum case, we can mention the law of transitivity, which, in its strong form, reads

$$i(a, b) \wedge i(b, c) \leqslant i(a, c).$$

Counterexamples for each of the choices i_1-i_5 are immediately found by considering, for instance, the proposition lattice of a spin-$\frac{1}{2}$ system.

---

*In particular, (20.3.4') is precisely the orthomodular law if referred to i_1 or i_2. To see this fact it suffices to express orthomodularity in the form $a \leqslant b \Rightarrow b \wedge (b^{\perp} \vee a) \leqslant a$ and apply it to the ordered pairs $(a \wedge b, a)$ and $(b, a \vee b)$.

Another classical law that is off limits for the quantum case is the "implicative condition"

$$a \wedge c \leq b \quad \text{if and only if} \quad c \leq i(a, b). \tag{20.3.5}$$

Lattices in which this condition holds true are often called implicative lattices,* and they are known to be distributive (see Exercise 1). Various weakenings of the implicative condition (20.3.5) have been considered that characterize the orthomodularity and single out a unique conditional. Consider the following series of conditions:

$$a \wedge c \leq b \quad \text{implies} \quad a^{\perp} \vee (a \wedge c) \leq i(a, b), \tag{20.3.6}$$

$$a \wedge (a^{\perp} \vee c) \leq b \quad \text{implies} \quad c \leq i(a, b) \text{ and } a \wedge b \leq i(a, b), \tag{20.3.7}$$

$$a \wedge c \leq b \quad \text{implies} \quad (a \wedge c) \vee (a^{\perp} \wedge c) \leq i(a, b), \tag{20.3.8}$$

$$a \wedge c \leq b \quad \text{implies} \quad c \leq i(a, b) \text{ if } (a, c)C. \tag{20.3.9}$$

Then we have the following result: [13,14,11]

THEOREM 20.3.1. *Every orthomodular lattice satisfies* (20.3.3′), (20.3.4′), *and* (20.3.6)–(20.3.9) *under the choice* i_1; *conversely, every lattice that satisfies* (20.3.3′),(20.3.4′) *and any one of* (20.3.6)–(20.3.9) *is orthomodular, and the i-function is necessarily* i_1.[†]

These facts show that orthomodularity corresponds to the survival, for the proposition lattice of a quantum system, of a notion of the logical conditional, which takes the place of the classical implication (20.3.2′) associated with Boolean algebras.

The physical interpretation of the proposition $i(a, b)$ can be clarified by looking at its certainly-yes domain. By recourse to the notion of ideal first-kind measurement developed in Chapter 16 we have (see Exercise 2)

$$\mathbb{S}_1^P(i_1(a, b)) = \mathbb{S}_1^P(a^{\perp}) \cup \{\alpha \in \mathbb{S}^P : G_a \alpha(b) = 1\}, \tag{20.3.10}$$

where $G_a \alpha$ is the state into which α is mapped by an ideal first-kind measurement of a [the condition $\alpha \in \mathfrak{D}(G_a)$ is obviously understood]. In words, the conditional $\langle \alpha, i_1(a, b) \rangle$ is true if either $\neg \langle \alpha, a \rangle$ is true or $\langle G_a \alpha, b \rangle$ is true.[‡]

Before ending this chapter let us add a few general comments. What is said in this and in the previous two sections is intended to provide a loose hint on how a discussion on quantum logic might start. As already mentioned it is not our purpose to give an account of what quantum logic

---

*An element $i(a, b)$ satisfying (20.3.5) is called, by G. Birkhoff, the *pseudocomplement* of a relative to b.

[†] The i_1-function is often called Mittelstaedt's conditional.

[‡] This recalls the idea of counterfactual conditional.[4,11,15]

is—or, perhaps, quantum logics are, the plural being motivated by the many branchings out on the way toward a logical calculus admitting proposition lattices of quantum systems as algebraic models. We only list, pell-mell, some questions the quantum-logic community wants to answer and has partially answered. What is the role of quantum logic? Is it a truly new logic? This calls into play a more general question: is logic empirical or *a priori*? If it is empirical, how can quantum theory, which is formulated in terms of classical logic, give rise to a new nonclassical logic? What is the relationship between quantum logic and classical logic? Consistency reasons would require that the metalogic of quantum logic be classical logic: can this be proved? If quantum logic has to supersede classical logic, as a more general logic, how do we justify the use of mathematics, wedded as it is to classical logic? Does quantum logic reduce to classical logic if restricted to those sentences that form the language of mathematics? If logic is an *a priori* structure, valid irrespective of experience, are there approaches to quantum logic that do not reduce to reading it off from the empirical laws of quantum mechanics? An answer is the dialogic approach of Mittelstaedt and his school,[16] a two-person game in which the truth of a given sentence is defined by a successful dialog about the sentence in question, and where the rules of the dialog game reflect precognitions of any scientific argumentation. Another answer might be the operational approach introduced by Foulis and Randall and their school.[17]

As a starting point for a bibliography on the complex issue of quantum logic we suggest the proceedings of recent meetings on the subject matter, for instance References 18 and 19.

Exercises

1. Prove that implicative lattices are distributive.
[*Hint.* Take $a, b_1, b_2 \in \mathcal{L}$, write $b = (a \wedge b_1) \vee (a \wedge b_2)$, use the inequalities $a \wedge b_1 \leq b$, $a \wedge b_2 \leq b$ to get from (20.3.5) $b_1 \vee b_2 \leq i(a, b)$; from (20.3.4′) [which is a consequence of (20.3.5)] deduce $a \wedge (b_1 \vee b_2) \leq b$, and note that the reversed inequality is obvious, whence $a \wedge (b_1 \vee b_2) = (a \wedge b_1) \vee (a \wedge b_2)$. The dual property, with meet and join interchanged, follows from the orthocomplementation of $\mathcal{L}$.]

2. Prove (20.3.10).
[*Hint.* Suppose $\alpha \in \mathcal{S}_1^P(i_1(a, b))$, and note that α is either in $\mathcal{S}_1^P(a^\perp)$ or in $\mathcal{D}(G_a)$, in which case use $s(\alpha) \leq i_1(a, b)$ to deduce $s(G_a \alpha) = [s(\alpha) \vee a]^\perp \wedge a \leq [i_1(a, b) \vee a^\perp] \wedge a = a \wedge b$, and hence $G_a \alpha(b) = 1$. Conversely, suppose α to be an element of the right-hand side of (20.3.10). If $\alpha(a^\perp) = 1$, then clearly $\alpha(i_1(a, b)) = 1$. If $\alpha \in \mathcal{D}(G_a)$ and $G_a \alpha(b) = 1$, then $s(G_a \alpha) \leq b$ and also $s(G_a \alpha) \leq a \wedge b$, whence $a^\perp \vee s(G_a \alpha) \leq i_1(a, b)$. But noticing that $(a^\perp, s(\alpha) \vee a^\perp, a)$ is a distributive triple, the left-hand side becomes $s(\alpha) \vee a^\perp$; therefore $s(\alpha) \leq i_1(a, b)$, that is, $\alpha \in \mathcal{S}_1^P(i_1(a, b))$.]

References

1. G. Birkhoff and J. von Neumann, *Ann. Math.* 37 (1936) 823.
2. H. Reichenbach, "Three-valued logic and interpretation of quantum mechanics", in *The Logico-Algebraic Approach to Quantum Mechanics*, C. A. Hooker, ed., Reidel, Dordrecht, 1979.
3. B. C. van Fraassen, "A formal approach to the philosophy of science", in *Paradigms and Paradoxes*, R. Colodny, ed., University of Pittsburgh Press, Pittsburgh, 1972.
4. B. C. van Fraassen, "Semantic analysis of quantum logic", in *Contemporary Research in the Foundations and Philosophy of Quantum Theory*, C. A. Hooker, ed., Reidel, Dordrecht, 1973.
5. B. C. van Fraassen, "The Einstein-Podolski-Rosen Paradox", in *Logic and Probability in Quantum Mechanics*, P. Suppes, ed., Reidel, Dordrecht, 1976.
6. M. L. Dalla Chiara, *J. Phil. Logic* 6 (1977) 391.
7. B. C. van Fraassen, "A modal interpretation of quantum mechanics", in *Current Issues in Quantum Logic*, E. G. Beltrametti and B. C. van Fraassen, eds., Plenum, New York, 1981.
8. P. Mittelstaedt, "Classification of different areas of work afferent to quantum logic", in *Current Issues in Quantum Logic*, E. G. Beltrametti and B. C. van Fraassen, eds., Plenum, New York, 1981.
9. M. Jammer, *The Philosophy of Quantum Mechanics*, Wiley, New York, 1974, Chapter 8.
10. J. Kotas, *Studia Logica* 21 (1967) 17.
11. G. M. Hardegree, *Z. Naturforsch.* 30A (1975) 1347; *Synthese* 29 (1974) 63.
12. G. Kalmbach, *Zeitschr. f. math. Logik und Grundlagen d. Math.* 20 (1974) 395.
13. P. Mittelstaedt, *Z. Naturforsch.* 25A (1970) 1773; *ibid.* 27A (1972) 1358.
14. L. Herman, E. L. Marsden, and R. Piziak, *Notre Dame J. Formal Logic* 16 (1975) 306.
15. R. Stalnaker and R. H. Thomason, *Theoria* 36 (1970) 23.
16. P. Mittelstaedt, *Quantum Logic*, Reidel, Dordrecht, 1978.
17. D. J. Foulis and C. Randall, *J. Math. Phys.* 13 (1972) 1167; *ibid.* 14 (1972) 1472; in *Current Issues in Quantum Logic*, E. G. Beltrametti and B. C. van Fraassen, eds., Plenum, New York, 1981.
18. G. Ludwig and H. Neumann, eds., *Interpretations and Foundations of Quantum Theories*, Bibliographisches Institut, München, 1980.
19. E. G. Beltrametti and B. C. van Fraassen, eds., *Current Issues in Quantum Logic*, Plenum, New York, 1981.

PART III

Reconstruction of Hilbert-Space Quantum Mechanics

This part is less self-contained than the preceding ones. This is because the reconstruction of Hilbert-space quantum mechanics from the simpler structures of Part II, which is here the guiding theme, involves too wide an array of mathematical techniques: from the theory of coordinatization of projective geometries to the theory of group representations, from number field theory to probability theory, etc. Thus we are often forced to mention mathematical results without proofs.

CHAPTER 21

The Coordinatization Problem

21.1 Lattice Approach

In Part I of this volume we sketched the usual Hilbert-space formulation of quantum mechanics, and in Part II we examined various simpler mathematical structures that underly quantum mechanics. In this chapter we explore to what extent these simpler mathematical structures imply the Hilbert-space formulation.

We have seen that the ordered structure of propositions plays a dominant role, being the quantum analogue of the phase space of classical systems. Thus it is not surprising that the program of recovering the Hilbert-space formulation of quantum theory takes a path that starts, in general, from that ordered structure.

The first mathematical problem one is faced with in building the bridge from the lattice of propositions to the Hilbert space takes the name of *lattice coordinatization*. By this is meant the problem of identifying the given lattice with some lattice of subspaces of a vector space. This is a classical mathematical technique, the main parts of which will be recapitulated in the sequel. But quantum mechanics requires us to identify the proposition lattice with the lattice of subspaces of a highly specialized vector space, a complex Hilbert space; as we shall see, the corresponding coordinatization procedure requires, for the proposition lattice $\mathcal{L}$, orthomodularity, completeness, atomicity, and the covering property. Scattered throughout Part II are various physical arguments in favor of this structure of $\mathcal{L}$; some summary will perhaps be useful:

(i) *Orthomodularity.* This is a natural outcome of Mackey's approach, as seen in Section 13.6. It is secured as soon as one agrees that the propositions

ENCYCLOPEDIA OF MATHEMATICS and Its Applications, Gian-Carlo Rota (ed.).
Vol. 15: E. G. Beltrametti and G. Cassinelli, The Logic of Quantum Mechanics

ISBN 0-201-13514-0

form a σ-orthocomplete poset endowed with a separating set of states (see Section 11.3). It results from the approach based on the state transformations induced by idealized measurements (see Section 16.4), and from the one based on transition probabilities (see Section 18.3). It is however, rather extraneous to the approaches considered in Chapters 17 and 19.

(ii) *Completeness.* To motivate this, it would be desirable to have a prescription which tells, given the experimental devices associated with $a, b \in \mathfrak{L}$, what experimental devices correspond to the meet $a \wedge b$ and to the join $a \vee b$: a partial answer is contained in Section 16.5. From a more formalized standpoint, the structure of complete lattice is reached for $\mathfrak{L}$ if one of the following procedures is adopted: (1) starting from on orthomodular poset with a set $\mathfrak{S}$ of probability measures, add appropriate regularity properties for $\mathfrak{S}$ (see Section 11.4); (2) impose the correspondence between $\mathfrak{L}$ and the lattice of the subsets of $\mathfrak{S}$ that are closed under superposition (see Section 11.5); (3) introduce the notion of state transformations induced by idealized measurements, as in Section 16.4; (4) taking the states as primitive entities, make the set they form a closure space (see Section 17.2) or a transition-probability space (see Section 18.3); (5) assume a convex structure of states, and interpret propositions as faces of the convex set (see Section 19.2).

(iii) *Atomicity.* Essentially, this amounts to the existence of pure states, or, in other terms, to the existence of complete sets of compatible observables (see, in particular, Section 11.6). Notice that in the approaches that adopt the states as primitive entities, the atomicity is almost automatic (see Chapters 17–19).

(iv) *Covering property.* In terms of state transformations caused by idealized measurements (see Chapter 16), this corresponds to the requirement that pure states be transformed into pure states (projection postulate in Lüders's form). In terms of quantum superpositions of states, it corresponds to a quite natural "exchange property" (see Sections 11.7 and 17.2).

21.2 Vector-Space Coordinatization

As a starting point we assume that the propositions of a quantum system form a complete lattice $\mathfrak{L}$ that is orthomodular, is atomic, and has the covering property. For the sake of simplicity, we shall here add the irreducibility of $\mathfrak{L}$; when that is not done one can refer to its irreducible addends, as will be discussed in Section 21.6. Recall that the first question we have to answer is whether $\mathfrak{L}$ is isomorphic to some lattice of subspaces of a vector space, or, as it is usually put, whether $\mathfrak{L}$ admits a *vector-space coordinatization*. This question is addressed in a series of rather difficult theorems, which have been fully presented in the literature: we refer in particular to the books of Maeda and Maeda,[1] Varadarajan,[2] and Piron.[3]

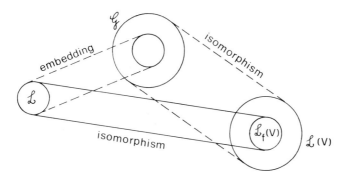

Figure 21.1. Steps of vector-space coordination.

We shall just outline, without proofs and details, the main steps of the procedure of vector-space coordinatization. They are essentially three. First, the search for an embedding of $\mathcal{L}$ into a larger lattice $\mathcal{G}$ that has modularity as a typical property, and that represents the lattice-theoretic counterpart of a projective geometry. Second, the identification of $\mathcal{G}$ with the lattice $\mathcal{L}(V)$ of all subspaces of some vector space V. Third, the determination of those elements of $\mathcal{L}(V)$ that are images of the elements of $\mathcal{L}$: they are recognized to form a lattice, say $\mathcal{L}_f(V)$, which is isomorphic to the original $\mathcal{L}$. This procedure is shown pictorially in Figure 21.1.

Let us proceed to say something more about the individual steps.

Step 1: *Embedding of $\mathcal{L}$ in a projective geometry.* This step works for any irreducible, complete, orthocomplemented AC lattice; orthomodularity is irrelevant.[1] Write $\mathcal{Q}(\mathcal{L})$ for the set of all atoms of $\mathcal{L}$, and define a *subspace of $\mathcal{Q}(\mathcal{L})$* to be any subset that contains every atom majorized by the join of any two of its elements. The empty set, the whole $\mathcal{Q}(\mathcal{L})$, and the singletons are obviously subspaces. A first important fact is that the set of all subspaces of $\mathcal{Q}(\mathcal{L})$ is a complete lattice with respect to set inclusion; we denote it by $\mathcal{L}(\mathcal{Q}(\mathcal{L}))$. The meet is set intersection, and the join is

$$\bigvee_i \omega_i = \left\{ p \in \mathcal{Q}(\mathcal{L}): p \leqslant \bigvee_i q_i, q_i \in \omega_i \right\}, \qquad \omega_i \in \mathcal{L}(\mathcal{Q}(\mathcal{L}))$$

Another relevant fact is that $\mathcal{L}(\mathcal{Q}(\mathcal{L}))$ is

(1) irreducible;
(2) modular;
(3) AC, its atoms being the singletons;
(4) compactly atomistic, that is, if ω is an atom of $\mathcal{L}(\mathcal{Q}(\mathcal{L}))$ and L a set of atoms of $\mathcal{L}(\mathcal{Q}(\mathcal{L}))$ such that $\omega \leqslant \bigvee(\omega_i: \omega_i \in L)$, then there exists a finite subset $\omega_1, \ldots, \omega_n$ of L such that $\omega \leqslant \omega_1 \vee \cdots \vee \omega_n$.

One summarizes these facts by saying that $\mathcal{L}(\mathcal{Q}(\mathcal{L}))$ is a *modular geometric lattice*, or a *projective lattice*.

The last relevant fact is that the mapping $a \mapsto \omega(a) = \{p \in \mathcal{Q}(\mathcal{L}): p \leq a\}$ that embeds the proposition lattice $\mathcal{L}$ into $\mathcal{L}(\mathcal{Q}(\mathcal{L}))$ has the following properties:

(i) it is injective and order-preserving, that is, $a \leq b$ if and only if $\omega(a) \leq \omega(b)$,

(ii) it is meet-preserving, that is, $\omega(\bigwedge_i a_i) = \bigwedge_i \omega(a_i)$,

(iii) it is bijective between the finite elements of $\mathcal{L}$ and the finite elements of $\mathcal{L}(\mathcal{Q}(\mathcal{L}))$.

Notice that $\mathcal{L}(\mathcal{Q}(\mathcal{L}))$ is precisely what has been denoted by $\mathcal{G}$ in Figure 21.1.

Step 2: Vector-space coordinatization of projective lattices. We define the *length* of an AC lattice to be the number of elements of a maximal chain from 0 to 1, endings included (see Section 17.3). This step applies to any projective lattice of length ≥ 4. The main fact is as follows:[4] for any such lattice, say $\mathcal{G}$, there exist a field K and a vector space V over K such that $\mathcal{G}$ is isomorphic to the lattice $\mathcal{L}(V)$ of all linear subspaces of V. Here the ordering of $\mathcal{L}(V)$ is meant to be set-theoretic inclusion, with meet and join given respectively by set intersection and by the usual algebraic sum of subspaces of a vector space. Notice that by the term "field" we do not imply commutativity: at this stage, K may be commutative or not. Should it be noncommutative, the term "division ring" (that is, a ring with identity and without divisors of zero) might be preferred. The mentioned result is a classical one, usually referred to as "coordinatization of projective geometries", the name being motivated by the fact that a projective lattice can be viewed, in geometrical language, as a projective geometry.*

It is worth stressing that the existence of the field K is asserted, but its explicit nature is not given. Only in case $\mathcal{G}$ contains a finite number of elements, thus making K finite too, is the nature of K determined: indeed, a finite field is necessarily a Galois field[5] GF (k) over k elements, and k must be the nth power (n integer) of some prime ($n+1$ is the length of $\mathcal{G}$).

Notice that combining step 1 with step 2, one gets an embedding of $\mathcal{L}$ (provided it is of length ≥ 4) into $\mathcal{L}(V)$. Since the mapping $a \mapsto \omega(a)$ of $\mathcal{L}$ into $\mathcal{L}(\mathcal{Q}(\mathcal{L})$ is bijective between finite elements, the embedding becomes an isomorphism when $\mathcal{L}$ has finite length (and *a fortiori* when $\mathcal{L}$ has a finite number of elements). In this case the problem of vector-space coordinatization is over. Otherwise one must go to the third step.

Step 3: Determination of the subspaces of $\mathcal{L}$ (V) that correspond to the elements of $\mathcal{L}$. The procedure of picking up the elements of $\mathcal{L}(V)$ that are

---

*The geometrical standpoint is fully developed in Piron's book.[3]

images of the elements of $\mathfrak{L}$ under the embedding discussed above rests on a fundamental theorem of Birkhoff and von Neumann.[6] We shall only give the final result, but before doing that we need some definitions:

(a) An *involution* (short for involutive automorphism) of a field K is a mapping $\lambda \mapsto \lambda^*$ of K into itself such that for every $\lambda, \mu \in K$,

$$(\lambda + \mu)^* = \lambda^* + \mu^*, \qquad (\lambda\mu)^* = \mu^*\lambda^*, \qquad \lambda^{**} = \lambda.$$

(b) A mapping f from $V \times V$ into K is called a *hermitian form* on V whenever, for every $v, w, v_i, w_i \in V$ and $\lambda, \mu \in K$,

$$f(\lambda v_1 + \mu v_2, w) = \lambda f(v_1, w) + \mu f(v_2, w),$$
$$f(v, \lambda w_1 + \mu w_2) = f(v, w_1)\lambda^* + f(v, w_2)\mu^*,$$
$$f(v, w) = f^*(w, v),$$
$$f(v, v) = 0 \quad \text{if and only if} \quad v = 0.$$

(c) Given a subspace M of V, set

$$M^\circ = \{v \in V : f(v, w) = 0 \text{ for every } w \in M\}:$$

M° is still a subspace of V, and the mapping $M \mapsto M^{\circ\circ}$ is a closure operation in $\mathfrak{L}(V)$, to be called the *f-closure*. M is called *f-closed* if it is invariant under this operation, namely, if $M = M^{\circ\circ}$.

Now, the relevant fact is that, under the hypotheses of steps 1 and 2, V is endowed with a hermitian form f, and the subspaces of $\mathfrak{L}(V)$ that are f-closed are precisely the images in $\mathfrak{L}(V)$ of the elements of $\mathfrak{L}$. It is worth remarking that the set $\mathfrak{L}_f(V)$ of all f-closed elements of $\mathfrak{L}(V)$ is a complete lattice ordered by set-theoretic inclusion, with meet and join given by

$$\bigwedge_i M_i = \bigcap_i M_i, \qquad \bigvee_i M_i = (\Sigma M_i)^{\circ\circ}, \qquad M_i \in \mathfrak{L}_f(V);$$

moreover $\mathfrak{L}_f(V)$ is irreducible, orthocomplemented, and AC, the orthocomplementation being $M \mapsto M^\circ$.

The final result that summarizes the vector-space coordinatization of the original proposition lattice $\mathfrak{L}$ reads as follows: If $\mathfrak{L}$ is a complete, irreducible, orthocomplemented, AC lattice of length ≥ 4, then there exist a field K with an involution $\lambda \mapsto \lambda^*$, and a vector space V over K endowed with a hermitian form f, having the additional property $f(w, w) = 1$ for some $w \in V$, such that $\mathfrak{L}$ and $\mathfrak{L}_f(V)$ are isomorphic and the isomorphism preserves the orthocomplementations.

Notice that a finite-dimensional subspace of V is always f-closed; therefore, if V has finite dimension [that is, if $\mathcal{L}(V)$ has finite length], then $\mathcal{L}_f(V)$ and $\mathcal{L}(V)$ coincide.

21.3 Choice of the Number Field

As has already been said, the vector-space coordinatization of the proposition lattice $\mathcal{L}$ ensures the existence of the field K on which the vector space V is defined, but the nature of K is not specified. Given $\mathcal{L}$, what K to use? Notice that also orthomodularity and separability are attributes of our proposition lattice, but neither the former nor the latter is required by the procedure of vector-space coordinatization. Then the question is: do orthomodularity and separability restrict the choice of K? Certainly, they cannot force a unique choice of K, for we know that a projection lattice $\mathcal{P}(\mathcal{H})$ on a separable Hilbert space has all the desired properties no matter whether $\mathcal{H}$ is over the complex field $\mathbb{C}$, or the real field $\mathbb{R}$, or the quaternion field $\mathbb{Q}$ (the reader can refer to Section 10.3 for the complex case, the other two cases being quite analogous). Do perhaps the properties of the proposition lattice, notably orthomodularity and separability, restrict the choice of K to the three mentioned cases, that is $\mathbb{R}, \mathbb{C}, \mathbb{Q}$? No sharp answer is known, though there are conjectures that the answer might be yes. For a defence of these conjectures we refer in particular to a series of papers by Morash.[7]

Beyond conjectures, two possible strategies have been considered: a strategy based on exclusion criteria and a constructive strategy.

(1) *The exclusion strategy.* We mean the following. One takes a number field and a vector space V over it, and then one verifies the possibility of defining a hermitian form relative to some involution of the field; if such a hermitian form does not exist, then the field is excluded (compare with the preceding section). This exclusion procedure has been used to rule out finite fields as well as p-adic fields.* The following results have been obtained:

(i) If K is a Galois field (or an extension of it), and V is a vector space over it with $\dim V \geqslant 3$, then there does not exist any hermitian form on V, no matter which involution of K is chosen.[9, 10]

(ii) If K is a p-adic field (or an extension of it), and V is a vector space over it with $\dim V \geqslant 5$, then there is no hermitian form on V with respect to continuous involutions of K,[11] while the noncontinuous case appears too exotic for physical purposes.

These results can be generalized to every power-series field and every local field.[12] A consistent indication in favor of restricting K to $\mathbb{R}, \mathbb{C}, \mathbb{Q}$ thus emerges.

*For definitions and properties of these fields we refer to textbooks.[8]

(2) *The constructive strategy.* Generally speaking, the idea is to endow the proposition lattice $\mathfrak{L}$ with structural properties that are sufficient to force K into $\mathbb{R}$ or into its extensions $\mathbb{C}, \mathbb{Q}$. This can be done by postulating suitable compactness conditions and the existence of a continuous function from a real segment into a segment of $\mathfrak{L}$ under a finite element,[13] or by postulating a topology in $\mathfrak{L}$ defined via the pure states,[14] or by assuming the existence of an observable on $\mathfrak{L}$ having special "smoothness" properties.[15,16]

Let us end this section with the following remark. Suppose that $\mathfrak{L}$ has a finite number of elements but has length $\geqslant 4$. By step 2 of the last section we know that $\mathfrak{L}$ should be coordinatized by a vector space V over a Galois field, and by step 3 we know that there should exist a hermitian form on V, which, on the other hand, would contradict what was said under item (ii) above. We are forced to conclude that there is no complete, irreducible, orthocomplemented AC lattice with length $\geqslant 4$ and containing a finite number of elements, a conclusion independently known in a purely lattice-theoretic context.

21.4 Hilbert-Space Coordinatization

What was said in the last section suggests that the proposition lattices that describe quantum systems favor, in the procedure of vector-space coordinatization, one of the three fields $\mathbb{R}, \mathbb{C}, \mathbb{Q}$. In this section we show the following fact: if we actually choose $\mathbb{R}$, $\mathbb{C}$, or $\mathbb{Q}$, and we add a requirement of continuity of the involution of the field, then the usual Hilbert space of quantum mechanics is at hand.

The procedure consists of two steps. The first is to show that the continuity requirement forces the hermitian form f to be a scalar product. The second is a theorem of Amemiya and Araki,[17] which makes V a Hilbert space.

We start specifying all continuous involutions of $\mathbb{R}, \mathbb{C}, \mathbb{Q}$.

(a) *The real case.* $\mathbb{R}$ admits just one involution: the identity, which is obviously continuous.

(b) *The complex case.* $\mathbb{C}$ has various involutions, but just two of them are continuous: the identity and the usual complex conjugation.

(c) *The quaternionic case.* Every involution $\lambda \mapsto \lambda^*$ in $\mathbb{Q}$ is of the form $\lambda^* = \mu \bar{\lambda} \mu^{-1}$, where $\bar{\lambda}$ is the canonical conjugate of λ (in standard notation, if $\lambda = q_0 \mathbf{1} + \vec{q} \cdot \vec{j}$, then $\lambda^* = q_0 \mathbf{1} - \vec{q} \cdot \vec{j}$) and μ is some fixed (nonzero) element of $\mathbb{Q}$.[2] In other words, $\mathbb{Q}$ has only inner involutions, and they are continuous.

Now we can analyse the impact of the continuity of the involution on the hermitian forms in vector spaces over $\mathbb{R}$, $\mathbb{C}$, or $\mathbb{Q}$. The following fact will emerge: the continuity of the involution implies that the quadratic form $f(v, v)$ associated with the hermitian form f is real-valued and positive-definite, so that $f(v, w)$ is recognized to be a scalar product. Let us see this fact for each of the three cases.

(a) *The real case.* Let V be a vector space over $\mathbb{R}$, and f a hermitian form on V (relative to the identical involution). By linearity $f(v, v)$ is a continuous* function of v, and it vanishes only at $v=0$. If $\dim V \geqslant 2$, any nonzero vector can be reached continuously from any other nonzero vector without going through the zero vector; hence the sign of $f(v, v)$ is constant, and recalling that f has the further property $f(w, w)=1$ for some $w \in V$, we conclude that $f(v, v)$ is positive for all $v \neq 0$. Thus, $f(v, w)$ has all the properties of a scalar product between v and w. Notice that the case $\dim V=1$ has little interest but, anyhow, admits the same result by *ad hoc* reasoning.

(b) *The complex case.* Let V be a vector space over $\mathbb{C}$. There is no hermitian form on V relative to the identical involution (if $\dim V \geqslant 2$); in fact it would always be possible[2] to choose two nonzero vectors u_1, u_2 such that $f(u_1, u_2)=0$, $f(u_1, u_1)=1$, $f(u_2, u_2)=1$, whence the contradiction $f(u_1 +iu_2, u_1+iu_2)=0$. We are left with the possibility of taking hermitian forms relative to the usual complex conjugation. In this case $f(v, v)$ is real-valued and we fall back on the same arguments used for the real case: $f(v, v)$ is positive-definite and $f(v, w)$ is a scalar product.

(c) *The quaternionic case.* Among the inner involutions $\lambda \mapsto \mu\bar{\lambda}\mu^{-1}$, just one allows the construction of a hermitian form. It corresponds to taking $\mu = \pm 1$, and thus reduces to the canonical involution $\lambda \mapsto \bar{\lambda}$. Again, $f(v, v)$ is real-valued and we fall back on the same arguments as before, with the conclusion that $f(v, w)$ is a scalar product.

Summing up, and referring to Section 21.2, we have the following: restricting the field to $\mathbb{R}$, $\mathbb{C}$, or $\mathbb{Q}$ and adopting the continuity requirement on the involution, the proposition lattice $\mathscr{L}$ is made isomorphic to the lattice $\mathscr{L}_f(V)$ of all f-closed subspaces of a pre-Hilbert space V, in which $f(\cdot, \cdot)$ is the scalar product.

We come now to the second and last step of the Hilbert-space coordinatization procedure. To make V a Hilbert space, only one property has still to be achieved: the completeness. This property is however at hand, for it is implied by the orthomodulatity of $\mathscr{L}$ according to the following sharp theorem of Amemiya and Araki:[17]

---

*To use continuity arguments freely, V should be finite-dimensional. However, inspection of the proof of the relevant theorems on vector-space coordinatization reveals that in the infinite-dimensional case, the hermitian form is built up from a form on a finite-dimensional subspace.[1] Hence the continuity arguments used here hold true in general.

THEOREM 21.4.1. *Let V be a pre-Hilbert space with respect to the scalar product $f(\cdot, \cdot)$; then $\mathcal{L}_f(V)$ is orthomodular if and only if V is Hilbert.*

Notice that once V is a Hilbert space (denoted as usual by $\mathcal{H}$), the f-closed subspaces are the closed subspaces with respect to the topology of $\mathcal{H}$; hence $\mathcal{L}_f(\mathcal{H})$ is precisely what we have throughout denoted by $\mathcal{P}(\mathcal{H})$. In case $\mathcal{L}$ is separable, the isomorphism between $\mathcal{L}$ and $\mathcal{P}(\mathcal{H})$ makes $\mathcal{H}$ separable too.

This concludes the Hilbert-space coordinatization of $\mathcal{L}$. As is seen, the orthomodularity of $\mathcal{L}$ is crucially responsible for the Hilbert-space structure of the coordinatizing vector space.

21.5 States and Physical Quantities

As seen above, once one agrees that the proposition lattice $\mathcal{L}$ of a quantum physical system is a complete, irreducible, separable, orthomodular AC lattice with length $\geqslant 4$, and once one agrees to consider vector spaces over $\mathbb{R}, \mathbb{C}, \mathbb{Q}$ endowed with continuous involutions, $\mathcal{L}$ can be identified with $\mathcal{P}(\mathcal{H})$ for some separable Hilbert space over $\mathbb{R}, \mathbb{C}, \mathbb{Q}$. Thus, the first basic rule of Hilbert-space quantum mechanics is achieved: to every quantum system a separable Hilbert space is associated. The purpose of this section is to outline that the other basic rules concerning states and physical quantities are also achieved.

Consider first the states. In our context they are defined as probability measures on $\mathcal{L}$, and hence on $\mathcal{P}(\mathcal{H})$. Now, Gleason's theorem states, as seen in Section 11.2, that the set of all probability measures on $\mathcal{P}(\mathcal{H})$ is in one-to-one correspondence with the set of all density operators on $\mathcal{H}$. Notice that Gleason's theorem holds true provided the dimension of $\mathcal{H}$ is at least 3, and this is precisely the condition that $\mathcal{P}(\mathcal{H})$ and hence $\mathcal{L}$ have length $\geqslant 4$, the condition to be presupposed in order to work out the vector-space coordinatization. Though in Section 11.2 Gleason's theorem referred to a Hilbert space over $\mathbb{C}$, it applies unchanged also in case $\mathcal{H}$ is over $\mathbb{R}$ or $\mathbb{Q}$.[2] Thus, no matter whether we choose $\mathbb{R}$, or $\mathbb{C}$, or $\mathbb{Q}$, we recover the second basic rule (in the list used in Section 1.1) of Hilbert-space quantum mechanics: states are identified with density operators.

Consider now the physical quantities. In our context they are the $\mathcal{L}$-valued probability measures on the real line: as soon as $\mathcal{L}$ is recognized as $\mathcal{P}(\mathcal{H})$, they become $\mathcal{P}(\mathcal{H})$-valued probability measures. But now the spectral theorem, which holds unchanged no matter whether $\mathcal{H}$ is over $\mathbb{R}$, $\mathbb{C}$, or $\mathbb{Q}$, ensures that they are in one-to-one correspondence with the self-adjoint operators on $\mathcal{H}$. This is precisely the third basic rule of Hilbert-space quantum mechanics (see Section 1.1). The rule giving the probability distribution of physical quantities in a given state is now automatically embodied.

In this way, the starting ingredients of the Hilbert-space formulation of quantum mechanics are recovered. Hence, all that follows from these ingredients, in particular what has been examined in Chapters 2–4, is similarly recovered.

Those parts of Hilbert-space quantum mechanics that fall beyond the above scheme will be considered in the sequel. No special problem arises with the possible occurrence of superselection rules, as we shall see in next section. More problematic is the recovery of the rules about the dynamical evolution and about the description of compound systems; this will be examined in Chapters 23 and 24.

21.6 Reducibility and Superselection Rules

The coordinatization procedure examined up to now requires also the irreducibility of the proposition lattice $\mathcal{L}$, and leads to recovering the familiar rules for physical systems with unlimited quantum behavior. We have found, in particular, that the pure states are in one-to-one correspondence with one-dimensional subspaces of $\mathcal{H}$, and that physical quantities are in one-to-one correspondence with the self-adjoint operators on $\mathcal{H}$. On the other hand we know that there are physical systems with limited quantum behavior (as discussed in Chapter 5), and they need a more flexible formalism, that of superselection rules. The purpose of this section is to show, by use of what was said in Section 12.5, that the usual pattern of superselection rules corresponds to abandoning the requirement of irreducibility of the proposition lattice.

When $\mathcal{L}$ is reducible, we know from Chapter 12 that its center $Z(\mathcal{L})$ is nontrivial: it is a Boolean σ-algebra generated by the (countable) set $\{z_i\}$ of its atoms. $\mathcal{L}$ is known to be the direct sum of the segments $\mathcal{L}[0, z_i]$, each of them being irreducible and having all the relevant properties — orthomodularity, atomicity, and the covering property. Thus the coordinatization procedure applies to every segment, and we are allowed to write $\mathcal{L}[0, z_i] = \mathcal{P}(\mathcal{H}_i)$; hence

$$\mathcal{L} = \cup \, \mathcal{P}(\mathcal{H}_i), \qquad\qquad (21.6.1)$$

where each $\mathcal{H}_i$ is a separable Hilbert space over $\mathbb{R}$, $\mathbb{C}$, or $\mathbb{Q}$. Now, in order to get a formalism that properly generalizes the one relative to strictly quantum systems, it is still desirable to represent states and physical quantities associated with $\mathcal{L}$ as density operators and self-adjoint operators pertaining to some Hilbert space $\mathcal{H}$, even if we must be prepared to find that not every density operator in $\mathcal{H}$ is a state of the physical system and not every self-adjoint operator on $\mathcal{H}$ is a physical quantity. The lattice (21.6.1) is precisely the one described in Example 3 of Section 12.5: it is a sublattice of

$\mathcal{P}(\mathcal{H})$, where

$$\mathcal{H} = \bigoplus_i \mathcal{H}_i. \tag{21.6.2}$$

The structure of pure states of $\mathcal{L}$ is obtained by use of the results contained in Section 14.8 (in particular Theorems 14.8.5–14.8.10): only those unit vectors of $\mathcal{H}$ that belong to some subspace $\mathcal{H}_i$ correspond to pure states on $\mathcal{L}$. Thus we are back to the pattern of superselection rules described, in the framework of Hilbert space, in Chapter 5.

Of course the structure of the set of observables is also affected by the reducibility of $\mathcal{L}$. It still holds true that every observable associated with $\mathcal{L}$ is a projection-valued measure and determines a self-adjoint operator on $\mathcal{H}$, but a self-adjoint operator on $\mathcal{H}$ now represents an observable associated with $\mathcal{L}$ if and only if all its spectral projections are in $\mathcal{L}$. Again we are back to the discussion in Chapter 5.

21.7 Other Ways of Recovering the Hilbert-Space Representation

The question we consider in this section is whether there are alternatives to the procedure of Hilbert-space representation examined in the rest of this chapter. We can say at once that, though a number of ideas have been advanced, no result exists comparable, as regards power and completeness, to the standard representation theorems. Our aim here is just to supply some bibliography with brief remarks.

In the framework of the traditional $(\mathcal{L}, \mathcal{S})$ structure with the added assumption that the pure states form a transition-probability space (see Chapter 18) with respect to the transition probability $\langle \alpha | \beta \rangle = \alpha(s(\beta))$,* Deliyannis[18] was able to construct explicitly a real or complex vector space endowed with a nonsingular bilinear form induced by the transition probability. Then $\mathcal{L}$ can be identified with a certain set of subspaces of that vector space. Unfortunately, this procedure, though direct and simple, has some unsatisfactory features: for instance, if one applies it to $\mathcal{L} = \mathcal{P}(\mathcal{H})$, one ends up with a vector space different from the original $\mathcal{H}$.[18] Thus, this vector-space coordinatization of $\mathcal{L}$ cannot be regarded as a true substitute for the coordinatization procedures of the foregoing sections, though it opens a way to be further exploited. Still in the same framework, Deliyannis[19] has also considered the possibility of assigning from the outset an explicit form of the superposition of two pure states. This gives the possibility of constructing a real Hilbert space $\mathcal{H}$ such that $\mathcal{P}(\mathcal{H})$ is isomorphic to the original

---

*Of course this demands the nonobvious condition $\alpha(s(\beta)) = \beta(s(\alpha))$.

proposition lattice $\mathcal{L}$. This result constitutes an actual representation theorem, and hence an alternative to the procedure of Hilbert-space representation of the preceding sections. The limitation to real Hilbert spaces comes from the fact that Deliyannis's formula for superpositions of pure states involves only real parameters.

To put this idea in more explicit terms, consider a real Hilbert space $\mathcal{H}$ and, in it, a state vector $\psi = c_1\varphi_1 + c_2\varphi_2$, a superposition of the state vectors φ_1, φ_2. For $P \in \mathcal{P}(\mathcal{H})$ we have

$$(\psi, P\psi) = c_1^2(\varphi_1, P\varphi_1) + c_2^2(\varphi_2, P\varphi_2)$$
$$+ c_1c_2(\varphi_1, P\varphi_2) + c_1c_2(\varphi_2, P\varphi_1).$$

Noticing that $\mathcal{H}$ is a transition-probability space with respect to $\langle\varphi|\varphi'\rangle = (\varphi, \varphi')^2$, and that we have also $\langle\varphi|\varphi'\rangle = (\varphi, s(\varphi')\varphi)$, we can rewrite the foregoing formula as

$$(\psi, P\psi) = c_1^2(\varphi_1, P\varphi_1) + c_2^2(\varphi_2, P\varphi_2) + c_1c_2(\varphi_2, P\varphi_2)(\varphi_1, s(G_P\varphi_2)\varphi_1)^{1/2}$$
$$+ c_1c_2(\varphi_1, P\varphi_1)(\varphi_2, s(G_P\varphi_1)\varphi_2)^{1/2},$$

where $G_P\varphi_i$ stands for the unit vector along $P\varphi_i$, and can also be viewed as the state in which φ_i is mapped by an idealized measurement of P (in the sense of von Neumann's projection postulate of Section 8.1). This expression allows an immediate translation into purely $(\mathcal{L}, \mathcal{S})$ terms; by reference to the results of Chapter 16, it reads as follows: if $\alpha \in \mathcal{S}$ is a quantum superposition of $\alpha_1, \alpha_2 \in \mathcal{S}$, then there exist $c_1, c_2 \in \mathbb{R}$, $c_1^2 + c_2^2 = 1$, such that for all $a \in \mathcal{L}$,

$$\alpha(a) = c_1^2\alpha_1(a) + c_2^2\alpha_2(a) + c_1c_2\alpha_2(a)\left[\alpha_1([s(\alpha_2)\vee a^\perp]\wedge a)\right]^{1/2}$$
$$+ c_1c_2\alpha_1(a)\left[\alpha_2([s(\alpha_1)\vee a^\perp]\wedge a)\right]^{1/2}. \tag{21.7.1}$$

Now, the question is: if, starting from the general $(\mathcal{L}, \mathcal{S})$ structure of a quantum system, without presupposing any Hilbert-space structure, we assume that the equality (21.7.1) characterizes the superpositions of α_1 and α_2, can we conclude $\mathcal{L} = \mathcal{P}(\mathcal{H})$ for some real Hilbert space $\mathcal{H}$? The answer is yes, for the explicit expression (21.7.1) fulfills the condition of Deliyannis. But notice that it is quite a strong assumption to require that (21.7.1) be a probability measure on $\mathcal{L}$: the additivity on orthogonal propositions of the last two terms is nonobvious.

Another idea for bringing the Hilbert space forward has been advanced by Gudder.[20] He takes an $(\mathcal{L}, \mathcal{S})$ structure with rather minimal requirements (such as the structure arising from Mackey's axioms) and adds other physical structures such as the physical space in which the system resides, the symmetry properties of this space, and the position observables. A complex Hilbert space $\mathcal{H}$ is then constructed and a partial Hilbert-space coordinatization of $\mathcal{L}$ is achieved. Unpleasant aspects are however encountered: the correspondence between $\mathcal{L}$ and $\mathcal{P}(\mathcal{H})$ is not an isomorphism, and the representation of $\mathcal{L}$ erases the distinction between pure and mixed states of $\mathcal{L}$, in the sense that both are mapped into pure states of $\mathcal{P}(\mathcal{H})$.

The possibility of taking advantage of the remarkable connection between orthomodular lattices and Baer *-semigroups (see Section 16.1) has also been explored.[21,22] The idea is to characterize the orthomodular lattices that can be embedded in Hilbert space via the characterization of the Baer *-semigroups that are representable in Hilbert space.

All the alternatives mentioned above start from an $(\mathcal{L}, \mathcal{S})$ structure, in which the elements of $\mathcal{L}$ (the propositions) are primitive elements, and the elements of $\mathcal{S}$ (the states) are understood as probability measures on $\mathcal{L}$. We can ask whether there are ways toward the Hilbert space that start from a different point of view: that of taking states as primitive entities, as in the approaches discussed in Chapters 17–19. Two choices can be imagined. The first is to construct, out of $\mathcal{S}$, an ordered structure $\mathcal{L}$ (see Sections 17.2, 18.3, 19.3) and then apply to it, whenever possible, the available representation procedures. The second is to look for direct representation procedures that, starting from the various structures $\mathcal{S}$ can be endowed with (closure space, transition-probability space, convex space), provide a Hilbert -space representation of $\mathcal{S}$. For instance it would be nice to characterize the transition-probability spaces that can be identified with the set of all rays, or at least with a distinguished subset of rays, in a Hilbert space, or the convex structures that can be identified with the set of all density operators in a Hilbert space.[23] To the extent of our knowledge such a program is more a hope than an immediate possibility.

References

1. F. Maeda and S. Maeda, *Theory of Symmetric Lattices*, Springer, Berlin, Heidelberg, New York, 1970 (in particular Chapter 7).
2. V. S. Varadarajan, *Geometry of Quantum Theory*, Vol. I, Van Nostrand, Princeton, N.J., 1968 (in particular Chapters 3, 4, 5, and Section 5 of Chapter 7).
3. C. Piron, *Foundations of Quantum Physics*, Benjamin, Advanced Book Program, Reading, Mass., 1976 (in particular Chapter 3).
4. R. Baer, *Linear Algebra and Projective Geometry*, Academic Press, New York, 1952.
5. See, e.g., L. E. Dickson, *Linear Groups with an Exposition of the Galois Field Theory*, Dover, New York, 1958.
6. G. D. Birkhoff and J. von Neumann, *Ann. Math.* 37 (1936) 823.
7. R. P. Morash, *Proc. Amer. Math. Soc.* 24 (1970) 716; *ibid.* 27 (1971) 446; *ibid* 36 (1972) 63; *ibid* 43 (1974) 42; *Canad. J. Math.* 25 (1973) 261; *Glasnik Matematički* 10 (1975) 107.
8. See, e.g., G. Bachmann, *Introduction to p-adic Numbers and Valuation Theory*, Academic Press, New York, London, 1964.
9. J. P. Eckmann and P. Zabey, *Helv. Phys. Acta* 42 (1969) 420.
10. P. A. Ivert and T. Sjödin, *Helv. Phys. Acta* 51 (1978) 635.

11. E. G. Beltrametti and G. Cassinelli, *Found. of Phys.* 2 (1972) 1.
12. N. Schaefer, Thesis, University of Clausthal, Germany, 1980.
13. N. Zierler, *Pacific J. Math.* 11 (1961) 1151.
14. R. Cirelli and P. Cotta-Ramusino, *Int. J. Theor. Phys.* 8 (1973) 11.
15. S. P. Gudder and C. Piron, *J. Math. Phys.* 12 (1971) 1583.
16. M. J. Maczynski, *J. Math. Phys.* 14 (1973) 1469.
17. I. Amemiya and H. Araki, *Publications Research Inst. Math. Sci., Kyoto Univ.* A2 (1966) 423.
18. P. C. Deliyannis, *J. Math. Phys.* 14 (1973) 249.
19. P. C. Deliyannis, *J. Math. Phys.* 17 (1976) 248.
20. S. P. Gudder, *Rep. Math. Phys.* 4 (1973) 193.
21. S. P. Gudder and J. Michel, *Proc. Amer. Math. Soc.* 81 (1981) 157.
22. S. P. Gudder, in *Current Issues in Quantum Logic*, E. G. Beltrametti and B. C. van Fraassen, eds., Plenum, New York, 1981.
23. S. P. Gudder, *Commun. Math. Phys.* 29 (1975) 249.

CHAPTER 22

Use of Real and Quaternionic Hilbert Spaces: A Simple Example

22.1 Quantum Description of a Free Particle

The usual quantum descriptions of physical quantities in terms of self-adjoint operators, and of states in terms of density operators, are based respectively on the spectral theorem and on Gleason's theorem. In fact, when we identify the logic $\mathcal{L}$ of a quantum system with $\mathcal{P}(\mathcal{H})$, the $\mathcal{P}(\mathcal{H})$-valued measures on the Borel sets of $\mathbb{R}$ are in one-to-one correspondence with the self-adjoint operators on $\mathcal{H}$ by the projection-valued measure form of the spectral theorem, while the probability measures on $\mathcal{P}(\mathcal{H})$ are in one-to-one correspondence with the density operators on $\mathcal{H}$ by Gleason's theorem, and both theorems retain their validity also in real and quaternionic Hilbert spaces. This means that the basic structure of quantum mechanics could be formulated also in terms of Hilbert spaces over $\mathbb{R}$ or $\mathbb{Q}$. In the next two sections we shall test this possibility on the concrete example consisting of a spin-zero free particle, and we shall point out the main differences that arise with respect to the usual description based on complex Hilbert spaces. As sketched in Section 4.5, we shall derive the quantum description of the system from the structure and the symmetries of the physical space where the system resides. We associate to the particle a separable Hilbert space $\mathcal{H}$ (not necessarily complex, now), and we assume that the particle is free in the three-dimensional space $\mathbb{R}^3$. The description of the particle is then based on:

(1) a projection-valued measure $P(\cdot)\colon \mathcal{B}(\mathbb{R}^3)\to\mathcal{P}(\mathcal{H})$, which represents the "localization" of the particle in a (Borel) subset of $\mathbb{R}^3$,

ENCYCLOPEDIA OF MATHEMATICS and Its Applications, Gian-Carlo Rota (ed.). Vol. 15: E. G. Beltrametti and G. Cassinelli, The Logic of Quantum Mechanics

ISBN 0-201-13514-0

(2) a projective, unitary, strongly continuous representation, in the Hilbert space $\mathcal{H}$, of the symmetry group of $\mathbb{R}^3$ (i.e. the Euclidean group $\mathbb{G}$), which we denote by $g \mapsto U_g$, $g \in \mathbb{G}$.

$P(\cdot)$ and U_g are linked together by the fundamental relation (imprimitivity system)

$$U_g P(\Delta) U_g^{-1} = P(g[\Delta]) \qquad \text{for all} \quad \Delta \in \mathcal{B}(\mathbb{R}^3) \text{ and } g \in \mathbb{G}$$

$$(22.1.1)$$

(where $g[\Delta]$ stands for the transform of the set Δ under g), which expresses the invariance with respect to the Euclidean group, and hence the fact that the particle is localizable and all positions in space are unitarily equivalent. Moreover, if the dynamics of the system is described by a one-parameter group $t \mapsto V_t$ of unitary operators on $\mathcal{H}$ (see Section 6.1), then V_t is linked to U_g by the second fundamental relation

$$U_g V_t = V_t U_g \qquad \text{for all} \quad g \in \mathbb{G}, \quad t \in \mathbb{R}, \qquad (22.1.2)$$

which expresses the dynamical invariance and hence the fact that the system is isolated (first acting with g and then letting the system evolve for a time t is the same as first letting the system evolve for a time t and then acting with g).

At this stage the problem is to find the solutions of (22.1.1) and (22.1.2), that is, the explicit form of $\mathcal{H}$ (as a function space), V_t, $P(\Delta)$, and U_g, as well as the form of the self-adjoint operators on $\mathcal{H}$ that represent the interesting physical quantities of the particle (position, momentum, energy, etc.). These self-adjoint operators will appear as generators of subgroups of U_g.

In this chapter we freely use some concepts of the theory of induced representations of noncompact groups; they are rather far from the rest of this volume, and the reader is referred to Mackey[1] and Varadarajan.[2] The approach we shall outline gives results comparable with the ones obtained, by quite different techniques, by Stueckelberg[3] for the real case and by Emch[4] for the quaternionic case.

22.2 Mackey's Analysis of the Real Case

In this section we briefly review the results of G. W. Mackey[5] on the possibility of describing a spin-zero free particle by means of a real Hilbert space $\mathcal{H}$. As in the complex case, we have that a key role is played by the irreducible representation of the rotation group associated with the intrinsic spin of the particle. Recall that in the complex case the pertinent complex-valued representation was denoted D^j, it acted in a $(2j+1)$-dimensional

Hilbert space $\mathfrak{K}(D^j)$, and j was recognized as the intrinsic spin of the particle: in particular, the representation D^0 associated with a zero-spin particle was one-dimensional. In the real case, we have to look for irreducible real-valued representations of the rotation group, and the dimensionality of the Hilbert space in which the representation acts is no longer related to the intrinsic spin of the particle by the same rule as in the complex case. Let Ξ be the real-valued representation of the rotation group associated with spin zero (that is, the representation of lowest order), and write $\mathfrak{K}(\Xi)$ for the real Hilbert space in which it acts; as we shall say later, $\mathfrak{K}(\Xi)$ is not one-dimensional, it is two-dimensional.

Starting from the relation (22.1.1) and using the imprimitivity theorem, one gets, in analogy to the complex case, the following facts:

(i) the real Hilbert space $\mathfrak{K}$ associated with the spin-zero particle can be realized as

$$\mathcal{L}^2\big(\mathbb{R}^3, \mathfrak{K}(\Xi)\big),$$

the real Hilbert space of (equivalence classes of) square-integrable functions defined in $\mathbb{R}^3$ with values in $\mathfrak{K}(\Xi)$;

(ii) $(P(\Delta)f)(x_1, x_2, x_3) = \chi_\Delta(x_1, x_2, x_3)f(x_1, x_2, x_3)$, $f \in \mathcal{L}^2(\mathbb{R}^3, \mathfrak{K}(\Xi))$, where χ_Δ is the characteristic function of Δ;

(iii) $(U_g f)(x_1, x_2, x_3) = \Xi_{\tilde{g}}f(g^{-1}(x_1, x_2, x_3))$, $f \in \mathcal{L}^2(\mathbb{R}^3, \mathfrak{K}(\Xi))$, where $\Xi_{\tilde{g}}$ is the unitary real (orthogonal) operator on $\mathfrak{K}(\Xi)$ that corresponds, in the representation Ξ, to the rotation component $\tilde{g}$ of the element g of $\mathbf{G}$.

From (ii) it follows at once that the operator representing the ith component of the particle position is the multiplication by x_i. From (iii) one gets, by appropriate choices of g, the operators that represent other physical quantities of interest. Suppose, for instance, that we are interested in the first component of the linear momentum. Consider the subgroup of $\mathbf{G}$ consisting of all translations along the x_1-axis, and denote by g_1^λ the translation of λ units. Then $U_{g_1^\lambda}$, as λ varies in $\mathbb{R}$, is a one-parameter group of unitary operators on $\mathfrak{K}$ which is strongly continuous. In the complex case, at this stage, one can use Stone's theorem, thus getting

$$U_{g_1^\lambda} = e^{-i\lambda P_1}, \qquad \lambda \in \mathbb{R},$$

where P_1 is the self-adjoint operator representing the first component of the linear momentum. Of course we cannot have such a result in the real case, because we have not at our disposal the imaginary unit i. The analogue of Stone's theorem for real Hilbert spaces gives

$$U_{g_1^\lambda} = e^{-\lambda R_1}, \qquad \lambda \in \mathbb{R},$$

where R_1 is an anti-self-adjoint operator ($R^* = -R$), and we cannot resort to a self-adjoint operator P_1 by setting $R_1 = iP_1$, for we are confined to the reals. A similar difficulty, related to a different form of Stone's theorem, is found when we introduce the Hamiltonian operator. In fact, while in the complex case the Hamiltonian H is defined as the unique self-adjoint operator such that the evolution operator V_t takes the form

$$V_t = e^{-itH}, \qquad t \in \mathbb{R},$$

in the real case we have only

$$V_t = e^{-tW}, \qquad t \in \mathbb{R},$$

where W is anti-self-adjoint.

Thus we see that, though every physical quantity is represented by a self-adjoint operator in $\mathfrak{L}^2(\mathbb{R}^3, \mathcal{H}(\Xi))$, Stone's theorem cannot provide the explicit form of these operators. A way to overcome this difficulty comes from the consideration that the real Hilbert space $\mathcal{H}(\Xi)$ that carries the spin-zero irreducible representation of the rotation group is two-dimensional.[5] Thus if we fix a basis in $\mathcal{H}(\Xi)$, every element f of $\mathfrak{L}^2(\mathbb{R}^3, \mathcal{H}(\Xi))$ can be written as pair $\langle f_1, f_2 \rangle$ of real functions. In the example of the first component of the momentum, the action of R_1 turns out to be, in proper units,

$$R_1 f = R_1 \langle f_1, f_2 \rangle = \left\langle \frac{\partial f_2}{\partial x_1}, \frac{\partial f_1}{\partial x_1} \right\rangle$$

(of course one should be careful about domains of the differential operators).

The elements of $\mathfrak{L}^2(\mathbb{R}^3, \mathcal{H}(\Xi))$ can be identified with the elements of $\mathfrak{L}^2(\mathbb{R}^3, \mathbb{C})$, the space of complex-valued square-integrable functions defined in $\mathbb{R}^3$, by putting

$$\langle f_1, f_2 \rangle = f_1 + if_2.$$

In this way we recognize that R_1 becomes the usual operator $-i(\partial/\partial x_1)$. What has been said for R_1 can be repeated for all operators representing physical quantities of interest (momentum, energy, orbital angular momentum), and all of them recover their familiar form in $\mathfrak{L}^2(\mathbb{R}^3, \mathbb{C})$.

Though the description of the spin-zero free particle in a real Hilbert space induces in a natural way a description in a complex Hilbert space, there is however a difference. In the original real Hilbert space the physical quantities are represented by self-adjoint real linear operators, while in the associated complex space, not only the self-adjoint complex linear operators represent physical quantities, but also the real-linear ones. In a loose sense we have a larger number of operators representing physical quantities.

This result can, perhaps, be better understood if we recall that our original hypothesis was that the logic $\mathfrak{L}$ of the system was isomorphic to $\mathcal{P}(\mathcal{H})$ for some real Hilbert space $\mathcal{H}$. We have succeeded in finding a complex Hilbert space $\mathcal{H}'$ where we are able to describe the system; nevertheless, $\mathcal{L}$ is not isomorphic to $\mathcal{P}(\mathcal{H}')$. Therefore the set of observables associated with $\mathcal{P}(\mathcal{H})$ cannot coincide with the set of observables associated with $\mathcal{P}(\mathcal{H}')$.

22.3 Quaternionic Hilbert Spaces

In this section we assume as a starting point that the logic of our system is isomorphic to the lattice of projections of a quaternionic Hilbert space. We shall treat this case much in the same way as Mackey treated the real case; we shall sketch only the main results without entering into details, for which we refer to the literature.[6]

As in the real case, the main difficulty in getting the explicit form of operators representing physical quantities is the different form of Stone's theorem. As an example, for the dynamical group one finds

$$V_t = e^{-t\mathcal{J}H}, \qquad t \in \mathbb{R},$$

where $\mathcal{J}$ is any unitary anti-self-adjoint operator that commutes with V_t for all t, and H is a self-adjoint operator that depends on the choice of $\mathcal{J}$ (such a $\mathcal{J}$ surely exists, but it is not uniquely defined by the requirement of commuting with V_t for all t).

The problem of the choice of $\mathcal{J}$, and then of H, can be solved by considering some additional physical requirements. Since we want to interpret H as the Hamiltonian of a free particle with nontrivial dynamics ($V_t \neq I$), we require that H be a strictly positive definite operator. It turns out[6] that this physical requirement is sufficient to fix $\mathcal{J}$ uniquely. We shall denote this particular operator by $\mathcal{J}_H$; it commutes not only with every V_t, but also with every operator that commutes with every V_t.

To determine the form of the operators that represent other physical quantities, we consider, as usual, the imprimitivity relation (22.1.2). We confine ourselves again to the case of a spin-zero particle, and we have that:

(i) the Hilbert space $\mathcal{H}$ can be identified with the space of all quaternion-valued functions, defined in $\mathbb{R}^3$, square-integrable with respect to the Lebesgue measure;

(ii) $(P(\Delta)f)(x_1, x_2, x_3) = \chi_\Delta(x_1, x_2, x_3)f(x_1, x_2, x_3)$, $f \in \mathcal{H}$;

(iii) $(U_g f)(x_1, x_2, x_3) = f(g^{-1}(x_1, x_2, x_3))$, $g \in \mathbf{G}, f \in \mathcal{H}$.*

From (ii) we get at once that the operator that represents the ith component of the position of the particle is the multiplication by x_i. By repeated use of (iii), identifying g with translations along and rotations

---

*Note that in the corresponding formula for the real case the operator $\Xi_{\bar{g}}$ operates on f: this is because in the real case the representation for spin zero is two-dimensional, while in the quaternionic case it is one-dimensional.

about the three coordinate axes, we get, by use of the quaternionic Stone's theorem, the following expressions for the self-adjoint operators that represent components of momentum and orbital angular momentum:

$$P_1 = -\oint_{P_1} \frac{\partial}{\partial x_1}, \qquad P_2 = -\oint_{P_2} \frac{\partial}{\partial x_2}, \qquad P_3 = -\oint_{P_3} \frac{\partial}{\partial x_3},$$

$$L_3 = -\oint_{L_3} \left(x_1 \frac{\partial}{\partial x_2} - x_2 \frac{\partial}{\partial x_1} \right), \dots$$

(as before, one should be careful about domains of differential operators). The main problem is that each operator is defined in terms of a unitary anti-self-adjoint operator $\oint$ whose choice is quite arbitrary.

At this stage we are faced with the problem of "unifying" the various operators $\oint_{P_1}, \oint_{P_2}, \oint_{P_3}, \oint_{L_1}, \oint_{L_2}, \oint_{L_3}$ by adopting a unique operator of this kind that commutes with every $U_g, V_t, P(\Delta)$ and hence with every operator representing a physical quantity. The candidate is $\oint_H$, which was uniquely determined by the physical hypothesis of positivity of the Hamiltonian, and which indeed commutes with every U_g and V_t [also[6] with every $P(\Delta)$]. Thus we define the operators corresponding to the momentum and angular momentum in the following way:

$$P_1 = -\oint_H \frac{\partial}{\partial x_1}, \dots, \qquad L_3 = -\oint_H \left(x_1 \frac{\partial}{\partial x_2} - x_2 \frac{\partial}{\partial x_1} \right).$$

By means of the operator $\oint_H$ it is possible to decompose $\mathcal{H}$ into the direct sum of two complex Hilbert spaces $\mathcal{H}_1$ and $\mathcal{H}_2$ (the "symplectic decomposition" of $\mathcal{H}$), and it is possible to restrict the operators $U_g, V_t, P(\Delta)$ to the spaces $\mathcal{H}_1$ and $\mathcal{H}_2$.[6] Using the symplectic decomposition, the explicit form of the Hamiltonian can be determined. In each of the spaces $\mathcal{H}_1$ and $\mathcal{H}_2$ the restriction of the "quaternionic Hamiltonian" coincides with the usual Hamiltonian that we have met for the complex case. The unitary anti-self-adjoint operator $\oint_H$, which is the formal analogue in the quaternionic case of the imaginary unit i in $\mathbb{C}$, allows the reduction of the quantum quaternionic theory to the complex one.

References

1. G. W. Mackey, *Induced Representations of Groups and Quantum Mechanics*, Benjamin, Advanced Book Program, Reading, Mass., 1968.
2. V. S. Varadarajan, *Geometry of Quantum Theory*, Vol. II, Van Nostrand, Princeton, N.J., 1970.
3. E. C. G. Stueckelberg, *Helv. Phys. Acta* 33 (1960) 727.
4. G. C. Emch, *Helv. Phys. Acta* 36 (1963) 739; *ibid.* 36 (1963) 770.
5. G. W. Mackey, *Unitary Group Representations in Physics, Probability and Number Theory*, Benjamin Cummings, Advanced Book Program, Reading, Mass., 1978.
6. G. Cassinelli and P. Truini, preprint, Duke University, 1980.

CHAPTER 23

Dynamics

23.1 General Description of Dynamical Evolution

In Chapter 21 we have discussed the connection between the Hilbert-space description of a quantum system and its description in terms of the $(\mathcal{L}, \mathcal{S})$ structure, but only static properties of the system were considered. Here we discuss the dynamical evolution. We anticipate that within the $(\mathcal{L}, \mathcal{S})$ structure a satisfactory characterization of quantum dynamics seems to be out of reach. This is why the discussion of the dynamics remains somewhat marginal to the aim, pursued in this volume, of focusing attention on those aspects of quantum theory that can be founded on simple mathematical substructures underlying the Hilbert-space formulation of the theory. Actually, the studies on quantum dynamics have usually been carried out by full use of the mathematical apparatus of Hilbert spaces; the results of these studies fill a rich literature whose discussion might need a volume of its own.

The problem of describing the dynamical evolution of a physical system in terms of a proposition-state structure $(\mathcal{L}, \mathcal{S})$ amounts to giving the time dependence of the probabilities of yes outcomes of propositions. In principle, this can be done by introducing a time evolution of states, or of propositions, or of both. But the choice that is closer to what actually goes on in a laboratory is to describe the dynamical evolution of the system in terms of the time evolution of its states, leaving fixed in time the propositions, i.e. the measuring instruments. This is the so-called Schrödinger picture.

Causality puts a first restriction on the time dependence: the state at time t cannot depend on what will happen to the system at later times, but only

ENCYCLOPEDIA OF MATHEMATICS and Its Applications, Gian-Carlo Rota (ed.). Vol. 15: E. G. Beltrametti and G. Cassinelli, The Logic of Quantum Mechanics

ISBN 0-201-13514-0

on what happened in the past. Thus, the state at time t has to be a function of the states and of the interactions the system has experienced at times before t. But it is an empirical fact that there are many physical systems that admit a simpler scheme of time evolution—systems that have, so to speak, only a partial memory of their past.[1] For such systems the state at time t is completely determined by the knowledge of the state at an arbitrary fixed time $t_0 < t$ and of the interactions the system has experienced in the interval between t_0 and t. Such time evolutions are generally referred to as Markovian, after the notion of Markov process in ordinary probability theory.

Formally, if we write α_t for the state of the system at time t, a Markovian dynamics is given by a two-parameter family of mappings V_{t,t_0} of the set $\mathbb{S}$ of states into itself, such that

$$\alpha_t = V_{t,t_0}\alpha_{t_0}, \qquad t > t_0.$$

Notice that if $t_0 < t_1 < t_2$ we have

$$\alpha_{t_2} = V_{t_2,t_1}\alpha_{t_1} = V_{t_2,t_1}V_{t_1,t_0}\alpha_{t_0}$$

and also

$$\alpha_{t_2} = V_{t_2,t_0}\alpha_{t_0},$$

thus concluding

$$V_{t_2,t_1}\cdot V_{t_1,t_0} = V_{t_2,t_0}, \tag{23.1.1}$$

where $\cdot$ is map composition.

A further remarkable simplification occurs when V_{t,t_0} actually depends only on the difference $t - t_0$: in this case the time evolution is specified by a one-parameter family of mappings V_τ ($\tau \geq 0$) such that if α_{t_0} is the state at time t_0, the state τ units of time after t_0 will be

$$\alpha_{t_0+\tau} = V_\tau\alpha_{t_0}. \tag{23.1.2}$$

The occurrence of dynamics of this kind corresponds to those situations in which there is homogeneity in time, or, in other words, to those situations in which no absolute time enters, but only relative time is relevant. Interactions of a physical system with nonstationary surroundings are here excluded. On the other hand, every dynamical evolution of an isolated (closed) system is included. According to the terminology of Markov processes in ordinary probability theory,[2] we shall refer to (23.1.2) as the characterization of stationary Markovian dynamics.

It must be noticed that for a stationary Markovian dynamics, the relation (23.1.1) is replaced by

$$V_\tau \cdot V_{\tau'} = V_{\tau+\tau'}, \qquad (23.1.3)$$

showing that the mathematical structure that now carries the dynamical evolution is a one-parameter semigroup of mappings of the set $\mathcal{S}$ of states into itself.

Up to now we have not considered a requirement that appears quite natural from the physical standpoint: continuity in time. In a structure $(\mathcal{L}, \mathcal{S})$ of propositions and states with a stationary Markovian evolution, this amounts to requiring that the real function

$$\tau \mapsto \alpha_{t_0+\tau}(a) = V_\tau \alpha_{t_0}(a)$$

be continuous for any fixed $\alpha_{t_0} \in \mathcal{S}$ and any fixed $a \in \mathcal{L}$. Small changes in time produce small changes in probabilities.*

Another requirement that has, in many circumstances, natural physical motivations is that the dynamical evolution be an affine transformation, i.e., it preserves the convex structures of the set $\mathcal{S}$ of states. Still referring to a stationary Markovian evolution, this means that the dynamical operator V_τ has the property

$$V_\tau(w_1\alpha_1 + w_2\alpha_2) = w_1 V_\tau\alpha_1 + w_2 V_\tau\alpha_2 \qquad (23.1.4)$$

for any positive τ, for arbitrary $\alpha_1, \alpha_2 \in \mathcal{S}$, and for arbitrary weights w_1, w_2.†

All the requirements considered heretofore are unable to separate the reversible from the irreversible dynamical evolutions. Reversibility means, in the language of stationary Markovian dynamics, that to every V_τ ($\tau \geqslant 0$) that maps α_{t_0} into $\alpha_{t_0+\tau}$ [see (23.1.2)] there corresponds, in the family of all mappings that describe the dynamics of the system, another element that "brings back" the state $\alpha_{t_0+\tau}$ into α_{t_0}. It is natural to write V_τ^{-1} for such a mapping, so that

$$V_\tau^{-1}\alpha_{t_0+\tau} = \alpha_{t_0}.$$

Thus we see that, if reversibility holds, the dynamical semigroup is indeed something more: it is a one-parameter group, with V_τ^{-1} acting as group-theoretical inverse of V_τ ($V_\tau \cdot V_\tau^{-1} = V_\tau^{-1} \cdot V_\tau =$ the identity map in $\mathcal{S}$).

---

*When $\mathcal{L} = \mathcal{P}(\mathcal{H})$, the class of continuous stationary Markovian dynamics is structured enough to allow an interesting analysis,[3] and it is seen to contain examples of nonlinear evolutions.

†Cases of nonaffine dynamical evolution can however be conceived: e.g. the dynamics induced by a compound system on its subsystems, as discussed in Section 7.5.

Conversely, if the dynamical mappings form a group, then reversibility holds. Thus we see that reversibility and group structure of the dynamical mappings are the same. It can also be noticed that reversibility forces every V_τ to be a bijection of S onto itself, for if not, it would be impossible to define the inverse of V_τ.

23.2 From Automorphism to the Hamiltonian

In this section we assume from the outset that the proposition lattice $\mathcal{L}$ is the projection lattice $\mathcal{P}(\mathcal{H})$ of a separable complex Hilbert space and that the set S of states is the set $S(\mathcal{H})$ of density operators on $\mathcal{H}$. We can then sketch the logical steps that lead to the usual Hamiltonian description of the dynamical evolution. Recalling what was said in Sections 6.1 and 6.2, the door to Hamiltonians is open as soon as the evolution map V_τ is recognized to be of the form $V_\tau D = U_\tau D U_\tau^*$ for every $D \in S(\mathcal{H})$, where $\tau \mapsto U_\tau$ is a one-parameter weakly continuous group of unitary operators on $\mathcal{H}$.

There are two alternative points of departure. The first, mostly related to works of Mackey and Kadison, is to assume that

(1) V_τ is a bijection of $S(\mathcal{H})$ onto $S(\mathcal{H})$;
(2) V_τ is affine, that is, it preserves the convex structure of $S(\mathcal{H})$ [see (23.1.4)];
(3) $\tau \mapsto \mathrm{tr}(V_\tau DP)$ is continuous for every $D \in S(\mathcal{H})$ and every $P \in \mathcal{P}(\mathcal{H})$, which expresses the physical fact that small changes in time produce small changes of probabilities of propositions.

The second point of departure, mostly related to works of Wigner, makes reference only to the set $S^P(\mathcal{H})$ of one-dimensional projectors (the pure states) and amounts to assuming that

(1') V_τ is a bijection of $S^P(\mathcal{H})$ onto $S^P(\mathcal{H})$;
(2') V_τ preserves the transition probabilities, that is,

$$\mathrm{tr}\left(P^{\hat{\varphi}} P^{\hat{\psi}} \right) = \mathrm{tr}\left(V_\tau P^{\hat{\varphi}} V_\tau P^{\hat{\psi}} \right)$$

for any two vector states φ, ψ — in other words, the time evolution preserves the modulus of scalar products;*
(3') $\tau \mapsto \mathrm{tr}(V_\tau P^{\hat{\varphi}} V_\tau P^{\hat{\psi}})$ is continuous for every $P^{\hat{\varphi}}, P^{\hat{\psi}} \in S^P(\mathcal{H})$

Both sets of requirements have physical naturalness, and both have the power of leading to a unitary dynamical group. Let us mention the main steps.

---

*Should one assume the preservation of the scalar products themselves, the unitarity would be ensured.

Step 1.   Every automorphism V_τ of $\mathcal{S}(\mathcal{K})$ that satisfies (1)–(3) must take the form[4,5]

$$V_\tau D = U_\tau D U_\tau^* \qquad \text{for every} \quad D \in \mathcal{S}(\mathcal{K}), \qquad (23.2.1)$$

where U_τ is a unitary or an antiunitary operator on $\mathcal{K}$, determined up to a phase factor. Similarly, every automorphism V_τ of $\mathcal{S}^P(\mathcal{K})$ that satisfies (1′)–(3′) is, by Wigner's theorem,[6] of the form

$$V_\tau P^{\hat{\psi}} = U_\tau P^{\hat{\psi}} U_\tau^* \qquad \text{for every} \quad P^{\hat{\psi}} \in \mathcal{S}^P(\mathcal{K}), \qquad (23.2.1')$$

where, again, U_τ is a unitary or an antiunitary operator on $\mathcal{K}$, determined up to a phase factor.

Step 2.   The result (23.2.1), or (23.2.1′), comes uniquely from the properties (1)–(3),or (1′)–(3′). According to the discussion of the preceding section, it is however understood that the evolution map V_τ satisfies also the semigroup property (23.1.3) (hence also the group property, for we are here assuming that V_τ is a bijection). The consequence of this semigroup structure is that in (23.2.1), or (23.2.1′), we can restrict ourselves to unitary weakly continuous* operators U_τ.[7] Thus, combining (23.2.1), or (23.2.1′), with (23.1.3), we have the further property

$$U_\tau U_{\tau'} = \omega(\tau, \tau') U_{\tau + \tau'} \qquad (23.2.2)$$

for some complex-valued function $\omega(\tau, \tau')$ of modulus 1.

Step 3.   This step reduces to a property of $\mathbb{R}$ as a Borel group. (23.2.2) is a projective representation (weakly continuous and hence measurable) of the additive group of the real line: it is therefore known that the phase of the U_τ's can always be chosen in such a way that $\omega(\tau, \tau') = 1$, leaving V_τ unaffected. This result is due to Wigner[7] (for more general groups the analogous problem is called the multiplier problem [8–10]). Thus we get the group property $U_\tau U_{\tau'} = U_{\tau + \tau'}$.

The three steps above leave us with all the premises about the dynamical evolution that we have considered in Sections 6.1 and 6.2: if D_t is the state at time t, the state at time $t + \tau$ is given by

$$D_{t+\tau} = U_\tau D_t U_\tau^*, \qquad (23.2.3)$$

where $\tau \mapsto U_\tau$ is a weakly continuous one-parameter group of unitary operators. The passage to the Hamiltonian is now achieved by Stone's theorem, and Schrödinger's equation is at hand (see Section 6.2).

---

*Recall that the operator U_τ is called weakly continuous when $\tau \mapsto (\varphi, U_\tau \psi)$ is continuous for every $\varphi, \psi \in \mathcal{K}$.

As a general reference for the program sketched in this section we mention Simon's overview of the subject matter.[11]

Notice that if the hypotheses of this section are weakened by giving up the reversibility of the time evolution [that is, giving up condition (1) or (1')], the linearity of V_τ is still preserved. Indeed,[12] V_τ can be uniquely extended to a linear operator in a vector (Banach) space generated by $\mathcal{S}(\mathcal{H})$; the corresponding class of dynamical evolutions includes the ones based on Kossakowski's axiomatics,[13] which in turn include cases of interest for quantum systems.

Let us finally remark that the idea of characterizing quantum dynamics as a noncommutative Markov process has been successfully worked out by Accardi[14] in the context of the algebraic approach to quantum mechanics based on C^*-algebras. This is an important result, an account of which, however, would go beyond the purposes of the present volume.

23.3 On Heisenberg's Picture

In Section 23.1 we have described the dynamical evolution of the system by letting the states evolve and keeping the propositions fixed. In other words, we have adopted the Schrödinger picture. The question whether other pictures can give equivalent descriptions, and notably whether the Heisenberg picture, in which states are kept fixed while propositions evolve with time, is possible, can be put in the following terms: given the state automorphism (in Schrödinger's picture)

$$\alpha \mapsto V_\tau \alpha, \qquad \alpha \in \mathcal{S},$$

is it possible to determine a one-parameter family of automorphisms of $\mathcal{L}$, say $\tilde{V}_\tau$, such that

$$\alpha(\tilde{V}_\tau a) = (V_\tau \alpha)(a) \qquad (23.3.1)$$

for all $a \in \mathcal{L}$? In case of an affirmative answer the state automorphism V_τ is said to be *implemented by* $\tilde{V}_\tau$, and the Schrödinger picture of time evolution admits an equivalent Heisenberg picture.*

In a general $(\mathcal{L}, \mathcal{S})$ scheme a characterization of the state automorphisms that admit an implementation is not available, and examples are known of state automorphisms that cannot be implemented.[16] These facts indicate that, within an $(\mathcal{L}, \mathcal{S})$ scheme, even the study of the reversible dynamics of a physical system has problematic aspects.[17]

---

*The notion of implementability of a state automorphism (and, vice versa, of a lattice automorphism) is the basic tool for the problem of symmetries in a propositions-states structure.[15]

Notice that, according to the discussion in Section 6.3, the class of dynamical evolutions that, in the Hilbert-space framework, admit a Heisenberg picture is precisely the one in which D_t evolves, at time $t + \tau$, into $D_{t+\tau}$ given by (23.2.3).

References

1. A. Frigerio, V. Gorini, A. Kossakowski, E. C. G. Sudarshan, and M. Verri, *Rep. Math. Phys.* 13 (1978) 149.
2. L. Breiman, *Probability*, Addison-Wesley, Reading, Mass., 1968.
3. W. Guz, *Rep. Math. Phys.* 8 (1975) 49.
4. G. W. Mackey, *Mathematical Foundations of Quantum Mechanics*, Benjamin Advanced Book Program, Reading, Mass., 1963.
5. R. Kadison, *Topology* 3, Suppl. 2 (1965) 177; *Ann. Math.* 54 (1951) 325.
6. E. P. Wigner, *Group Theory and its Applications to the Quantum Mechanics of Atomic Spectra*, Academic Press, New York, 1959.
7. E. P. Wigner, *Ann. Math.* 40 (1939) 149.
8. V. S. Varadarajan, *Geometry of Quantum Theory*, Vol. II, Van Nostrand, Princeton, N.J., 1970.
9. V. Bargmann, *Ann. Math.* 59 (1954) 1.
10. G. W. Mackey, *Acta Math.* 99 (1958) 265.
11. B. Simon, in *Studies in Mathematical Physics—Essays in Honour of Valentine Bargmann*, E. H. Lieb, B. Simon, and A. S. Wightman, eds., Princeton University Press, Princeton, N.J., 1976.
12. W. Guz, *Rep. Math. Phys.* 6 (1974) 455.
13. A. Kossakowski, *Rep. Math. Phys.* 3 (1972) 247.
14. L. Accardi, *Adv. Math.* 20 (1976) 329.
15. T. A. Cook and G. T. Ruttimann, "Symmetries on quantum logics", preprint, University of Massachusetts, 1975.
16. S. P. Gudder, *Int. J. Theor. Phys.* 6 (1973) 205.
17. V. Gorini and A. Zecca, *J. Math. Phys.* 16 (1975) 667.

CHAPTER 24

Composition of Physical Systems

24.1 Formalization of the Notion of Composition

In Chapter 7 the rules that specify how to describe a compound system in terms of its component subsystems (and vice versa) were examined in the framework of Hilbert-space quantum mechanics. There we saw that given the Hilbert spaces $\mathcal{H}_1, \mathcal{H}_2$ associated with the component subsystems $\mathbb{S}_1, \mathbb{S}_2$, the arena in which the compound system $\mathbb{S}_1 + \mathbb{S}_2$ has to be described is the tensor product $\mathcal{H}_1 \otimes \mathcal{H}_2$ or some subspace of it. In case $\mathbb{S}_1$ and $\mathbb{S}_2$ are identical we saw that characteristic subspaces of $\mathcal{H}_1 \otimes \mathcal{H}_2$ are called into play: either the symmetric or the antisymmetric tensor products, according as the spin of $\mathbb{S}_1$ (and of $\mathbb{S}_2$) is integer or half-integer.

Why all that? Is there the possibility of motivating those rules on a more fundamental level? To discuss these questions, briefly and partially, is the purpose of the present chapter. Briefly and partially, for there are still open problems, and current research is active on the subject.

Hellwig and Krauser,[1] Zecca,[2] and especially Aerts[3-6] have tackled the problem by approaches that recognize proposition lattices as basic carriers of the description of physical systems.* It is to these attempts that we shall mainly refer in the sequel; in particular we shall refer to Aerts's work. We shall limit ourselves to the composition of nonidentical subsystems, for this is the case where the most promising results are available.

We write $(\mathcal{L}_1, \mathcal{S}_1)$, $(\mathcal{L}_2, \mathcal{S}_2)$, and $(\mathcal{L}, \mathcal{S})$ for the proposition-state structures associated with $\mathbb{S}_1$, $\mathbb{S}_2$, and $\mathbb{S}$ respectively, and we look for a way to

---

*The problem has also been the object of other approaches, notably by Foulis and Randall within their operational framework.

ENCYCLOPEDIA OF MATHEMATICS and Its Applications, Gian-Carlo Rota (ed.).
Vol. 15: E. G. Beltrametti and G. Cassinelli, The Logic of Quantum Mechanics

ISBN 0-201-13514-0

formalize the fact that S is a physical system composed of the two subsystems S_1 and S_2. We shall adopt for these proposition-state structures the standard requirements encountered in Chapter 14. We are going to use the same symbols for the lattice operations ($\wedge, \vee, \perp$) of $\mathcal{L}_1, \mathcal{L}_2$, and $\mathcal{L}$; the elements of $\mathcal{L}_1$ (and S_1) will have the index 1, the elements of $\mathcal{L}_2$ (and S_2) the index 2, and the elements of $\mathcal{L}$ (and S) no index.

It is implicit in the assertion that S is composed of two different subsystems S_1 and S_2 that each of S_1 and S_2 preserves its own identity when thought of as part of the bigger physical system S. In other words, the interactions that may be present between S_1 and S_2 when they form the system S are implicitly assumed to be gentle enough to preserve unchanged the internal structure of S_1 and of S_2 thought of as isolated noninteracting systems. The translation of this basic idea is, roughly speaking, that $\mathcal{L}$ has to contain a part that resembles $\mathcal{L}_1$ and another part that resembles $\mathcal{L}_2$. Technically, and referring to the terminology of Section 12.5, there must be as a minimal requirement an injective morphism, say h_1, from $\mathcal{L}_1$ into $\mathcal{L}$, and an injective morphism, say h_2, from $\mathcal{L}_2$ into $\mathcal{L}$. We shall refer to h_1 and h_2 as the recognition maps, for they allow one to recognize S_1 and S_2 as parts of S. The injectivity, we recall, means that different elements are mapped into different elements; an injective morphism preserves ordering (in both directions) as well as the join and meet operations, it maps the least element into the least element, it preserves the orthogonality, and more specifically, it maps the orthocomplement into the "relative orthocomplement". In our case,

$$h_1(a_1^\perp)=h_1(a_1)^\perp \wedge h_1(1_1), \qquad h_2(a_2^\perp)=h_2(a_2)^\perp \wedge h_2(1_2).$$

There is a natural physical interpretation of the recognition maps: $h_1(a_1)$ represents the proposition, relative to the compound system S, that consists of questioning the proposition a_1 for the subsystem S_1 without questioning anything about the subsystem S_2 but the fact that it is present, and similarly for $h_2(a_2)$. Then we see that the proposition $h_1(1_1)$, as well as $h_2(1_2)$, simply amounts to saying that both S_1 and S_2 are present as parts of S. Now we have at hand the way to express the fact that S is composed of nothing else than S_1 and S_2: the assertion that both S_1 and S_2 are present has to be the same as the assertion that S is present; otherwise there would be in S some third piece, besides S_1 and S_2. Hence we come to the requirement

AXIOM 24.1.1. $\qquad\qquad h_1(1_1)=h_2(1_2)=1,$

namely, that h_1 and h_2 are unitary morphisms. This makes h_1 and h_2 orthocomplement-preserving:

$$h_1(a_1^\perp)=h_1(a_1)^\perp, \qquad h_2(a_2^\perp)=h_2(a_2)^\perp. \qquad (24.1.1)$$

Notice that, instead of saying that there are two unitary injective morphisms h_1 and h_2 from $\mathcal{L}_1$ and from $\mathcal{L}_2$, respectively, into $\mathcal{L}$, we might say that there are in $\mathcal{L}$ two sublattices, both containing 0 and 1, which are isomorphic as lattices to $\mathcal{L}_1$ and to $\mathcal{L}_2$, respectively.

There is another requirement that looks basic from the physical point of view and that reflects the assumed distinguishability of the two subsystems $\mathcal{S}_1$ and $\mathcal{S}_2$: every proposition on $\mathcal{S}$ that actually refers only to $\mathcal{S}_1$, thus being of the form $h_1(a_1)$ for some $a_1 \in \mathcal{L}_1$, has to be compatible with every proposition on $\mathcal{S}$ that actually refers only to $\mathcal{S}_2$, thus being of the form $h_2(a_2)$ for some $a_2 \in \mathcal{L}_2$. Then we have

AXIOM 24.1.2. $(h_1(a_1), h_2(a_2))C$ for every $a_1 \in \mathcal{L}_1$ and every $a_2 \in \mathcal{L}_2$,

where the commutativity is understood in its lattice-theoretic meaning (see Section 12.1).

One more fact has to be acknowledged in order to recognize $\mathcal{S}$ as a quantum system composed of $\mathcal{S}_1$ and $\mathcal{S}_2$: maximum information about $\mathcal{S}_1$ together with maximum information about $\mathcal{S}_2$ means maximum information about $\mathcal{S}$ (the converse is not true, as emphasized in Section 7.2). In terms of states, this amounts to saying that if $\mathcal{S}_1$ is in a pure state and $\mathcal{S}_2$ is in a pure state, then $\mathcal{S}$ has to be in a pure state too. But recalling that, in our context, pure states correspond to atoms of the proposition lattice, we can further rephrase this requirement by saying that a proposition about $\mathcal{S}$ that consists of questioning an atomic proposition p_1 about $\mathcal{S}_1$ and an atomic proposition p_2 about $\mathcal{S}_2$ must be atomic in $\mathcal{L}$. Formally:

AXIOM 24.1.3. $h_1(p_1) \wedge h_2(p_2)$ is an atom of $\mathcal{L}$ if p_1 is an atom of $\mathcal{L}_1$ and p_2 an atom of $\mathcal{L}_2$.

(Again we notice that, according to Section 7.2, not every atom of $\mathcal{L}$ will be of that form unless we deal with classical systems.)

For brevity, we shall define the *composition conditions* as the conditions of the existence of two injective morphisms h_1 from $\mathcal{L}_1$ into $\mathcal{L}$ and h_2 from $\mathcal{L}_2$ into $\mathcal{L}$, satisfying Axioms 24.1.1–24.1.3. Given the physical systems $\mathcal{S}_1$ and $\mathcal{S}_2$ to be composed, or given the proposition lattices $\mathcal{L}_1$ and $\mathcal{L}_2$, a solution of the composition conditions will then consist of the triple $(\mathcal{L}, h_1, h_2)$.

Now a problem arises in a natural way: to what extent do $\mathcal{L}_1$ and $\mathcal{L}_2$ determine the solution of the composition conditions? The mathematical translation of this question is what we consider in the next section.

24.2 A Question of Uniqueness

Suppose $\mathcal{L}_1$ and $\mathcal{L}_2$ are given, and let $(\mathcal{L}, h_1, h_2)$ be a solution of the composition conditions. Suppose now that $\mathcal{L}'$ is another lattice with the same structural properties (orthomodularity, atomicity, and so on) of $\mathcal{L}$, and

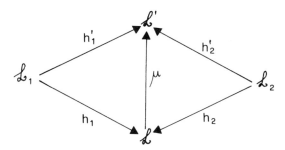

Figure 24.1. The commutative diagram associated with the coproduct of two proposition lattices.

suppose that there exists a unitary injective morphism μ from $\mathcal{L}$ into $\mathcal{L}'$ that maps atoms into atoms. It is then a simple exercise to check that the triple $(\mathcal{L}', \mu \cdot h_1, \mu \cdot h_2)$ is also a solution of the composition conditions.* But this new solution adds nothing interesting to the original one, for it is either equivalent to $(\mathcal{L}, h_1, h_2)$ in case μ is an isomorphism, thus making $\mathcal{L}'$ a replica of $\mathcal{L}$, or less economic than $(\mathcal{L}, h_1, h_2)$ in case $\mathcal{L}'$ is, so to speak, bigger than $\mathcal{L}$. Therefore, we can take the solution $(\mathcal{L}, h_1, h_2)$ as representative of the whole class of solutions that have the form $(\mathcal{L}', \mu \cdot h_1, \mu \cdot h_2)$. In technical words, the triple $(\mathcal{L}, h_1, h_2)$ can be taken as representative of any other triple, say $(\mathcal{L}', h_1', h_2')$, such that the diagram in Figure 24.1 is commutative for a unitary, injective morphism μ that maps atoms into atoms. Thus we see that in order to get a solution $(\mathcal{L}', h_1', h_2')$ genuinely different from $(\mathcal{L}, h_1, h_2)$, we must require that no commutative diagram of the form shown in Figure 24.1 can connect them.

It may happen that a solution $(\mathcal{L}, h_1, h_2)$ of the composition conditions is "universal", in the sense that every other solution is of the kind $(\mathcal{L}', h_1', h_2')$ depicted in Figure 24.1. In this case the solution $(\mathcal{L}, h_1, h_2)$ is called the coproduct of $\mathcal{L}_1$ and $\mathcal{L}_2$, in the terminology of category theory. Thus we can say that the solution $(\mathcal{L}, h_1, h_2)$ is essentially unique, or that $(\mathcal{L}, h_1, h_2)$ is the solution of the composition conditions, if it is the coproduct of $\mathcal{L}_1$ and $\mathcal{L}_2$ (within the category of lattices we are interested in).

From the physical point of view, the knowledge of the physical systems $\mathbb{S}_1$ and $\mathbb{S}_2$ should determine the nature of the compound system they form: correspondingly, either we should fall into the situation in which the composition conditions determine a coproduct, or we should find physical criteria to rule out all but one solution of the composition conditions.

Up to now we have considered the problem of describing a compound system in a general, even generic, way: we have only discussed a characterization of the composition conditions, without entering into how to construct solutions in explicit cases. As a matter of fact, it seems that there is no

---

*Of course, $\mu \cdot h_1$ denotes the mapping from $\mathcal{L}_1$ into $\mathcal{L}'$ defined by $(\mu \cdot h_1)(a_1) = \mu(h_1(a_1))$, and similarly for $\mu \cdot h_2$.

general recipe to construct the solution(s) of the composition conditions as long as the proposition-state structures involved are as general as above. To make the problem more definite, Zecca[2] has considered the idea of relating the superselection rules of the subsystems $\mathbb{S}_1$ and $\mathbb{S}_2$ to those of the compound system $\mathbb{S}$—in other words, of relating, by means of simple axioms, the degree of reducibility of $\mathcal{L}_1$ and $\mathcal{L}_2$ to the degree of reducibility of $\mathcal{L}$, and hence relating the center of $\mathcal{L}$ to the centers of $\mathcal{L}_1$ and $\mathcal{L}_2$. Aerts* has considered the problem by assuming more specific structures for the proposition lattices; we are going to mention, without proofs, his main results.

24.3 Composition of a Classical with a Quantum System

Consider the case in which $\mathbb{S}_1$ is a classical system, with the usual description in terms of a phase space Ω, and with $\mathcal{L}_1$ becoming the Boolean algebra $\mathcal{B}$ of all subsets of Ω. To each element (or point) ω of Ω let us associate a replica of $\mathcal{L}_2$, say $\mathcal{L}_2^{(\omega)}$. By $a_2^{(\omega)}$ we shall denote the image in $\mathcal{L}_2^{(\omega)}$ of $a_2 \in \mathcal{L}_2$. Then consider the direct product $\prod_{\omega \in \Omega} \mathcal{L}_2^{(\omega)}$, ω running over Ω, as defined in Section 12.5, and recall that an element of the direct product can be thought of as a collection obtained by picking an element from each factor $\mathcal{L}_2^{(\omega)}$. Write for short $\mathcal{L} = \prod_{\omega \in \Omega} \mathcal{L}_2^{(\omega)}$.

Define the recognition map h_1 from $\mathcal{B}$ into $\mathcal{L}$ by

$$h_1(a_1) = \left\{ \delta^{(\omega)}(a_1) : \omega \in \Omega \right\}.$$

where $\delta^{(\omega)}(a_1)$ denotes the least element of $\mathcal{L}_2^{(\omega)}$ if the point ω does not belong to the subset a_1 of Ω, and the greatest element of $\mathcal{L}_2^{(\omega)}$ if ω belongs to a_1. Define the recognition map h_2 from $\mathcal{L}_2$ into $\mathcal{L}$ by

$$h_2(a_2) = \left\{ a_2^{(\omega)} : \omega \in \Omega \right\}.$$

Now we can state the result:[†] the triple $(\mathcal{L}, h_1, h_2)$ defined above is a solution of the composition conditions and constitutes the coproduct of $\mathcal{B}$ and $\mathcal{L}_2$. Thus, according to the discussion of the last section, we can look at $(\mathcal{L}, h_1, h_2)$ as the unique solution of the composition conditions. This means that, given a classical system $\mathbb{S}_1$ and a quantum system $\mathbb{S}_2$, the compound system they form is determined and corresponds to the proposition lattice $\mathcal{L}$. As seen in Section 12.5, the direct product of orthomodular lattices is the same as the direct sum; hence in our case we might equivalently write $\mathcal{L} = \bigcup_{\omega \in \Omega} \mathcal{L}_2^{(\omega)}$.

---

*For a review of Aerts's work we refer especially to Reference 6.
†See Theorem 5.3.1 of Reference 5.

Suppose, for simplicity, that $\mathbb{S}_2$ is a purely quantum system, namely that $\mathcal{L}_2$ is irreducible. Then $\mathcal{L}$ is the sum of irreducible addends so that we come to the conclusion that composing a purely classical with a purely quantum system we get a system with limited quantum behavior: it has superselection rules labeled by the points of the phase space of the classical subsystem (compare also References 1 and 2). This is precisely the pattern that was assumed in Sections 8.2 and 8.3; it was seen to give consistency to the measurement problem in quantum mechanics, resolving at the same time certain supposed paradoxes of the theory.

Before ending this section let us mention one more result:[3] in case not only $\mathbb{S}_1$ but also $\mathbb{S}_2$ is classical, so that not only $\mathcal{L}_1$ but also $\mathcal{L}_2$ is a Boolean algebra, one gets the expected result that the proposition lattice of the compound system becomes the Boolean algebra of all subsets of the cartesian product $\Omega_1 \times \Omega_2$, where Ω_1 and Ω_2 are respectively the phase spaces of $\mathbb{S}_1$ and of $\mathbb{S}_2$ (compare also Reference 2).

24.4 Composition of Two Quantum Systems

We tackle the problem of composing two quantum systems in a context more specialized than above: the proposition lattices coming into play will be assumed to belong to the class of lattices that admit the vector-space coordinatization, discussed in Section 21.2. The covering property and irreducibility have thus to be added to the structural properties assumed in the preceding sections. Notice, in particular, that once irreducibility is assumed we are confined to the case of pure quantum systems: superselection rules are ruled out from the outset.

The lattices $\mathcal{L}_1, \mathcal{L}_2, \mathcal{L}$ associated with $\mathbb{S}_1, \mathbb{S}_2, \mathbb{S}$, will now be lattices of closed subspaces of vector spaces. Since we have to handle three vector spaces over possibly different fields, it is worth modifying a bit the notation used in Section 21.2.

Instead of using an asterisk to denote the involution in the field, say K, we shall use the letter ξ and write $\xi(\lambda)$ in place of λ^*; moreover, instead of writing $\mathcal{L}_f(V)$ to denote the lattice of all f-closed subspaces of V, we shall write $\mathcal{L}(V, K, f, \xi)$, thus making apparent the reference to the underlying field and involution. Then our problem reads as follows: given the lattices $\mathcal{L}(V_1, K_1, f_1, \xi_1)$ and $\mathcal{L}(V_2, K_2, f_2, \xi_2)$ associated with $\mathbb{S}_1$ and $\mathbb{S}_2$, are there a lattice $\mathcal{L}(V, K, f, \xi)$ and two recognition maps h_1 and h_2 such that the triple $(\mathcal{L}(V, K, f, \xi), h_1, h_2)$ is a solution of the composition conditions? Aerts* proves the following result:

THEOREM 24.4.1. $(\mathcal{L}(V, K, f, \xi), h_1, h_2)$ is a solution of the composition conditions if and only if:

*See Theorems 2.21 and 2.22 of Reference 4.

(i) *there exist a morphism k_1 from K_1 into K and a morphism k_2 of K_2 into K such that $k_1(\xi_1(\lambda_1)) = \xi k_1(\lambda_1)$, $k_2(\xi_2(\lambda_2)) = \xi k_2(\lambda_2)$, and $k_1(\lambda_1)k_2(\lambda_2) = k_2(\lambda_2)k_1(\lambda_1)$ for all $\lambda_1 \in K_1$ and $\lambda_2 \in K_2$;*

(ii) *there exists a mapping Θ from $V_1 \times V_2$ into V such that for all $v_1, w_1 \in V_1$, $v_2, w_2 \in V_2$, $\lambda_1 \in K_1$, $\lambda_2 \in K_2$,*

$$\Theta(v_1 + w_1, v_2) = \Theta(v_1, v_2) + \Theta(w_1, v_2),$$

$$\Theta(v_1, v_2 + w_2) = \Theta(v_1, v_2) + \Theta(v_1, w_2),$$

$$\Theta(\lambda_1 v_1, v_2) = k_1(\lambda_1)\Theta(v_1, v_2),$$

$$\Theta(v_1, \lambda_2 v_2) = k_2(\lambda_2)\Theta(v_1, v_2),$$

$$f(\Theta(v_1, v_2), \Theta(w_1, w_2)) = k_1(f_1(v_1, w_1))k_2(f_2(v_2, w_2)) \cdot \lambda$$

for some $\lambda \in K$,

$$\{\Theta(v_1, v_2): v_1 \in M_1^{\circ\circ}, v_2 \in M_2^{\circ\circ}\} \subseteq \{\Theta(v_1, v_2): v_1 \in M_1, v_2 \in M_2\}^{\circ\circ}$$

for arbitrary $M_1 \subseteq V_1$, $M_2 \subseteq V_2$,

$$\{\Theta(v_1, v_2): v_1 \in V_1, v_2 \in V_2\}^{\circ\circ} = V;$$

and

(iii) *the recognition maps h_1, h_2 are related to Θ by*

$$h_1(a_1) = \{\Theta(v_1, v_2): v_1 \in a_1, v_2 \in V_2\}^{\circ\circ},$$

$$h_2(a_2) = \{\Theta(v_1, v_2): v_1 \in V_1, v_2 \in a_2\}^{\circ\circ}$$

(*where the orthogonal operation $^\circ$ is the one defined in Section 21.2*).

A nice feature of this result is that it transfers the problem of solving the composition conditions into a problem of properties of the fields K_1, K_2, K [see item (i)] and of the vector spaces V_1, V_2, V [see item (ii)]. But on the other hand this translation of the problem into terms of fields and vector spaces outlines its difficulty, for no general characterizations are known for the existence conditions occurring in items (i) and (ii). Thus, in the context here considered, the problem of composing two quantum systems has unsolved aspects.

However, the physically minded reader will find the context we have used exceedingly general: he may well find it needlessly exotic to use different number fields underlying the vector spaces associated with the physical systems $\mathbb{S}_1, \mathbb{S}_2, \mathbb{S}$. Thus one might re-pose the problem under the hypothesis that $K_1 = K_2 = K$. Notice however that this does not entail $\xi_1 = \xi_2 = \xi$, for K may possess more than a single involution. If we look at item (i) of Theorem 24.4.1, we now have that k_1 and k_2 are two (possibly different) automorphisms of K, and the condition

$$k_1(\lambda_1)k_2(\lambda_2) = k_2(\lambda_2)k_1(\lambda_1) \qquad \text{for all} \quad \lambda_1, \lambda_2 \in K \quad (24.4.1)$$

represents some sort of commutativity for the field K (it would be precisely the commutativity of K if k_1 and k_2 were the identity mappings). We can expect that (24.4.1) will be enough to rule out some fields: according to Aerts,* it cannot be satisfied when K is the field $\mathbb{Q}$ of quaternions.

This last fact modifies the situation emerging from the discussion of Chapter 21, where we saw that the whole vector-space representation machinery leaves on the same footing the reals, the complexes, and the quaternions, a situation essentially confirmed by the more explicit study of the one-particle system given in Chapter 22. Now we have that, once we come to compound systems, $\mathbb{R}$, $\mathbb{C}$, and $\mathbb{Q}$ are no longer on the same footing, and we are left with arguments against $\mathbb{Q}$.

Let us now briefly see what we can do about the reals and the complexes.

The case $K_1 = K_2 = K = \mathbb{R}$. There is only the identical involution in $\mathbb{R}$; thus $\xi_1 = \xi_2 = \xi = $ identity. Moreover, the automorphisms k_1 and k_2 are necessarily the identity.† According to Section 21.4 the vector spaces V_1, V_2, V become Hilbert spaces, say $\mathcal{H}_1, \mathcal{H}_2, \mathcal{H}$. The properties under item (ii) of Theorem 24.4.1 ensure that Θ is a bilinear mapping of $\mathcal{H}_1 \otimes \mathcal{H}_2$ into $\mathcal{H}$ and, more precisely, that $\mathcal{H}$ coincides with the tensor product $\mathcal{H}_1 \otimes \mathcal{H}_2$. This shows that there is a solution of the composition conditions and that the Hilbert space associated with the compound system is indeed the tensor product of the Hilbert spaces of the subsystems. This reproduces the familiar prescription of Hilbert-space quantum mechanics. It might also be shown that $(\mathcal{L}(\mathcal{H}_1 \otimes \mathcal{H}_2), h_1, h_2)$ is the coproduct of $\mathcal{L}(\mathcal{H}_1)$ and $\mathcal{L}(\mathcal{H}_2)$; hence we have uniqueness, in the sense of Section 24.2 above, of the solution of the composition conditions.

The case $K_1 = K_2 = K = \mathbb{C}$. $\mathbb{C}$ admits various involutions, but, as seen in Section 21.4, only by choosing complex conjugation for the involution can one reach the Hilbert-space structure. Thus assume $\xi_1 = \xi_2 = \xi = $ complex conjugation. As for the automorphisms k_1, k_2 occurring in item (i) of Theorem 24.4.1, each of them can be either the identity or the complex conjugation. Consider first the simplest choice, $k_1 = k_2 = $ identity. Again we have that the properties under item (ii) of Theorem 24.4.1 ensure that Θ is a bilinear mapping from $\mathcal{H}_1 \otimes \mathcal{H}_2$ into $\mathcal{H}$ and, more precisely, that $\mathcal{H}$ is exactly the tensor product $\mathcal{H}_1 \otimes \mathcal{H}_2$. The other choices, namely (1) $k_1 = $ identity, $k_2 = $ complex conjugation, (2) $k_1 = $ complex conjugation, $k_2 = $ identity, (3) $k_1 = k_2 = $ complex conjugation, lead similarly to the following results: (1) $\mathcal{H} = \mathcal{H}_1 \otimes \mathcal{H}_2^*$, (2) $\mathcal{H} = \mathcal{H}_1^* \otimes \mathcal{H}_2$, (3) $\mathcal{H} = \mathcal{H}_1^* \otimes \mathcal{H}_2^*$, where the asterisk denotes the dual. Now, the projection lattices $\mathcal{L}(\mathcal{H}_1 \otimes \mathcal{H}_2)$, $\mathcal{L}(\mathcal{H}_1 \otimes \mathcal{H}_2^*)$, $\mathcal{L}(\mathcal{H}_1^* \otimes \mathcal{H}_2)$,

---

*Result announced in Reference 6.
†The choice $k_1(\lambda) = -\lambda$ [$k_2(\lambda) = -\lambda$] is compatible with item (i) of Theorem 24.4.1 but is excluded by item (ii), for we would have for Θ the null mapping.

and $\mathfrak{L}(\mathcal{H}_1^* \otimes \mathcal{H}_2^*)$ are mutually isomorphic, but nevertheless the associated solutions of the composition conditions (which involve also the recognition maps) are not always mutually related by a commutative diagram like the one of Figure 24.1.* There are essentially two distinct solutions: the one associated with $\mathfrak{L}(\mathcal{H}_1 \otimes \mathcal{H}_2)$ or $\mathfrak{L}(\mathcal{H}_1^* \otimes \mathcal{H}_2^*)$, and the one associated with $\mathfrak{L}(\mathcal{H}_1 \otimes \mathcal{H}_2^*)$ or $\mathfrak{L}(\mathcal{H}_1^* \otimes \mathcal{H}_2)$. This means that, technically, there is no coproduct of $\mathfrak{L}(\mathcal{H}_1)$ and $\mathfrak{L}(\mathcal{H}_2)$. However, from the physical standpoint, the choices involving duals of Hilbert spaces do not appear as genuinely different, but rather as unnecessary complications [as said above, $\mathfrak{L}(\mathcal{H}_1 \otimes \mathcal{H}_2)$, $\mathfrak{L}(\mathcal{H}_1 \otimes \mathcal{H}_2^*)$, $\mathfrak{L}(\mathcal{H}_1^* \otimes \mathcal{H}_2)$, and $\mathfrak{L}(\mathcal{H}_1^* \otimes \mathcal{H}_2^*)$ are isomorphic as lattices]. Thus we come to motivating the use of tensor products to describe compound systems also in the case of complex Hilbert spaces.

References

1. K. E. Hellwig and D. Krauser, *Int. J. Theor. Phys.* 16 (1977) 775.
2. A. Zecca, *J. Math. Phys.* 19 (1978) 1482; in *Current Issues in Quantum Logic*, E. G. Beltrametti and B. C. van Fraassen, eds., Plenum, New York, 1981.
3. D. Aerts and I. Daubechies, *Helv. Phys. Acta* 51 (1978) 661.
4. D. Aerts, *J. Math. Phys.*, 21 (1980) 778.
5. D. Aerts, "Construction of a Structure Which Makes it Possible to Describe the Joint System of a Classical System and a Quantum System," *Rep. Math. Phys.*, to appear.
6. D. Aerts, in *Current Issues in Quantum Logic*, E. G. Beltrametti and B. C. van Fraassen, eds., Plenum, New York, 1981.

*See Theorem 5.2.2 of Reference 5.

CHAPTER 25

Hidden - Variable Theories and Gleason's Theorem

25.1 Noncontextual Hidden-Variable Theories in Hilbert Spaces

In Sections 15.2 and 15.3 we have considered, in a purely lattice-theoretical framework, the noncontextual hidden-variable theories, that is, those hidden-variable theories for which the completed states are thought of as dispersion-free probability measures on the whole logic $\mathcal{L}$ of the system, and we have seen that no-go theorems are encountered if the completed states are required to share typical properties of quantum states such as sufficiency and the property (15.3.1).

If we go to noncontextual hidden-variable theories in Hilbert space, Gleason's theorem makes things more immediate and allows much sharper no-go theorems. Indeed, if the dimension of the Hilbert space is greater than two, Gleason's theorem secures that every probability measure on $\mathcal{P}(\mathcal{H})$ arises from a density operator (and vice versa), with the pure probability measures arising from the projectors onto one-dimensional subspaces of $\mathcal{H}$ (see Section 11.2), so that the existence question for noncontextual hidden-variables theories takes the simple form: can a one-dimensional projector on $\mathcal{H}$ determine a dispersion-free probability measure on $\mathcal{P}(\mathcal{H})$? The negative answer to this question is immediate: writing α_φ for the probability measure arising from the projector $P^{\hat{\varphi}}$ (onto the subspace spanned by the unit vector φ), the probability assigned to an element $P^{\hat{\psi}}$ of $\mathcal{P}(\mathcal{H})$ reduces to

$$\alpha_\varphi(P^{\hat{\psi}}) = |(\varphi, \psi)|^2,$$

so that, whenever ψ is neither orthogonal to nor coincident with φ, we get

ENCYCLOPEDIA OF MATHEMATICS and Its Applications, Gian-Carlo Rota (ed.). Vol. 15: E. G. Beltrametti and G. Cassinelli, The Logic of Quantum Mechanics

ISBN 0-201-13514-0

$\alpha_\varphi(P^{\hat\psi})\neq 0, 1$, showing that it is not dispersion-free. Thus the conclusion that there is no place for noncontextual hidden-variable theories in the framework of Hilbert space when the dimension of the space is (strictly) greater than two.

A negative result of this kind was anticipated by von Neumann[1] many years before Gleason's theorem. Let us recall that his approach was characterized by the following premises:

(a) bounded physical quantities are represented by bounded self-adjoint operators on $\mathcal{H}$ (denoted by $A, B,\ldots$),

(b) states are represented by real-valued functions m on the bounded self-adjoint operators such that

$$m(I)=1, \qquad m(A^2)\geq 0,$$

$$m(\lambda_1 A +\lambda_2 B)=\lambda_1 m(A)+\lambda_2 m(B), \qquad \lambda_1, \lambda_2 \in \mathbb{C},$$

$$m(f(A))=f(m(A)) \qquad \text{for any real-valued Borel function } f.$$

Recall also that the main effort behind his no-go theorem for hidden-variable theories was to establish that states must have the form

$$m(A)=\text{tr}(DA)$$

for some density operator D. To rule out dispersion-free states was then an obvious conclusion.

Seemingly, this was the first time that the physical notion of state was dressed with a precise mathematical definition, an accomplishment that led to important developments. The impact of von Neumann's theorem on hidden-variable theories was quite strong for many years, and it remained the chief result until Gleason's theorem. Advocates of hidden variables strove to find physical weaknesses in the premises: in particular, the linearity condition $m(\lambda_1 A +\lambda_2 B)=\lambda_1 m(A)+\lambda_2 m(B)$ was questioned, at least in the case of noncompatible physical quantities. Of course, this criticism is overcome by Gleason's theorem.

But the production of no-go theorems for noncontextual hidden-variable theories in Hilbert space continued even after Gleason's theorem. In 1966 Bell[2] reproduced the impossibility result without using the full power of Gleason's theorem. In 1967 Kochen and Specker[3] formulated a new no-go theorem, which had common roots with Gleason's theorem and the same domain of application (dimension of the Hilbert space greater than two).

Kochen and Specker framed their theorem in a general scheme that contained some novelty: in place of the traditional proposition lattice they took a partial Boolean algebra, a mathematical object that generalizes the notion of orthomodular lattice. Essentially, a partial Boolean algebra is a poset endowed with a binary reflexive symmetric relation, to be interpreted as a compatibility relation, such that every maximal subset in which all pairs belong to that relation is a Boolean algebra. Orthomodular lattices are examples of partial Boolean algebras, provided we choose as binary relation the usual commutativity relation of orthomodular lattices: the maximal subsets formed by pairwise commuting elements are indeed Boolean algebras with respect to the original lattice operations. Hence, also the projection lattices of the form $\mathcal{P}(\mathcal{H})$ can be thought of as partial Boolean algebras. Let us stress that in talking about lattices the emphasis is on their operational structure (meet, join, orthocomplementation), whereas in talking about partial Boolean algebras the emphasis is on their relational structure (the compatibility relation).

Within this scheme, Kochen and Specker interpret the hidden-variables issue as that of the existence of an embedding of the partial Boolean algebra X of propositions into a Boolean algebra. They show that the existence of such an embedding is fully equivalent to the existence, for every pair of distinct elements a, b of X, of a two-valued homomorphism h (depending upon the pair a, b) from X into the Boolean algebra Z_2 formed by just two elements (0 and 1) such that $h(a) \neq h(b)$. Notice that a two-valued homomorphism of the partial Boolean algebra X is the counterpart of a dispersion-free state on the orthomodular lattice of propositions; hence the analogy between Kochen and Specker's idea and the usual notion of noncontextual hidden-variable theory.

Now, the main point in Kochen and Specker's no-go theorem is to exhibit a finite partial Boolean algebra Y (which is part of the lattice of linear subspaces of the 3-dimensional real Euclidean space; see Section 10.2) that cannot have two-valued homomorphisms, and to show that whenever Y can be embedded into $\mathcal{P}(\mathcal{H})$, thought of as a partial Boolean algebra rather than a lattice, then $\mathcal{P}(\mathcal{H})$ inherits the impossibility of having two-valued homomorphisms. Since Y can be embedded in $\mathcal{P}(\mathcal{H})$ if $\mathcal{H}$ is at least three-dimensional, Kochen and Specker get their no-go theorem for (non-contextual) hidden-variable theories in Hilbert spaces of dimension (strictly) greater than two. As already mentioned, this result is a by-product of Gleason's theorem; nevertheless, it shows that the ruling out of hidden variables can be traced back to properties of a small specimen of partial Boolean algebras.*

---

*Some remarks and improvements on the mathematical side of Kochen and Specker's theorem have been put forward by Czelakowski[4] and Latzer.[5] The behavior of the spin-1 system should provide, in Kochen and Specker's opinion, an intuitive physical account of their no-go theorem.

25.2 Models of Noncontextual Hidden-Variable Theories; Exceptional States

In the sixties, some explicit models of noncontextual hidden-variable theories appeared. At first sight, they appeared as a challenge to the no-go theorems, but on the contrary they finally had a clarifying influence, for they were recognized to lie outside the domain of those theorems, and served to outline the essential role of each of the hypotheses in the theorems. As soon as a hypothesis is dropped, an explicit model of hidden-variable theory can be found.

Here we mainly refer to the models proposed by Bell[2] and by Kochen and Specker.[3] Both models refer to the spin-$\frac{1}{2}$ system, namely, to a quantum system whose Hilbert space is two-dimensional. They do not contradict the no-go theorems based on Gleason's theorem, for it applies to Hilbert spaces of dimension not smaller than three, and they do not contradict the no-go theorems quoted in Section 15.3, for one of the hypotheses is missed, as we shall see.

25.2.1 *Bell's Model*

Consider the complex two-dimensional Hilbert space $\mathbb{C}^2$ associated with the spin-$\frac{1}{2}$ system, and recall (see Section 4.2) that the nontrivial elements of $\mathcal{P}(\mathbb{C}^2)$, that is, the projectors onto one-dimensional subspaces, have the form

$$\frac{1}{2}(I+\vec{r}\cdot\vec{\sigma}),\qquad(25.2.1)$$

where $\vec{r}\cdot\vec{\sigma}$ is the abbreviation for $r_1\sigma_1+r_2\sigma_2+r_3\sigma_3$, with r_1, r_2, r_3 real numbers such that $r_1^2+r_2^2+r_3^2=1$, and $\sigma_1, \sigma_2, \sigma_3$ the usual Pauli matrices.

The projectors of the form (25.2.1) are the quantum pure states of the spin-$\frac{1}{2}$ system: if we write $\alpha_{\vec{s}}$ for the probability measure on $\mathcal{P}(\mathbb{C}^2)$ induced by $\frac{1}{2}(I+\vec{s}\cdot\vec{\sigma})$, we have

$$\alpha_{\vec{s}}(0)=0,\qquad \alpha_{\vec{s}}(I)=1,$$

$$\alpha_{\vec{s}}\left(\tfrac{1}{2}(I+\vec{r}\cdot\vec{\sigma})\right)=\mathrm{tr}\left(\left[\tfrac{1}{2}(I+\vec{s}\cdot\vec{\sigma})\right]\left[\tfrac{1}{2}(I+\vec{r}\cdot\vec{\sigma})\right]\right)=\tfrac{1}{2}(I+\vec{s}\cdot\vec{r}).\quad(25.2.2)$$

Clearly, $\alpha_{\vec{s}}$ is not dispersion-free on $\mathcal{P}(\mathbb{C}^2)$.

In order to construct completed states, take for the space Ω of the hidden variables (see Section 15.2) the real segment $-\frac{1}{2}\leqslant\omega\leqslant\frac{1}{2}$ with the probability measure $d\omega$. Then, for each $\alpha_{\vec{s}}$ and each ω, consider the function $\alpha_{\vec{s},\omega}$ on $\mathcal{P}(\mathbb{C}^2)$ defined by

$$\alpha_{\vec{s},\omega}(0)=0,\qquad \alpha_{\vec{s},\omega}(I)=1,$$

$$\alpha_{\vec{s},\omega}\left(\tfrac{1}{2}(I+\vec{r}\cdot\vec{\sigma})\right)=\tfrac{1}{2}\left[1+\mathrm{sign}\left(\omega+\tfrac{1}{2}|\vec{r}\cdot\vec{s}|\right)\mathrm{sign}(\vec{r}\cdot\vec{s})\right]\qquad(25.2.3)$$

where the "sign function" is given by $\text{sign}(x)=1$ if $x \geqslant 0$ and $\text{sign}(x)=-1$ if $x<0$. It is easily checked that $\alpha_{\vec{s},\omega}$ is indeed a probability measure on $\mathscr{P}(\mathbf{C}^2)$,* and is dispersion-free by construction (the only values it can take are 0 and 1); moreover it satisfies the requirement

$$\alpha_{\vec{s}}(P)=\int_{-1/2}^{1/2}\alpha_{\vec{s},\omega}(P)\,d\omega \qquad \text{for all} \quad P\in\mathscr{P}(\mathbf{C}^2),$$

which ensures that the quantum probabilities are obtained by averaging over the hidden variables [see (15.2.1)], as is seen by use of (25.2.2) and (25.2.3). Thus $\alpha_{\vec{s},\omega}$ behaves precisely as a completed state in the sense of noncontextual hidden-variable theories.

Now, what we want to emphasize is that the probability measure $\alpha_{\vec{s},\omega}$ cannot arise from a density operator, and in that sense deserves the name of "exceptional state". To show this fact it is sufficient to prove that the (unique) extension of $\alpha_{\vec{s},\omega}$ to the set $\mathbb{B}(\mathbf{C}^2)$ of all operators on $\mathbf{C}^2$ is not linear: indeed, if we had $\alpha_{\vec{s},\omega}(P)=\text{tr}(DP)$ for some density operator D and for all $P\in\mathscr{P}(\mathbf{C}^2)$, then the extension of $\alpha_{\vec{s},\omega}$ to $\mathbb{B}(\mathbf{C}^2)$ would obviously be linear. Take a hermitian operator A in $\mathbb{B}(\mathbf{C}^2)$, recall that it has the form (see Section 4.2)

$$A=\vec{\rho}\cdot\vec{\sigma}+\rho_4 I, \qquad \rho_i\in\mathbb{R},$$

and has the spectral decomposition

$$A=\lambda_+P_+ +\lambda_- P_- \qquad \text{with} \quad \lambda_\pm=\rho_4\pm|\vec{\rho}| \text{ and } P_\pm=\frac{1}{2}\left(I\pm\frac{\vec{\rho}}{|\vec{\rho}|}\cdot\vec{\sigma}\right);$$

$$(25.2.4)$$

then notice that the extension, say $\tilde{\alpha}_{\vec{s},\omega}$, of $\alpha_{\vec{s},\omega}$ to $\mathbb{B}(\mathbf{C}^2)$ would give, by use of (25.2.3) and (25.2.4),

$$\tilde{\alpha}_{\vec{s},\omega}(A)=\lambda_+\alpha_{\vec{s},\omega}(P_+)+\lambda_-\alpha_{\vec{s},\omega}(P_-)$$
$$=\rho_4+|\vec{\rho}|\,\text{sign}\big(\omega|\vec{\rho}|+\tfrac{1}{2}|\vec{\rho}\cdot\vec{s}|\big)\,\text{sign}(\vec{\rho}\cdot\vec{s}),$$

and the presence of the term $|\vec{\rho}|$ would exclude the linearity of $\tilde{\alpha}_{\vec{s},\omega}$.

Therefore we see that Bell's model makes use of completed states, the $\alpha_{\vec{s},\omega}$'s, which do not arise from density operators. This is possible just in the case of two-dimensional Hilbert space, where Gleason's theorem does not apply. All no-go theorems based on Gleason's theorem have nothing to say against Bell's model of hidden variables.

---

*To show that $\alpha_{\vec{s},\omega}$ is additive on orthogonal projectors, notice that the only projector orthogonal to $\frac{1}{2}(I+\vec{r}\cdot\vec{\sigma})$ is $\frac{1}{2}(I-\vec{r}\cdot\vec{\sigma})$.

To show that even the no-go theorems considered in Section 15.3 are not contradicted by Bell's model, it is sufficient to show that the completed state $\alpha_{\vec{s},\omega}$ fails to satisfy the property

$$\alpha_{\vec{s},\omega}(P_1)=1 \text{ and } \alpha_{\vec{s},\omega}(P_2)=1 \quad \text{imply} \quad \alpha_{\vec{s},\omega}(P_1\wedge P_2)=1,$$
$$P_1, P_2 \in \mathcal{P}(\mathbb{C}^2), \quad (25.2.5)$$

which was presupposed by those theorems [see (15.3.1)]. Actually we have, more generally, that no dispersion-free state on $\mathcal{P}(\mathbb{C}^2)$ can satisfy the condition (25.2.5). In fact, given any projector on $\mathcal{P}(\mathbb{C}^2)$, a dispersion-free state assigns the value 1 either to that projector or to its orthocomplement; therefore, it is always possible to find two nontrivial projectors that take the value 1 in the dispersion-free state but whose meet is 0, thus violating (25.2.5). [Notice that for the quantum state $\alpha_{\vec{s}}$ it is not possible, as is evident from (25.2.2), to find two nontrivial projectors P_1, P_2 such that $\alpha_{\vec{s}}(P_1)=1$ and $\alpha_{\vec{s}}(P_2)=1$.]

25.2.2 Kochen and Specker's Model

As a starting point we still have the quantum pure states of the spin-$\frac{1}{2}$ system, namely the states $\alpha_{\vec{s}}$ given in (25.2.2), but the space Ω of the hidden variables is now the surface of the unit sphere in $\mathbb{R}^3$, whose elements will be denoted by unit vectors $\vec{\omega}$. The probability measure on Ω is parametrized by the quantum state $\alpha_{\vec{s}}$ according to

$$d\mu_{\vec{s}}(\vec{\omega})=\begin{cases} \dfrac{1}{\pi}\vec{s}\cdot\vec{\omega}\,dv & \text{if } \vec{s}\cdot\vec{\omega}\geqslant 0, \\ 0 & \text{if } \vec{s}\cdot\vec{\omega}<0, \end{cases} \qquad (25.2.6)$$

where dv denotes the invariant measure on the sphere surface.

For each $\alpha_{\vec{s}}$ and each $\vec{\omega}$, the completed state, denoted $\alpha_{\vec{s},\vec{\omega}}$, is constructed as follows:

$$\alpha_{\vec{s},\vec{\omega}}(0)=0, \qquad \alpha_{\vec{s},\vec{\omega}}(I)=1,$$

$$\alpha_{\vec{s},\vec{\omega}}(\tfrac{1}{2}(I+\vec{r}\cdot\vec{\sigma}))=\begin{cases} 1 & \text{if } \vec{\omega}\cdot\vec{r}\geqslant 0, \\ 0 & \text{if } \vec{\omega}\cdot\vec{r}<0. \end{cases} \qquad (25.2.7)$$

The function $\alpha_{\vec{s},\vec{\omega}}$ so defined is indeed a probability measure on $\mathcal{P}(\mathbb{C}^2)$, it is dispersion-free, and it satisfies the requirement of consistency with quantum mechanics [compare (15.2.1)],

$$\alpha_{\vec{s}}(P)=\int_{\Omega}\alpha_{\vec{s},\vec{\omega}}(P)\,d\mu_{\vec{s}}(\vec{\omega}) \qquad \text{for all } P\in\mathcal{P}(\mathbb{C}^2),$$

as seen by explicit use of (25.2.2), (25.2.7), and (25.2.6).*

---

*The calculation of the integral is simplified by choosing polar coordinates on Ω so that $\vec{s}$ and $\vec{r}$ are equatorial and $\vec{s}$ has zero azimuth.

Similarly to what happened with Bell's model, the states $\alpha_{\vec{s}, \vec{\omega}}$ are exceptional, in the sense that they do not arise from density operators. The argument that shows this fact is exactly the same as the one used for the completed states of Bell's model: the extension $\tilde{\alpha}_{\vec{s}, \vec{\omega}}$ of $\alpha_{\vec{s}, \vec{\omega}}$ to $\mathbb{B}(\mathbb{C}^2)$ is not linear. Indeed, if $A = \vec{\rho} \cdot \vec{\sigma} + \rho_4 I$ is a self-adjoint element of $\mathbb{B}(\mathbb{C}^2)$, we have, by use of (25.2.4) and (25.2.7),

$$\tilde{\alpha}_{\vec{s}, \vec{\omega}}(A) = \begin{cases} \rho_4 + |\vec{\rho}| & \text{if} \quad \vec{\omega} \cdot \vec{\rho} \geq 0, \\ \rho_4 - |\vec{\rho}| & \text{if} \quad \vec{\omega} \cdot \vec{\rho} < 0; \end{cases}$$

hence the nonlinearity carried by $|\vec{\rho}|$.

All this makes clear that Kochen and Specker's model of hidden variables, like Bell's model, makes recourse to the only way out left open by Gleason's theorem: the exceptional states that come out when the Hilbert space is two-dimensional. The no-go theorems coming from Gleason's theorem are irrelevant to the model, and the same is true for the no-go theorems discussed in Section 15.3, for we know that dispersion-free states on $\mathcal{P}(\mathbb{C}^2)$ fail to satisfy one of the premises of those theorems [namely the property (15.3.1), as already noticed].

No empirical evidence has emerged in favor of Bell's or Kochen and Specker's model: their interest rests more on the mathematical side. They supply a deficiency of the no-go theorems for noncontextual hidden-variable theories.

25.3 Local Contextual Hidden-Variable Theories; Bell's Inequalities

The particular class of hidden-variable theories considered in this section refers to composite systems and rests on a certain notion of locality that can be introduced for such physical systems.

Take a physical system composed of two nonidentical subsystems $\mathbb{S}_1$ and $\mathbb{S}_2$; let $\mathcal{H}_1$ and $\mathcal{H}_2$ be the Hilbert spaces associated with $\mathbb{S}_1$ and $\mathbb{S}_2$, so that the Hilbert space associated with the compound system $\mathbb{S}_1 + \mathbb{S}_2$ will be the tensor product $\mathcal{H}_1 \otimes \mathcal{H}_2$ (see Section 7.1). Let Φ represent a pure state of $\mathbb{S}_1 + \mathbb{S}_2$, and let A_1 and A_2 be two self-adjoint operators on $\mathcal{H}_1$ and on $\mathcal{H}_2$, respectively. For simplicity, suppose A_1 and A_2 to have simple spectra (see Section 3.2), thus representing physical quantities of $\mathbb{S}_1$ and of $\mathbb{S}_2$, respectively, which form by themselves complete sets of compatible physical quantities. Then we know (see Section 7.1) that $\{A_1 \otimes I_2, I_1 \otimes A_2\}$ is a complete set of compatible physical quantities for $\mathbb{S}_1 + \mathbb{S}_2$ (i.e., it is a complete set of commuting self-adjoint operators on $\mathcal{H}_1 \otimes \mathcal{H}_2$), so that it determines a maximal Boolean σ-algebra $\mathcal{B}$ in $\mathcal{P}(\mathcal{H}_1 \otimes \mathcal{H}_2)$, the algebra generated by projectors contained in the spectral decompositions of $A_1 \otimes I_2$

and $I_1 \otimes A_2$. The reader will notice that in what follows A_1 and A_2 have to be considered as fixed elements, for they determine $\mathcal{B}$, which is, so to speak, the environment of the theory under discussion.

At this stage, within the framework of a contextual hidden-variable theory (see Section 15.2) for the compound system $\mathbb{S}_1 + \mathbb{S}_2$, consider the completed state $(\Phi, \omega_{\mathcal{B}})$, defined on the Boolean σ-algebra $\mathcal{B}$ and dispersion-free on it. This state assigns a definite numerical value to every physical quantity represented by an operator whose spectral decomposition is entirely contained in $\mathcal{B}$. In particular, it assigns definite values to the physical quantities of the form $A_1 \otimes I_2$, $I_1 \otimes A_2$, $A_1 \otimes A_2$, and also to the ones of the form $C_1 \otimes C_2$ with C_1 commuting with A_1 and C_2 with A_2; for the sake of uniformity with previous notations (see, e.g., Section 3.1), we write, for these values, $\mathcal{E}(A_1 \otimes I_2, (\Phi, \omega_{\mathcal{B}}))$, $\mathcal{E}(I_1 \otimes A_2, (\Phi, \omega_{\mathcal{B}}))$, and so on, even if the reference to mean values is made superfluous by the fact that $(\Phi, \omega_{\mathcal{B}})$ is dispersion-free so that mean values are the same as actual values.

The hidden-variable theory is said to be *local* if whenever, in the quantum state Φ, the two subsystems $\mathbb{S}_1$ and $\mathbb{S}_2$ are spatially far apart, so that there is no mutual interaction, the value $\mathcal{E}(C_1 \otimes I_2, (\Phi, \omega_{\mathcal{B}}))$ of every physical quantity of the form $C_1 \otimes I_2$ (with C_1 commuting with A_1) becomes independent of every physical quantity of the form $I_1 \otimes C_2$ (with C_2 commuting with A_2), and vice versa. This notion of locality reflects a notion of causality and conforms to common sense.* Bell[7] considers a hidden-variable theory of this type, $\mathbb{S}_1$ and $\mathbb{S}_2$ being two spin-$\frac{1}{2}$ systems (say an electron and a positron). For the quantum state Φ the "singlet state" is taken, that is,

$$\Phi = \tfrac{1}{2}(\varphi_+ \otimes \psi_- - \varphi_- \otimes \psi_+), \qquad (25.3.1)$$

where φ_+ and φ_- (respectively ψ_+ and ψ_-) are the spin-up and spin-down states of $\mathbb{S}_1$ (respectively of $\mathbb{S}_2$).[†] Notice that only the spin-dependent part of the state vector Φ is considered: the dependence on position and momentum coordinates need not be made explicit, but it is understood that Φ describes $\mathbb{S}_1$ and $\mathbb{S}_2$ far apart. For the physical quantities A_1 and A_2, the spin components along given axes are taken: thus $A_1 = \vec{\sigma}^{(1)} \cdot \vec{r}$ and $A_2 = \vec{\sigma}^{(2)} \cdot \vec{s}$, where $\vec{\sigma}^{(1)}$ and $\vec{\sigma}^{(2)}$ denote the Pauli matrices for $\mathbb{S}_1$ and $\mathbb{S}_2$, respectively, while $\vec{r}$ and $\vec{s}$ are unit vectors in ordinary space. In other words, A_1 represents the spin of $\mathbb{S}_1$ along $\vec{r}$, and A_2 the spin of $\mathbb{S}_2$ along $\vec{s}$.

Then Bell points to the correlation between A_1 and A_2 in the quantum state Φ as a crucial criterion on which the predictions of quantum mechanics cannot agree with those of a local hidden variables theory. Denoting

---

*In Einstein's words:[6] "But on one supposition we should, in my opinion, absolutely hold fast: the real factual situation of the system $\mathbb{S}_1$ is independent of what is done with the system $\mathbb{S}_2$, which is spatially separated from the former".

†Φ is an eigenstate, with zero eigenvalue, of the spin operator of the compound system $\mathbb{S}_1 + \mathbb{S}_2$; in this state only one value (zero) is accessible to the third component of the spin, whence the name "singlet".

this correlation by $\rho(\vec{r}, \vec{s})$, the general rule of quantum mechanics discussed in Section 7.3 gives (expressing variance in terms of expectations)

$$\rho(\vec{r}, \vec{s}) = \frac{\xi}{\eta},$$

where

$$\xi = \mathcal{E}(\vec{\sigma}^{(1)} \cdot \vec{r} \otimes \vec{\sigma}^{(2)} \cdot \vec{s}, \Phi)$$
$$- \mathcal{E}(\vec{\sigma}^{(1)} \cdot \vec{r} \otimes I_2, \Phi) \mathcal{E}(I_1 \otimes \vec{\sigma}^{(2)} \cdot \vec{s}, \Phi),$$

$$\eta = \left[\mathcal{E}((\vec{\sigma}^{(1)} \cdot \vec{r} \otimes I_2)^2, \Phi) - \mathcal{E}^2(\vec{\sigma}^{(1)} \cdot \vec{r} \otimes I_2, \Phi) \right]^{1/2}$$
$$\cdot \left[\mathcal{E}((I_1 \otimes \vec{\sigma}^{(2)} \cdot \vec{s})^2, \Phi) - \mathcal{E}^2(I_1 \otimes \vec{\sigma}^{(2)} \cdot \vec{s}, \Phi) \right]^{1/2},$$

and by a straightforward calculation

$$\mathcal{E}(\vec{\sigma}^{(1)} \cdot \vec{r} \otimes \vec{\sigma}^{(2)} \cdot \vec{s}, \Phi) = -\vec{r} \cdot \vec{s},$$

$$\mathcal{E}(\vec{\sigma}^{(1)} \cdot \vec{r} \otimes I_2, \Phi) = \mathcal{E}(I_1 \otimes \vec{\sigma}^{(2)} \cdot \vec{s}, \Phi) = 0,$$

$$\mathcal{E}((\vec{\sigma}^{(1)} \cdot \vec{r} \otimes I_2)^2, \Phi) = \mathcal{E}((I_1 \otimes \vec{\sigma}^{(2)} \cdot \vec{s})^2, \Phi) = 1,$$

so that

$$\rho(\vec{r}, \vec{s}) = -\vec{r} \cdot \vec{s}. \tag{25.3.2}$$

The problem now arises of how the believer on the local hidden-variable theory will calculate the correlation coefficient, which, in view of possible disagreements with the result (25.3.2), we shall denote $\rho_{\text{H.V.}}(\vec{r}, \vec{s})$. Setting, for short,

$$\mathcal{E}(\vec{\sigma}^{(1)} \cdot \vec{r} \otimes I_2, (\Phi, \omega_{\mathcal{B}})) = \mathcal{E}_1,$$

$$\mathcal{E}(I_1 \otimes \vec{\sigma}^{(2)} \cdot \vec{s}, (\Phi, \omega_{\mathcal{B}})) = \mathcal{E}_2,$$

$$\mathcal{E}((\vec{\sigma}^{(1)} \cdot \vec{r} \otimes I_2)^2, (\Phi, \omega_{\mathcal{B}})) = \mathcal{E}_1',$$

$$\mathcal{E}((I_1 \otimes \vec{\sigma}^{(2)} \cdot \vec{s})^2, (\Phi, \omega_{\mathcal{B}})) = \mathcal{E}_2',$$

the starting point will be, according to the very definition of correlation coefficient,

$$\rho_{H.V.}(\vec{r}, \vec{s}) = \frac{\xi_{\text{H.V.}}}{\eta_{\text{H.V.}}}$$

where

$$\xi_{H.V.} = \int_{\Omega_\mathcal{B}} \mathcal{E}\big(\vec{\sigma}^{(1)} \cdot \vec{r} \otimes \vec{\sigma}^{(2)} \cdot \vec{s}, (\Phi, \omega_\mathcal{B})\big) \mu_\mathcal{B}(d\omega_\mathcal{B})$$

$$- \int_{\Omega_\mathcal{B}} \mathcal{E}_1 \mu_\mathcal{B}(d\omega_\mathcal{B}) \int_{\Omega_\mathcal{B}} \mathcal{E}_2 \mu_\mathcal{B}(d\omega_\mathcal{B}),$$

$$\eta_{H.V.} = \left[\int_{\Omega_\mathcal{B}} \mathcal{E}_1' \mu_\mathcal{B}(d\omega_\mathcal{B}) - \left(\int_{\Omega_\mathcal{B}} \mathcal{E}_1 \mu_\mathcal{B}(d\omega_\mathcal{B}) \right)^2 \right]^{1/2}$$

$$\left[\int_{\Omega_\mathcal{B}} \mathcal{E}_2' \mu_\mathcal{B}(d\omega_\mathcal{B}) - \left(\int_{\Omega_\mathcal{B}} \mathcal{E}_2 \mu_\mathcal{B}(d\omega_\mathcal{B}) \right)^2 \right]^{1/2}$$

At this stage the calculation must proceed along-the guidelines of the requirement of locality and the requirement that in averaging over the hidden variables the quantum expectations be recovered. And the very point raised by Bell is that these two requirements are sometimes mutually exclusive, as we shall see. The requirement of locality states a property of quantities like $\mathcal{E}_1, \mathcal{E}_2, \mathcal{E}_1', \mathcal{E}_2'$ [notice, for the last two, that $(\vec{\sigma}^{(1)} \cdot \vec{r} \otimes I_2)^2 = (\vec{\sigma}^{(1)} \cdot \vec{r})^2 \otimes I_2$ and $(I_1 \otimes \vec{\sigma}^{(2)} \cdot \vec{s})^2 = I_1 \otimes (\vec{\sigma}^{(2)} \cdot \vec{s})^2$]: it says that $\mathcal{E}_1$ and $\mathcal{E}_1'$ do not depend on the direction $\vec{s}$, and that $\mathcal{E}_2$ and $\mathcal{E}_2'$ do not depend on $\vec{r}$. This property raises no difficulty in satisfying the requirement that the averages of these quantities over the hidden variables coincide with the quantum expectations; thus we set

$$\int_{\Omega_\mathcal{B}} \mathcal{E}_1 \mu_\mathcal{B}(d\omega_\mathcal{B}) = \mathcal{E}\big(\vec{\sigma}^{(1)} \cdot \vec{r} \otimes I_2, \Phi\big) = 0,$$

$$\int_{\Omega_\mathcal{B}} \mathcal{E}_2 \mu_\mathcal{B}(d\omega_\mathcal{B}) = \mathcal{E}\big(I_1 \otimes \vec{\sigma}^{(2)} \cdot \vec{s}, \Phi\big) = 0,$$

$$\int_{\Omega_\mathcal{B}} \mathcal{E}_1' \mu_\mathcal{B}(d\omega_\mathcal{B}) = \mathcal{E}\big((\vec{\sigma}^{(1)} \cdot \vec{r} \otimes I_2)^2, \Phi\big) = 1,$$

$$\int_{\Omega_\mathcal{B}} \mathcal{E}_2' \mu_\mathcal{B}(d\omega_\mathcal{B}) = \mathcal{E}\big((I_1 \otimes \vec{\sigma}^{(2)} \cdot \vec{s})^2, \Phi\big) = 1,$$

and the correlation coefficient reduces to

$$\rho_{H.V.}(\vec{r}, \vec{s}) = \int_{\Omega_\mathcal{B}} \mathcal{E}\big(\vec{\sigma}^{(1)} \cdot \vec{r} \otimes \vec{\sigma}^{(2)} \cdot \vec{s}, (\Phi, \omega_\mathcal{B})\big) \mu_\mathcal{B}(d\omega_\mathcal{B}). \qquad (25.3.3)$$

Now, one cannot simply claim that this integral gives back the quantum expectation $\mathcal{E}(\vec{\sigma}^{(1)} \cdot \vec{r} \otimes \vec{\sigma}^{(2)} \cdot \vec{s}, \Phi)$ for it is not clear whether this is allowed by the locality requirements. Indeed, one has not to do with physical quantities of the form $C_1 \otimes I_2$ or $I_1 \otimes C_2$, the ones in terms of which locality is expressed. According to Bell, the believer in local hidden-variable theories takes the next step by writing

$$\rho_{\text{H.V.}}(\vec{r}, \vec{s}) = \int_{\Omega_{\mathcal{B}}} \mathcal{E}_1 \mathcal{E}_2 \mu_{\mathcal{B}}(d\omega_{\mathcal{B}}), \qquad (25.3.4)$$

and thus introducing, in the integrand, objects for which the locality requirement is formulated directly. Of course, the replacement of the integrand of (25.3.3) by that of (25.3.4) is physically natural, for the value of $\vec{\sigma}^{(1)} \cdot \vec{r} \otimes \vec{\sigma}^{(2)} \cdot \vec{s}$ in a dispersion-free state like $(\Phi, \omega_{\mathcal{B}})$ is indeed expected to be the product of the values of $\vec{\sigma}^{(1)} \cdot \vec{r} \otimes I_2$ and $I_1 \otimes \vec{\sigma}^{(2)} \cdot \vec{s}$.

Starting from the expression (25.3.4), Bell was able to show that the requirement of locality, which makes $\mathcal{E}_1$ independent of $\vec{s}$ and $\mathcal{E}_2$ independent of $\vec{r}$, carries an unavoidable disagreement between the correlation $\rho_{\text{H.V.}}(\vec{r}, \vec{s})$ calculated by the believer in local hidden-variable theories and the correlation (25.3.2) calculated according to the rules of quantum mechanics; this disagreement occurs no matter which particular explicit form the local hidden-variable theory takes, that is, no matter what explicit probability space is adopted for the hidden variables.* Bell's arguments consist in proving that for any three unit vectors $\vec{r}, \vec{s}, \vec{t}$ the following inequality must hold:

$$1 + \rho_{\text{H.V.}}(\vec{s}, \vec{t}) \geq |\rho_{\text{H.V.}}(\vec{r}, \vec{s}) - \rho_{\text{H.V.}}(\vec{r}, \vec{t})|, \qquad (25.3.5)$$

an inequality that is not satisfied if $\rho_{\text{H.V.}}$ is replaced by the quantum correlation (25.3.2).[†]

The discovery of the relation (25.3.5), generally known as Bell's inequality, had the great merit of bringing the discussion on hidden-variable theories, at least the local ones, from the rarefied atmosphere of go and no-go theorems onto the ground of experimental tests. The detection of correlation coefficients satisfying Bell's inequality would condemn quantum mechanics and would provide compelling evidence in favor of hidden variables. Unfortunately, the experimental study of Bell's inequality is rather difficult. This fact promoted the search for modified versions of Bell's inequality, involving spin correlations more accessible to experiment. Though the experimental difficulties remain quite severe, a number of experimental

---

*In Section 15.2 we have embodied in the very definition of hidden-variable theories the requirement that quantum expectations be recovered by averaging over the hidden variables [see (15.2.2)]. With this convention, Bell's result would mean that local hidden-variable theories do not exist.

[†] Notice that Bell's inequality has been recognized to be unstable against the introduction of any nonlocal term.[8]

tests have been performed and others are still in progress. A review of this kind of experiments[9, 10] goes beyond the purposes of this volume. We only stress that no evidence in favor of hidden variables has been obtained: the predictions of quantum mechanics have successfully passed the test of these experiments.

References

1. J. von Neumann, *Mathematical Foundations of Quantum Mechanics*, Princeton University Press, Princeton, N.J., 1955.
2. J. S. Bell, *Rev. Mod. Phys.* 38 (1966) 447.
3. S. Kochen and E. Specker, *J. Math. Mech.* 17 (1967) 59.
4. J. Czelakowski, *Studia Logica* 33 (1974) 371; *ibid.* 34 (1975) 69; *Colloq. Math.* 40 (1978) 13.
5. R. W. Latzer, *Synthèse* 29 (1974) 331.
6. A. Einstein, in *Albert Einstein: Philosopher-Scientist*, P. Schilpp, ed., Library of Living Philosophers, Evanston, Ill., 1949 (p. 85).
7. J. S. Bell, *Physics* 1 (1964) 195.
8. M. Flato, C. Piron, J. Grea, D. Sternheimer, and J. P. Vigier, *Helv. Phys. Acta* 48 (1975) 219.
9. M. Paty, in *Quantum Mechanics, a Half Century Later*, J. Leite Lopes and M. Paty, eds., Reidel, Dordrecht, 1977.
10. J. F. Clauser and A. Shimony, *Rep. Progr. Phys.* 41 (1978) 1881.

CHAPTER 26

Introduction to Quantum Probability Theory

26.1 Generalized Probability Spaces; the Hilbert-Space Model

In this chapter we only scratch the surface of quantum probability theory. In the first part of the volume we have presented quantum mechanics as a probabilistic theory, using concepts like distributions, means, and variances in the ordinary sense. We can go further: quantum mechanics is embedded in a theory that is a generalization of classical probability theory. We shall only introduce this subject, and present those aspects that are closest to quantum physics. For a lucid introduction to the world of quantum probability theory we refer to the book of S. P. Gudder,[1] which contains also an extensive bibliography on the subject.

Let α be a probability measure on an orthomodular lattice $\mathcal{L}$. The pair $\langle \mathcal{L}, \alpha \rangle$ is the natural generalization of the classical concept of probability space. Indeed, in Chapter 11 we have defined the notion of probability measure on an orthomodular lattice and we have seen that this definition is the exact analogue of the concept of probability measure on an ordinary probability space.

In Example C of Section 14.2 we have seen that the notion of an observable x associated with an orthomodular lattice $\mathcal{L}$ is the counterpart of the notion of a random variable f on a probability space $\langle \Omega, \Sigma, \mu \rangle$. (This is the standard notation for probability spaces: Ω is a set, Σ a Boolean σ-algebra of subsets of Ω, and μ a probability measure on Σ. This notation has already been used in Section 15.2. In Example C of Chapters 11 and 14 we used α in place of μ to emphasize the homogeneity of the subject matter. Here we return to the standard notation μ.) The probability measures on the

ENCYCLOPEDIA OF MATHEMATICS and Its Applications, Gian-Carlo Rota (ed.).
Vol. 15: E. G. Beltrametti and G. Cassinelli, The Logic of Quantum Mechanics
ISBN 0-201-13514-0

real line

$$E \mapsto \alpha(x(E)), \qquad E \in \mathcal{B}(\mathbb{R}),$$

$$E \mapsto \mu(f^{-1}(E)), \qquad E \in \mathcal{B}(\mathbb{R}),$$

are respectively the distribution of the (nonclassical) random variable x (with respect to the probability measure α) and the distribution of the (classical) random variable f (with respect to the probability measure μ). Led by the strict analogy, we call the pair $\langle \mathcal{L}, \alpha \rangle$ a *generalized probability space*, and in this chapter we shall see that some other concepts of ordinary probability theory can be generalized and translated into the $\langle \mathcal{L}, \alpha \rangle$ structure. This generalization has both mathematical and physical interest: some aspects of the behavior of quantum systems can be faced properly from a probabilistic standpoint only.

To avoid technical complications we consider only bounded random variables: in the classical case this means bounded measurable functions and, in the generalized space $\langle \mathcal{L}, \alpha \rangle$, $\mathcal{L}$-valued measures x such that there exists a real interval $[\lambda_1, \lambda_2]$ for which

$$x([\lambda_1, \lambda_2]) = I.$$

In the sequel, by the term random variable we shall always mean bounded random variable.

The expectation of a random variable has been defined (in the generalized and in the classical case) as the mean value of the probability distribution, that is,

$$\mathcal{E}(x, \alpha) = \int_{\mathbb{R}} \lambda \alpha(x(d\lambda)).$$

$$\mathcal{E}(f, \mu) = \int_{\mathbb{R}} \lambda \mu(f^{-1}(d\lambda)) = \int_{\Omega} f(\omega) \mu(d\omega),$$

At this stage we are faced with a branching point between ordinary and generalized probability spaces. The set of all random variables on a probability space $\langle \Omega, \Sigma, \mu \rangle$ has a natural linear structure, and the expectation is linear on this space. Moreover the restriction of the expectation to the random variables of the form $\chi_a(\omega)$, $a \in \Sigma$, coincides with $\mu(a)$. In other words the probability measure μ in a probability space is the restriction of a linear functional on the linear space of the random variables. In a generalized probability space, as we have seen in Section 14.5, the sum of random variables does not exist in general, and of course α is not the restriction of a linear functional on the set of all random variables associated with $\langle \mathcal{L}, \alpha \rangle$. This is a crucial point, and to develop a satisfactory theory of generalized probability spaces in the face of this fact seems a problem of a higher order of difficulty.[2, 3] This theory is in too early a stage of development and has too technical an aspect to be reported on in this volume.

What is more interesting is examining a particular concrete model of a generalized probability space, namely the physically relevant case where $\mathcal{L}$ is identified with $\mathscr{P}(\mathcal{H})$. We shall consider some probabilistic concepts in $\langle \mathscr{P}(\mathcal{H}), \alpha \rangle,$* for example conditional probabilities and expectations.

Gleason's theorem (see Section 11.2) ensures that if α is a probability measure on $\mathscr{P}(\mathcal{H})$, then it is the restriction of a unique linear functional on $\mathbb{B}(\mathcal{H})$; in fact, if D_α is the density operator of α, the functional

$$A \mapsto \text{tr}(D_\alpha A), \qquad A \in \mathbb{B}(\mathcal{H}),$$

is the unique linear functional on $\mathbb{B}(\mathcal{H})$ whose restriction to $\mathscr{P}(\mathcal{H})$ is α.

We can easily establish a "translation code" between classical probabilistic concepts and mathematical objects in the Hilbert space $\mathcal{H}$. We have

Event of Σ	$\leftrightarrow$	Projection operator of $\mathscr{P}(\mathcal{H})$
Involutive algebra of all complex-valued bounded measurable functions on Ω	$\leftrightarrow$	Involutive algebra $\mathbb{B}(\mathcal{H})$ of all bounded operators on $\mathcal{H}$
Real random variable	$\leftrightarrow$	Bounded projection-valued measure on $\mathscr{B}(\mathbf{R})$, i.e., bounded self-adjoint element of $\mathbb{B}(\mathcal{H})$
Probability measure $\mu(a)$	$\leftrightarrow$	$\alpha(P) = \text{tr}(D_\alpha P)$, $P \in \mathscr{P}(\mathcal{H})$
$\mathcal{E}(f, \mu) = \int_{\mathbf{R}} \lambda \mu(f^{-1}(d\lambda))$, f a real random variable	$\leftrightarrow$	$\mathcal{E}(A, \alpha) = \int_{\mathbf{R}} \lambda \alpha(P_A(d\lambda)) = \text{tr}(D_\alpha A)$, A a self-adjoint element of $\mathbb{B}(\mathcal{H})$

26.2 Conditioning with Respect to an Event

Let b be an event of a classical probability space $\langle \Omega, \Sigma, \mu \rangle$, such that $\mu(b) \neq 0$. It is a known result (see also Exercise 1) that there exists a unique functional, which we shall denote by $\mathbb{P}(\cdot | b)$, defined on Σ, with values in $[0, 1]$, such that:

$$\mathbb{P}(\cdot | b) \text{ is a probability measure on the event space } \langle \Omega, \Sigma \rangle, \quad (26.2.1)$$

$$\text{if} \quad a \leqslant b \quad \text{then} \quad \mathbb{P}(a | b) = \frac{\mu(a)}{\mu(b)}. \quad (26.2.2)$$

---

*In this chapter we shall always understand probability measures on $\mathscr{P}(\mathcal{H})$ to be defined by density operators.

Since $a \mapsto \mu(a \wedge b)/\mu(b)$ fulfills these conditions, we have

$$\mathbb{P}(\cdot \,|\, b) = \frac{\mu(\cdot \wedge b)}{\mu(b)}.$$

We call its value at a—that is, $\mathbb{P}(a\,|\,b)$—the *conditional probability* of a given b.

We define conditional probabilities in the generalized probability space $\langle \mathcal{P}(\mathcal{H}), \alpha \rangle$ in exactly the same way. First of all we have (see Exercise 2) that if $Q \in \mathcal{P}(\mathcal{H})$ is such that $\alpha(Q) \neq 0$, then there exists a unique probability measure $\mathbb{P}_\alpha(\cdot \,|\, Q)$ on $\mathcal{P}(\mathcal{H})$ with the property

$$\mathbb{P}_\alpha(P\,|\,Q) = \frac{\alpha(P)}{\alpha(Q)} \qquad \text{for} \quad P \leqslant Q. \tag{26.2.3}$$

Moreover the density operator of this probability measure, which we shall denote by D'_α, is given explicitly by (see Exercise 3)

$$D'_\alpha = \frac{Q D_\alpha Q}{\operatorname{tr}(D_\alpha Q)}; \tag{26.2.4}$$

hence

$$\mathbb{P}_\alpha(P\,|\,Q) = \frac{\operatorname{tr}(D_\alpha Q P Q)}{\operatorname{tr}(D_\alpha Q)}. \tag{26.2.5}$$

We shall call $\mathbb{P}_\alpha(P\,|\,Q)$ the conditional probability of the event P given the event Q (with respect to the "initial" probability measure α).*

Notice that when P and Q commute, the form of $\mathbb{P}_\alpha(P\,|\,Q)$ reduces to the classical one; in fact in this case

$$\mathbb{P}_\alpha(P\,|\,Q) = \frac{\operatorname{tr}(Q D_\alpha Q P)}{\operatorname{tr}(D_\alpha Q)} = \frac{\operatorname{tr}(D_\alpha Q P Q)}{\operatorname{tr}(D_\alpha Q)} = \frac{\alpha(P \wedge Q)}{\alpha(Q)}.$$

There is a deep difference between classical and quantum conditioning. It arises when we consider conditioning with respect to an event that is the sum of orthogonal elements.

---

*At first sight one might be tempted to define $\mathbb{P}_\alpha(P\,|\,Q) = \alpha(P \wedge Q)/\alpha(Q)$ in strict analogy with the classical case. However, this would not give a probability measure on $\mathcal{P}(\mathcal{H})$ (Exercise 4).

26.2.1 *Classical Case*

Let $b = +_i b_i$ with $\mu(b_i) \neq 0$ for all i, and consider the conditional probability $\mathbb{P}(a|b)$. As seen in Exercise 5, we have

$$\mathbb{P}(a|b) = \sum_i \frac{\mu(b_i)}{\mu(b)} \mathbb{P}(a|b_i); \qquad (26.2.6)$$

hence $\mathbb{P}(\cdot|b)$ has the structure of mixture of the probability measures $\mathbb{P}(\cdot|b_i)$ with weights $\mu(b_i)/\mu(b)$ (a pure probability measure concentrated at a point $\omega \in \Omega$ is obtained only in case b is the atom $\{\omega\}$).

26.2.2 *Quantum Case*

Suppose for simplicity that α is pure, whence $D_\alpha = P^{\hat{\psi}}$, and let $Q = +_i Q_i$ with $\|Q_i \psi\|^2 \neq 0$ for all i. From (26.2.5) it follows (Exercise 6) that

$$\mathbb{P}_\alpha(P|Q) = \left(\frac{Q\psi}{\|Q\psi\|}, P \frac{Q\psi}{\|Q\psi\|} \right) \qquad (26.2.7)$$

and, putting $Q_i\psi / \|Q_i\psi\| = \varphi_i$,

$$\mathbb{P}_\alpha(P|Q) = \sum_i \left(\frac{\|Q_i\psi\|}{\|Q\psi\|} \right)^2 \mathbb{P}_\alpha(P|Q_i)$$

$$+ \sum_{i \neq j} \frac{\|Q_i\psi\| \cdot \|Q_j\psi\|}{\|Q\psi\|^2} (\varphi_i, P\varphi_j). \qquad (26.2.8)$$

The expression (26.2.7) says that, contrary to the classical case, the conditioned state $\mathbb{P}_\alpha(\cdot|Q)$ is pure. More explicitly, (26.2.8) shows that $\mathbb{P}_\alpha(\cdot|Q)$ is the sum of two parts: the first part contains the diagonal terms and is the exact transcription of the classical form (26.2.6); the second part contains the off-diagonal terms, which are the typical quantum interference terms, and it is responsible of the fact that the state $\mathbb{P}_\alpha(\cdot|Q)$ is not a mixture. This quantum conditioning corresponds to concrete physical situations, for instance to the following one. Consider a spin-1 particle and two Stern-Gerlach devices that separate the possible values $-1, 0, +1$ of the spin component along the x- and the y-axis respectively. Let Q be the event "the x-component of the spin is 0 or 1", and let P be the event "the y-component is $+1$". The state α' shown in Figure 26.1 (d stands for a deflector superposing the $J_x = 1$ with the $J_x = 0$ channels, and we are neglecting possible differences in path lengths as well as coherence losses) is precisely $\mathbb{P}_\alpha(\cdot|Q)$, and $\mathbb{P}_\alpha(P|Q)$ is the probability of getting the system in the $J_y = 1$ channel. Notice that we have $Q = Q_1 + Q_2$, where Q_1 and Q_2 are the events "the x-component is $+1$" and "the x-component is 0", respectively.

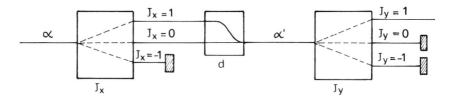

Figure 26.1. Conditioning with interfering channels.

One might wonder whether the first part of the right-hand side of (26.2.8), that is, the quantum transcription of the classical conditional probability (26.2.6), corresponds to some different conditioning device. In the example above we can refer to that of Figure 26.2, where the channels emerging from the first Stern-Gerlach apparatus are made totally independent. The probability measure

$$\left(\frac{\|Q_1\psi\|}{\|Q\psi\|}\right)^2 \mathbb{P}_\alpha(\cdot\,|Q_1) + \left(\frac{\|Q_2\psi\|}{\|Q\psi\|}\right)^2 \mathbb{P}_\alpha(\cdot\,|Q_2)$$

represents the statistical mixture of the $J_x = 1$ and $J_x = 0$ channels, and

$$\left(\frac{\|Q_1\psi\|}{\|Q\psi\|}\right)^2 \mathbb{P}_\alpha(P|Q_1) + \left(\frac{\|Q_2\psi\|}{\|Q_1\psi\|}\right)^2 \mathbb{P}_\alpha(P|Q_2)$$

is the probability of getting the particle in the $J_y = 1$ channel.

Note that the expression (26.2.4), which gives the density operator of the conditional probability with respect to an event, reproduces exactly Lüders's formulation of the projection postulate discussed in Section 8.1. This projection postulate gives the state transformation caused by an ideal first-kind measurement with yes outcome (see also Chapter 16) of the proposition represented by the projector Q. Let us stress that (26.2.4)

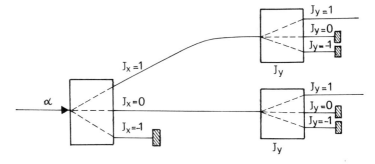

Figure 26.2. Conditioning with independent channels.

describes correctly physical situations both in case Q has the form $P_A(\{\lambda_i\})$ where λ_i is an eigenvalue of A, and in case Q has the form $P_A(E)$ where E is a set in the continuous spectrum of A [see (8.1.1)].

Notice also that the name of Lüders's projection postulate is often used in a different sense, namely, if $A = \sum_i \lambda_i Q_i$, the transformation

$$D_\alpha \to \frac{1}{\mathrm{tr}\left(D_\alpha \sum\limits_{i:\,\lambda_i \in E} Q_i\right)} \sum_{i:\,\lambda_i \in E} Q_i D_\alpha Q_i. \qquad (26.2.9)$$

This corresponds, when α is pure, to the diagonal part of the right-hand side of (26.2.8), and to the physical example of Figure 26.2. The discreteness of the spectrum of A is essential to get (26.2.9). In Section 26.4, (26.2.9) will appear as the result of conditioning with respect to an atomic Boolean sub-σ-algebra of $\mathcal{P}(\mathcal{H})$.

26.3 Two-slit Experiment

In this section we discuss the familiar two-slit experiment using the concept of conditional probabilities. We consider first the passage of a particle through a single slit.

Imagine a free particle traveling toward the screen S_1 in the direction of the x-axis with constant velocity $\vec{v}$ (Figure 26.3). The screen S_1 has a slit F, and after passage across F the particle reaches the screen S_2 (for example a photographic plate), where its position is recorded. We are interested in the probability distribution of the position of the particle on the screen S_2. A common simplifying approximation is to treat classically the motion along the x-axis. Let $t=0$ be the instant at which the particle reaches S_1, and $t=\tau$ the instant at which it reaches S_2; adopt an inertial frame of reference in which the particle has no velocity along the x-axis. Thus we have just to consider the position along the y axis: let $\mathcal{L}^2(\mathbb{R}, dy)$ be the Hilbert space that carries the description of this motion. The conditional probability that the y-coordinate of the particle has a value in the (Borel) set E on the screen S_2

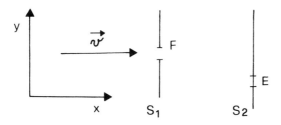

Figure 26.3. One-slit device.

at time $t=\tau$, given that it was localized in the interval F on the screen S_1 at time $t=0$, is

$$\mathbb{P}_\alpha(P(E), t=\tau \,|\, P(F), t=0), \qquad (26.3.1)$$

where α is the state of the incoming particle, and $P(\cdot)$ is the projection-valued measure that represents the position of the particle. We suppose that α is a pure state with density operator $P^{\hat\psi}$, and that the y-coordinate of the particle has a nonvanishing probability of having a value in F at time $t=0$:

$$(\psi, P(F)\psi) = \|P(F)\psi\|^2 \neq 0.$$

Using (26.2.4), we get (see Exercise 7)

$$\mathbb{P}_\alpha(P(E), t=\tau \,|\, P(F), t=0) = (U_\tau \psi_F, P(E) U_\tau \psi_F)$$

$$= \int_E |U_\tau \psi_F(y)|^2 \, dy, \qquad (26.3.2)$$

where U_τ is the free-evolution operator in $\mathcal{L}^2(\mathbb{R}, dy)$, and ψ_F is the "truncation" of ψ caused by F:

$$\psi_F = \frac{P(F)\psi}{\|P(F)\psi\|} = \frac{\chi_F(y)\psi(y)}{\left(\int_F |\psi(y)|^2 \, dy\right)^{1/2}}.$$

It is a well-known fact that $U_\tau \psi_F$ is spread over a support bigger than F, and we have "diffraction" of the particle.

We come now to the two-slit experiment. The situation is the same as before, but in this case the screen S_1 has two slits F_1 and F_2. We make the same simplifying assumptions about the motion along x; let $P^{\hat\psi}$ be the density operator of the initial pure state α, and let $\|P(F_i)\psi\|^2 \neq 0$, $i=1,2$. We are interested in the conditional probability that the y-coordinate of the particle has values in the (Borel) set E of the screen S_2 at time $t=\tau$, given that it was localized in the set $F_1 \cup F_2$ of the screen S_1 at time $t=0$. Writing

$$C_{F_i} = \frac{\|P(F_i)\psi\|}{\|P(F_1 \cup F_2)\psi\|}, \quad \psi_{F_i} = \frac{P(F_i)\psi}{\|P(F_i)\psi\|}, \qquad i=1,2,$$

this conditional probability is given by an expression like (26.2.8), which now takes the form (see Exercise 8)

$$\mathbb{P}_\alpha(P(E), t=\tau \,|\, P(F_1 \cup F_2), t=0) = \int_E |U_\tau(C_{F_1}\psi_{F_1} + C_{F_2}\psi_{F_2})|^2 \, dy.$$

$$(26.3.3)$$

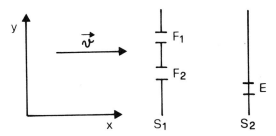

Figure 26.4. Two-slits device.

The interference term

$$2C_{F_1}C_{F_2}\,\mathrm{Re}\int_E \overline{U_\tau\psi_{F_1}(y)}\,U_\tau\psi_{F_2}(y)\,dy$$

is different from zero if $\tau\neq0$; in fact, though ψ_{F_1} and ψ_{F_2} have disjoint supports, U_τ spreads them over the whole y-axis. This term is a peculiar quantum effect, and it is responsible for the empirical fact that the probability of finding the particle in E is not the sum of the probabilities that one would have for each slit separately. The occurrence of the quantum superposition

$$C_{F_1}\psi_{F_1}+C_{F_2}\psi_{F_2}$$

in (26.3.3) is here a clear consequence of the formalism of conditioning with respect to events; it is often obscure in the teaching tradition of quantum mechanics.

26.4 Conditioning with Respect to Boolean Algebras

In a classical probability space $\langle\Omega,\Sigma,\mu\rangle$, let Σ_0 be a sub-σ-algebra of Σ. The conditional probability of the event $a\in\Sigma$ with respect to Σ_0, which we denote by $\mathbb{P}(a|\Sigma_0)$, is defined as the (μ-almost unique) random variable that satisfies the following two conditions:

$$\mathbb{P}(a|\Sigma_0)\text{ is measurable with respect to }\Sigma_0, \qquad (26.4.1)$$

$$\mu(a\wedge b)=\int_b \mathbb{P}(a|\Sigma_0)(\omega)\mu(d\omega) \qquad \text{for all } b\in\Sigma_0. \quad (26.4.2)$$

By use of the translation code of Section 26.1 this definition can be carried into the generalized probability space $\langle\mathcal{P}(\mathcal{H}),\alpha\rangle$,[4] or even into a generalized probability space $\langle\mathcal{L},\alpha\rangle$.[3] However it turns out that the analogues of (26.4.1) and (26.4.2) are too restrictive both from the physical and the mathematical point of view.[5] To avoid these difficulties we restrict ourselves to the conditioning with respect to atomic Boolean sub-σ-algebras of $\mathscr{P}(\mathfrak{K})$, and we proceed to find the explicit form of $\mathbb{P}(a|\Sigma_0)$ with Σ_0 atomic and countably generated. Then we find its generalization in $\langle \mathscr{P}(\mathfrak{K}), \alpha \rangle$ using our "translation code" of Section 26.1.

If $\{b_i\}$ is the countable set of atoms of Σ_0, then (see Exercises 9, 10)

$$\mathbb{P}(a|\Sigma_0) = \sum_{i:\,\mu(b_i)\neq 0} \mathbb{P}(a|b_i)\chi_{b_i}(\omega). \tag{26.4.3}$$

Suppose now that $\mathfrak{B}$ is an atomic Boolean sub-σ-algebra of $\mathscr{P}(\mathfrak{K})$. Since $\mathfrak{K}$ is separable, the set of atoms of $\mathfrak{B}$ [which is an orthogonal set in $\mathscr{P}(\mathfrak{K})$] is countable and $\mathfrak{B}$ is atomistic; we denote by $\{Q_i\}$ the set of atoms of $\mathfrak{B}$.

If P is an element of $\mathscr{P}(\mathfrak{K})$, we define the conditional probability of P with respect to $\mathfrak{B}$, denoted $\mathbb{P}_\alpha(P|\mathfrak{B})$, as the translation of (26.4.3), thus getting

$$\mathbb{P}_\alpha(P|\mathfrak{B}) = \sum_{i:\,\alpha(Q_i)\neq 0} \frac{\operatorname{tr}(Q_i D_\alpha Q_i P)}{\operatorname{tr}(D_\alpha Q_i)} Q_i. \tag{26.4.4}$$

Under our hypotheses it can be shown[6] that $\mathbb{P}_\alpha(P|\mathfrak{B})$ is a random variable in $\langle \mathscr{P}(\mathfrak{K}), \alpha \rangle$, and clearly all its spectral projections (the Q_i) belong to $\mathfrak{B}$; moreover it has the regularity properties[4] that mathematicians usually require.[7]*

Take now the expectation of $\mathbb{P}_\alpha(P|\mathfrak{B})$ in the state α; it is given by

$$\mathcal{E}(\mathbb{P}_\alpha(P|\mathfrak{B}), \alpha) = \sum_{i:\,\alpha(Q_i)\neq 0} \mathbb{P}_\alpha(P|Q_i)\alpha(Q_i).$$

The right-hand side represents the weighted average (over all Q_i's that have nonvanishing probability in the state α) of the conditional probabilities of the event P given the event Q_i. We can now define a new probability measure $\alpha_\mathfrak{B}$ on $\mathscr{P}(\mathfrak{K})$ as

$$\alpha_\mathfrak{B}(\cdot) = \sum_{i:\,\alpha(Q_i)\neq 0} \mathbb{P}_\alpha(\cdot|Q_i)\alpha(Q_i). \tag{26.4.5}$$

Since $Q_i D_\alpha = D_\alpha Q_i = 0$ if and only if $\alpha(Q_i) = 0$, we can also write

$$\alpha_\mathfrak{B}(\cdot) = \operatorname{tr}\left(\left(\sum_i Q_i D_\alpha Q_i\right)\cdot\right).$$

Conditioning with respect to an atomic Boolean sub-σ-algebra is thus

---

*Of course, $\mathbb{P}_\alpha(P|\mathfrak{B})$ will not fulfill, in general, the analogues in $\langle \mathscr{P}(\mathfrak{K}), \alpha \rangle$ of (26.4.1) and (26.4.2). It is a problem to define intrinsically a meaningful conditioning with respect to Boolean sub-σ-algebras, especially the continuous ones: we refer to References 1 and 8.

equivalent to going from the state α to a new state $\alpha_{\mathcal{B}}$ whose density operator is given by

$$D_{\alpha_{\mathcal{B}}} = \sum_i Q_i D_\alpha Q_i. \qquad (26.4.6)$$

This formula represents Luders's projection postulate (26.2.9) in the particular case $E = \mathbb{R}$. To get (26.2.9) exactly, one should generalize the formalism of the present section to the case in which the conditioning is done with respect to atomic Boolean σ-algebras that are subalgebras of segments of $\mathcal{P}(\mathcal{H})$ (see Exercise 13).

Remark. The *information* $I(\alpha)$ of the state α is defined as the Hilbert-Schmidt norm[9] of the density operator D_α:

$$I(\alpha) = \| D_\alpha \|_2 = \left[\mathrm{tr}\left(D_\alpha^2 \right) \right]^{1/2}$$

(a trace-class operator is also Hilbert-Schmidt[9]). The von Neumann entropy of the probability measure α is defined as

$$H(\alpha) = \mathrm{tr}(D_\alpha \ln D_\alpha),$$

and we have that $-H(\alpha)$ has the same extrema as $I(\alpha)$.

The density operator of a probability measure α on $\mathcal{P}(\mathcal{H})$ is an element of the Hilbert space of Hilbert-Schmidt operators,[9] which we denote by $\mathcal{I}_2(\mathcal{H})$. Let $\mathbf{A}$ be the von Neumann algebra $\{Q_i\}'$, where $\{Q_i\}$ is the set of atoms of $\mathcal{B}$. The set of all Hilbert-Schmidt operators whose spectral projections are contained in $\mathbf{A}$ is a closed subspace of $\mathcal{I}_2(\mathcal{H})$, and we denote it by $\mathcal{I}_2(\mathbf{A})$.

Now we have the following geometrical characterization of the state $\alpha_{\mathcal{B}}$ given in (26.4.5): the density operator $D_{\alpha_{\mathcal{B}}}$ is the projection of D_α, thought of as a vector of $\mathcal{I}_2(\mathcal{H})$, onto the closed subspace $\mathcal{I}_2(\mathbf{A})$.[4]

The state $\alpha_{\mathcal{B}}$ admits also a characterization in terms of its information content: it has the maximal information among all probability measures that coincide when restricted to $\mathcal{B}$.[4]

Exercises

1. Let $\langle \Omega, \Sigma, \mu \rangle$ be a probability space, $b \in \Sigma$ such that $\mu(b) \neq 0$. Then there exists a unique functional, $\mathbb{P}(\cdot | b): \Sigma \to [0, 1]$, that satisfies the conditions (26.2.1) (26.2.2).

[*Hint.* Suppose two such functionals, $\mathbb{P}(\cdot | b)$ and $\mathbb{P}'(\cdot | b)$, exist. Write any $a \in \Sigma$ as $a = (a \wedge b) + (a \wedge b^\perp)$. By (26.2.1)

$$\mathbb{P}(a | b) = \mathbb{P}(a \wedge b | b) + \mathbb{P}(a \wedge b^\perp | b),$$
$$\mathbb{P}'(a | b) = \mathbb{P}'(a \wedge b | b) + \mathbb{P}'(a \wedge b^\perp | b).$$

Having $\mathbb{P}(b|b)=\mathbb{P}'(b|b)=1$, we have $\mathbb{P}(b'|b)=\mathbb{P}'(b'|b)=0$; then $\mathbb{P}(a\wedge b'|b)=\mathbb{P}'(a\wedge b'|b)=0$. Moreover $\mathbb{P}(a\wedge b|b)=\mathbb{P}'(a\wedge b|b)$, because $a\wedge b\leqslant b$.]

2. Let α be a probability measure on $\mathcal{P}(\mathcal{H})$, and let $Q\in\mathcal{P}(\mathcal{H})$ be such that $\alpha(Q)\neq0$. Then there exists a unique probability measure on $\mathcal{P}(\mathcal{H})$, which we denote by $\mathbb{P}_\alpha(\cdot|Q)$, such that for all $P\leqslant Q$, $\mathbb{P}_\alpha(P|Q)=\alpha(P)/\alpha(Q)$.

[*Hint.* By Gleason's theorem $\mathbb{P}_\alpha(\cdot|Q)$ must be of the form $\mathrm{tr}(D\cdot)$ for some density operator D. Suppose D_1 and D_2 are two density operators for which $\mathrm{tr}(D_1P)=\mathrm{tr}(D_2P)=\alpha(P)/\alpha(Q)$ for all $P\leqslant Q$. Hence $\mathrm{tr}(D_1Q)=\mathrm{tr}(D_2Q)=1$ and $\mathrm{tr}(D_1Q^\perp)=\mathrm{tr}(D_2Q^\perp)=0$. The two last relations imply that $D_i\varphi=0$ $(i=1,2)$ for all vectors φ in the range of the projector $Q^\perp$. Denoting $\mathcal{Q}=Q\mathcal{H}$ and $\mathcal{Q}^\perp=Q^\perp\mathcal{H}$, we have $\mathcal{H}=\mathcal{Q}\oplus\mathcal{Q}^\perp$. Let R be any one-dimensional projector in $\mathcal{P}(\mathcal{H})$; then $\mathrm{tr}(D_iR)=(\varphi,D_i\varphi)$, $i=1,2$, where φ is any unit vector in the range of R. φ can be uniquely decomposed as $\varphi=\varphi_1+\varphi_2$ where $\varphi_1\in\mathcal{Q}$ and $\varphi_2\in\mathcal{Q}^\perp$. Then we have

$$(\varphi,D_i\varphi)=(\varphi_1,D_i\varphi_1)+(\varphi_2,D_i\varphi_2)+(\varphi_1,D_i\varphi_2)+(\varphi_2,D_i\varphi_1)$$
$$=(\varphi_1,D_i\varphi_1)+(\varphi_2,D_i\varphi_2)+2\,\mathrm{Re}(\varphi_1,D_i\varphi_2),\qquad i=1,2.$$

We have $D_i\varphi_2=0$; then $\mathrm{tr}(D_iR)=(\varphi_1,D_i\varphi_1)=\|\varphi_1\|^2\mathrm{tr}(D_iR_1)$, where R_1 is the projector onto the one-dimensional subspace generated by $\varphi_1(R_1\leqslant Q)$; hence $\mathrm{tr}(D_1R_1)=\mathrm{tr}(D_2R_1)$. This shows that $\mathrm{tr}(D_1R)=\mathrm{tr}(D_2R)$ for all one-dimensional projectors R; thus $D_1=D_2$.]

3. If α is a probability measure on $\mathcal{P}(\mathcal{H})$ and $Q\in\mathcal{P}(\mathcal{H})$ such that $\alpha(Q)\neq0$, then $QD_\alpha Q/\mathrm{tr}(D_\alpha Q)$ is a density operator and

$$\mathbb{P}_\alpha(P|Q)=\frac{\mathrm{tr}(QD_\alpha QP)}{\mathrm{tr}(D_\alpha Q)}=\frac{\mathrm{tr}(D_\alpha QPQ)}{\mathrm{tr}(D_\alpha Q)}.$$

4. Let φ_1 and φ_2 be two vectors in $\mathcal{H}$, and ψ an element of the subspace spanned by φ_1 and φ_2. Show that

$$\frac{\alpha\left((P^{\hat{\varphi}_1}\oplus P^{\hat{\varphi}_2})\wedge P^{\hat{\psi}}\right)}{\alpha(P^{\hat{\psi}})}\neq\frac{\alpha(P^{\hat{\varphi}_1}\wedge P^{\hat{\psi}})}{\alpha(P^{\hat{\psi}})}+\frac{\alpha(P^{\hat{\varphi}_2}\wedge P^{\hat{\psi}})}{\alpha(P^{\hat{\psi}})}.$$

[*Hint.* Note that $(P^{\hat{\varphi}_1}+P^{\hat{\varphi}_2})\wedge P^{\hat{\psi}}=P^{\hat{\psi}}$ and $P^{\hat{\varphi}_i}\wedge P^{\hat{\psi}}=0$, $i=1,2$.]

5. Suppose $\langle b_i\rangle$ is an orthogonal sequence of elements of Σ such that $\mu(b_i)\neq0$ for all i. Show that $\mathbb{P}(a|+b_i)=\Sigma_i[\mu(b_i)/\mu(+b_i)]P(a|b_i)$.
[*Hint.* $\mathbb{P}(a|+b_i)=\mu(a\wedge+b_i)/\mu(+b_i)$; then use distributivity).

6. Let $D_\alpha=P^{\hat{\psi}}$, and $\langle Q_i\rangle$ an orthonormal sequence of elements of $\mathcal{P}(\mathcal{H})$

such that $\|Q_i\psi\|^2 \neq 0$ for all i. Put $Q = +_i Q_i$; then show

(a) $\mathbb{P}_\alpha(P|Q) = \left(\dfrac{Q\psi}{\|Q\psi\|}, P\dfrac{Q\psi}{\|Q\psi\|} \right)$,

(b) $\mathbb{P}_\alpha(P|Q) = \sum_i \left(\dfrac{\|Q_i\psi\|}{\|Q\psi\|} \right)^2 \mathbb{P}_\alpha(P|Q_i) + \sum_{i \neq j} \dfrac{\|Q_i\psi\|\,\|Q_j\psi\|}{\|Q\psi\|^2}(\varphi_i, P\varphi_j)$,

where $\varphi_i = \dfrac{Q_i\psi}{\|Q_i\psi\|}$.

[*Hint*. (a): $\mathbb{P}_\alpha(P|Q) = \dfrac{\mathrm{tr}\left(QP^{\hat\psi}QP\right)}{\mathrm{tr}\left(P^{\hat\psi}Q\right)} = \dfrac{\mathrm{tr}(P\widehat{Q\psi}P)}{\|Q\psi\|^2} = \left(\dfrac{Q\psi}{\|Q\psi\|}, P\dfrac{Q\psi}{\|Q\psi\|} \right)$;

(b): by trivial computation.]

7. Prove (26.3.2).

[*Hint*. $\mathbb{P}_\alpha(P(E),\ t = \tau | P(F),\ t = 0) = \dfrac{\mathrm{tr}\left(P(F)P^{\hat\psi}P(F)U_\tau^{-1}P(E)U_\tau \right)}{\mathrm{tr}\left(P^{\hat\psi}P(F) \right)}$;

denoting $\psi_F = \dfrac{P(F)\psi}{\|P(F)\psi\|}$, we get $\mathrm{tr}(P^{\hat\psi_F}U_\tau^{-1}P(E)U_\tau)$; hence the result.]

8. Prove (26.3.3).
[*Hint*. Note $F_1 \cap F_2 = \varnothing$; hence $P(F_1 \cup F_2) = P(F_1) + P(F_2)$. Use (26.2.5) and proceed by analogy with Exercise 6.]

9. Let $\langle \Omega, \Sigma, \mu \rangle$ be a classical probability space and Σ_0 be a sub-σ-algebra of Σ that is atomic and generated by the countable set $\{b_i\}$ of its atoms. Show that any bounded function f defined in Ω and measurable with respect to Σ_0 has the form

$$f(\omega) = \sum_i \lambda_i \chi_{b_i}(\omega).$$

[*Hint*. Routine verification.]

10. Prove (26.4.3).
[*Hint*. By the previous exercise $\mathbb{P}_\alpha(a|\Sigma_0) = \Sigma_i \lambda_i \chi_{b_i}(\omega)$; using (26.4.2) and chosing $b = b_i$, $\mu(b_i) \neq 0$, we get the desired result.]

11. Find the form of $D_{\alpha_{\mathcal{B}}}$ when each Q_i is one-dimensional.
[*Hint*. If $Q_i = P^{\hat\varphi_i}$, by direct computation $D_{\alpha_{\mathcal{B}}} = \Sigma_i(\varphi_i, D_\alpha\varphi_i)P^{\hat\varphi_i}$.]

12. Find the form of $D_{\alpha_{\mathcal{B}}}$:

(a) when α is pure,

(b) when also $Q_i = P^{\hat\varphi_i}$.
[*Hint*. (a): $D_{\alpha_{\mathcal{B}}} = \displaystyle\sum_{\|Q_i\psi\| \neq 0} \|Q_i\psi\|^2 P^{\widehat{Q_i\psi/\|Q_i\psi\|}}$; (b): $D_{\alpha_{\mathcal{B}}} = \sum_i |(\psi, \varphi_i)|^2 P^{\hat\varphi_i}$.]

13. Show (26.2.9).

[*Hint.* Let $Q = +_i Q_i$, and consider the relatively orthocomplemented lattice $\mathcal{L}[0, Q]$; in this lattice $+_{i: \lambda_i \in E} Q_i = I$, and the set $\{Q_i : \lambda_i \in E\}$ generates an atomic Boolean sub-σ-algebra of this lattice. $\alpha(\cdot)/\alpha(Q)$ is a probability measure on $\mathcal{L}[0, Q]$. Then proceed as in Section 26.4.]

14. Let $\{P^{\hat{\psi}_i}\}_{i=1}^5$ be a set of orthogonal projectors in a 5-dimensional Hilbert space. Denote by $\mathcal{B}$ the atomic Boolean algebra they generate. Consider a vector ψ such that $\|P^{\hat{\psi}_i}\psi\|^2 = \frac{1}{5}$, $i = 1, \ldots, 5$. Consider also a projector $P^{\hat{\varphi}}$. Denote by α the pure state with density operator $P^{\hat{\psi}}$. Then calculate explicitly $\mathbb{P}_\alpha(P^{\hat{\varphi}} | P^{\hat{\psi}_1} + P^{\hat{\psi}_2})$ and $\mathbb{P}_\alpha(P^{\hat{\varphi}} | \mathcal{B})$. [*Answer.* $\mathbb{P}_\alpha(P^{\hat{\varphi}} | P^{\hat{\psi}_1} + P^{\hat{\psi}_2}) = \frac{1}{2}|(\varphi, \psi_1)|^2 + \frac{1}{2}|(\varphi, \psi_2)|^2 + \mathrm{Re}((\varphi, \psi_2)(\psi_1, \varphi))$; $\mathbb{P}_\alpha(P^{\hat{\varphi}} | \mathcal{B}) = \frac{1}{2}|(\varphi, \psi_1)|^2 + \frac{1}{2}|(\varphi, \psi_2)|^2.$]

References

1. S. P. Gudder, *Stochastic Methods in Quantum Mechanics*, North-Holland, New York, 1979.
2. S. P. Gudder, in *Probabilistic Methods in Applied Mathematics*, Vol. 2, A. T. Bharucha-Reid, ed., Academic Press, New York, 1970.
3. G. Cassinelli and P. Truini, *Rep. Math. Phys.*, to appear.
4. S. P. Gudder and J. P. Marchand, *J. Math. Phys.* 13 (1972) 799.
5. M. Takesaki, *J. Funct. Anal.* 9 (1972) 306.
6. S. P. Gudder and J. P. Marchand, *Rep. Math. Phys.* 12 (1977) 317.
7. H. Umegaki, *Tohoku Math. J.* 6 (1954) 177.
8. M. D. Srinivas, *Commun. Math. Phys.* 71 (1980) 131.
9. M. Reed and B. Simon, *Methods of Modern Mathematical Physics*, Vol. I, Academic Press, New York, 1972.

APPENDIX A

Trace-Class Operators

We collect in this appendix some properties of trace-class operators and of the trace functional. Almost everything we need about these topics can be found in Chapter VI of the book by M. Reed and B. Simon (Reference 1 of Chapter 2).

Let $\mathcal{H}$ be any separable, infinite-dimensional, complex Hilbert space. A bounded operator $A \in \mathbb{B}(\mathcal{H})$ is called *positive* when $(\varphi, A\varphi) \geqslant 0$ for all $\varphi \in \mathcal{H}$. For any positive operator A there exists a unique positive operator B such that $B^2 = A$. If A is any (not necessarily positive) bounded operator, then A^*A is positive, so that there exists a unique positive operator, denoted $|A|$, such that $|A|^2 = A^*A$.

Let $\langle \varphi_i \rangle_{i=1}^\infty$ be an orthonormal basis in $\mathcal{H}$; then for any positive operator $A \in \mathbb{B}(\mathcal{H})$ we define

$$\mathrm{tr}\, A = \sum_{i=1}^\infty (\varphi_i, A\varphi_i).$$

The number $\mathrm{tr}\, A$ is independent of the choice of the orthonormal basis $\langle \varphi_i \rangle$.

An operator $A \in \mathbb{B}(\mathcal{H})$ is called a *trace-class operator* when $\mathrm{tr}\,|A| < \infty$. The set of trace-class operators is denoted by $\mathcal{I}_1$. We give some properties of $\mathcal{I}_1$ that are used in this volume:

THEOREM A.1. *$\mathcal{I}_1$ is a vector space, that is, if $A, B \in \mathcal{I}_1$ then $\lambda_1 A + \lambda_2 B \in \mathcal{I}_1$, $\lambda_1, \lambda_2 \in \mathbb{C}$.*

THEOREM A.2. *If $A \in \mathcal{I}_1$ and $B \in \mathbb{B}(\mathcal{H})$, then $AB, BA \in \mathcal{I}_1$.*

ENCYCLOPEDIA OF MATHEMATICS and Its Applications, Gian-Carlo Rota (ed.).
Vol. 15: E. G. Beltrametti and G. Cassinelli, The Logic of Quantum Mechanics

ISBN 0-201-13514-0

THEOREM A.3. *If $A \in \mathcal{I}_1$ then $A^* \in \mathcal{I}_1$.*

THEOREM A.4. *Let $\|\cdot\|_1$ be defined in $\mathcal{I}_1$ by $\|A\|_1 = \mathrm{tr}|A|$. Then $\mathcal{I}_1$ is a Banach space with norm $\|\cdot\|_1$ and $\|A\| \leqslant \|A\|_1$.*

THEOREM A.5. *Every $A \in \mathcal{I}_1$ is compact.*

THEOREM A.6. *If $\langle A_n \rangle$ is a Cauchy sequence with respect to the norm $\|\cdot\|_1$, then its uniform limit A is trace-class and it is also the $\|\cdot\|_1$-limit of $\langle A_n \rangle$.*

THEOREM A.7. *If $A \in \mathcal{I}_1$ and $B \in \mathbb{B}(\mathcal{H})$, then $\|AB\|_1 \leqslant \|A\|_1 \|B\|$.*

THEOREM A.8. *$A \in \mathbb{B}(\mathcal{H})$ is trace-class if and only if, for at least one orthonormal basis $\langle \varphi_i \rangle$ of $\mathcal{H}$, the series $\sum_{i=1}^{\infty} \|A\varphi_i\|$ converges.*[*]

If $A \in \mathcal{I}_1$ and $\langle \varphi_i \rangle$ is any orthonormal basis of $\mathcal{H}$, then the series $\sum_{i=1}^{\infty} (\varphi_i, A\varphi_i)$ is absolutely convergent and its sum is independent of the choice of the basis. We call this sum the *trace* of A, and we denote it by $\mathrm{tr}\, A$.

The trace satisfies the following properties:

THEOREM A.9. $\mathrm{tr}(\cdot)$ *is linear;*

THEOREM A.10. *If $A \in \mathcal{I}_1$ and $B \in \mathbb{B}(\mathcal{H})$, then $\mathrm{tr}\, AB = \mathrm{tr}\, BA$;*

THEOREM A.11. *If $\langle P_i \rangle$ is a sequence of mutually orthogonal projectors, then[†] for any $A \in \mathcal{I}_1$*

$$\mathrm{tr}\, A \sum_i P_i = \sum_i \mathrm{tr}\, AP_i.$$

[*]See Exercise XV-7 of Reference 20 of Chapter 3.
[†]See Exercise 9, Section 1.4 of Reference 4 of Chapter 3.

APPENDIX B _____

The Spectral Theorem

This appendix is devoted to the spectral theorem for self-adjoint operators in Hilbert space. We recall here two among various equivalent formulations of this theorem which are often used throughout this volume. For
proofs, deepenings, and explanations see Reference 1 of Chapter 2.

B.1 The Multiplication-Operator Form of the Spectral Theorem

Let A be a self-adjoint operator, not necessarily bounded, defined in a
separable, complex Hilbert space $\mathcal{H}$, whose domain is $\mathcal{D}(A)$. Then there
exist a measure space $\langle M, \Sigma, \mu \rangle$, where μ is a finite measure; a unitary
operator $U: \mathcal{H} \to L^2(M, \mu)$; and a real-valued function f, defined on M,
which is finite almost everywhere with respect to μ, such that

(a) $\psi \in \mathcal{D}(A)$ if and only if $f(\lambda)(U\psi)(\lambda) \in L^2(M, \mu)$,
(b) if $\varphi \in \{U\psi : \psi \in \mathcal{D}(A)\}$ then $(UAU^{-1}\varphi)(\lambda) = f(\lambda)\varphi(\lambda)$.

All this means that it is always possible to find a Hilbert space of
square-integrable functions, on a suitable (finite) measure space, that is
isometrically isomorphic to $\mathcal{H}$ and in which A is simply a multiplication
operator. This particular realization of $\mathcal{H}$ and A is called the *spectral
representation* of A.

There is a special case, which is worthy of mention, for which the
statement of the preceding theorem can be made more precise. We say that

ENCYCLOPEDIA OF MATHEMATICS and Its Applications, Gian-Carlo Rota (ed.).
Vol. 15: E. G. Beltrametti and G. Cassinelli, The Logic of Quantum Mechanics

ISBN 0-201-13514-0

A is *simple* (or that A has a *simple spectrum*) if there exists $\psi \in \mathcal{D}(A)$ such that the vectors of the form $A^n \psi$, n a positive integer, belong to $\mathcal{D}(A)$, and their finite linear combinations form a set that is dense in $\mathcal{D}(A)$. Such a vector ψ is called a *cyclic vector* for A. The multiplication-operator form of the spectral theorem for simple self-adjoint operators can be stated as follows: Let A be a simple self-adjoint operator with domain $\mathcal{D}(A)$. Then there exist a finite measure μ on the Borel sets of the real line $\mathbb{R}$ that vanishes on the resolvent set of A, and a unitary operator $U: \mathcal{K} \to L^2(\mathbb{R}, \mu)$, such that

(a) $\psi \in \mathcal{D}(A)$ if and only if $\lambda(U\psi)(\lambda) \in L^2(\mathbb{R}, \mu)$,
(b) if $\varphi \in \{U\psi : \psi \in \mathcal{D}(A)\}$ then $(UAU^{-1}\varphi)(\lambda) = \lambda\varphi(\lambda)$.

Thus, in this case, the measure space $\langle M, \Sigma, \mu \rangle$ is simply $\langle \mathbb{R}, \mathcal{B}(\mathbb{R}), \mu \rangle$, and A is, in the function space $L^2(\mathbb{R}, \mu)$, the multiplication by λ, the independent variable of the functions $\varphi(\lambda) \in L^2(\mathbb{R}, \mu)$.

The measure μ that appears in the statements of the preceding theorems is not uniquely determined. We shall briefly discuss this lack of uniqueness in the case where A is simple. We can define another measure ν, possibly only σ-finite, on the Borel subsets of $\mathbb{R}$, by putting

$$\nu(E) = \int_E F(\lambda)\mu(d\lambda), \qquad E \in \mathcal{B}(\mathbb{R}),$$

where F is a Borel function that is positive and nonzero almost everywhere with respect to μ. Then the mapping $V: L^2(M, \mu) \to L^2(M, \nu)$, defined by $\varphi \mapsto V\varphi = (1/\sqrt{F})\varphi$, is unitary and commutes with multiplication by any function defined in $\mathbb{R}$, that is, $g(V\varphi) = V(g\varphi)$. This shows that if the measure μ and the unitary operator U define a spectral representation for A, then ν and $V \circ U: \mathcal{K} \to L^2(\mathbb{R}, \nu)$ define a spectral representation for A. We stress that in this case μ and ν have the same null sets, and this determines the full extent of the ambiguity: every measure (finite or only σ-finite) on $\mathcal{B}(\mathbb{R})$ having the same null sets as μ can be used in the spectral representation of A. As we shall see below, the null sets of μ are an intrinsic property of A.

B.2 The Projection-Valued Measure Form of the Spectral Theorem

A map $E \mapsto P(E)$, from the Borel sets $\mathcal{B}(\mathbb{R})$ into the set of projection operators of some separable complex Hilbert space $\mathcal{K}$, is called a *projection-valued measure* (p.v.m.) when the following conditions are satisfied:

(1) $P(\varnothing) = 0$, $P(\mathbb{R}) = I$.
(2) If $E = \cup_i E_i$, with $E_i \cap E_j = \varnothing$ if $i \neq j$, then $P(E) = \text{s-}\lim_{n \to \infty} \Sigma_{i=1}^n P(E_i)$.

If $E \mapsto P(E)$ is a p.v.m. and ψ is any vector in $\mathcal{H}$, then $E \mapsto (\psi, P(E)\psi)$ is a (finite) measure on the real Borel sets, and we can consider the integration with respect to this measure. If $f(\lambda)$ is a (not necessarily bounded) complex-valued Borel function, define

$$\mathcal{D}_f^P = \left\{ \psi \in \mathcal{H}: \int_{\mathbb{R}} |f(\lambda)|^2 (\psi, P(d\lambda)\psi) < \infty \right\}.$$

$\mathcal{D}_f^P$ is dense in $\mathcal{H}$, and there exists a unique operator A_f^P, with domain $\mathcal{D}_f^P$, such that

$$\left(\psi, A_f^P \psi \right) = \int_{\mathbb{R}} f(\lambda)(\psi, P(d\lambda)\psi) \qquad \text{for all} \quad \psi \in \mathcal{D}_f^P.$$

If f is real-valued, then A_f^P is self-adjoint on $\mathcal{D}_f^P$; if f is bounded, then A_f^P is bounded.

Conversely, if A is a self-adjoint operator with domain $\mathcal{D}(A)$, then there exists a uniquely determined p.v.m., which we denote by $E \mapsto P_A(E)$, such that

$$\mathcal{D}(A) = \left\{ \psi \in \mathcal{H}: \int_{\mathbb{R}} \lambda^2 (\psi, P_A(d\lambda)\psi) \right\}$$

and

$$(\psi, A\psi) = \int_{\mathbb{R}} \lambda(\psi, P_A(d\lambda)\psi) \qquad \text{for all} \quad \psi \in \mathcal{D}(A).$$

This shows that there is a one-to-one correspondence between the set of p.v.m. and the set of self-adjoint operators.

By this form of the spectral theorem we can give meaning to $f(A)$ whenever f is a (not necessarily bounded) complex-valued Borel function. We define $f(A)$ as $A_f^{P_A}$, with domain $\mathcal{D}_f^{P_A}$. If f is real-valued, then $f(A)$ is self-adjoint and

$$P_{f(A)}(E) = P_A\left(f^{-1}[E] \right) \qquad \text{for all} \quad E \in \mathcal{B}(\mathbb{R}).$$

The link between the two forms of the spectral theorem is particularly clear for simple operators. In this case the measure μ that appears in the statement of the multiplication-operator form of the spectral theorem is given by

$$\mu(E) = (\psi, P_A(E)\psi), \qquad E \in \mathscr{B}(\mathbb{R}),$$

where ψ is a cyclic vector for A. From this we see that $\mu(E) = 0$ if and only if $P_A(E) = 0$, and this is a property that depends only on A.

Proofs for Chapter 11

In this appendix we prove statements of Chapter 11. As a rule we shall not repeat the statements themselves, but refer only to their text numeration. We state explicitly, with appropriate numeration, only lemmas whose use is limited to this appendix.

C.1 Results of Section 11.4

Proof of Theorem 11.4.1. We have to show that $\mathcal{S}_1(a) \subseteq \mathcal{S}_1(b)$ implies $a \leqslant b$. We can assume without loss of generality that $a \neq 0$. Then $a = s(\alpha)$ for some $\alpha \in \mathcal{S}$, but $\alpha(s(\alpha)) = 1 \Rightarrow \alpha(b) = 1$; hence $b \geqslant s(\alpha) = a$, by Definition 11.4.3.

LEMMA C.1.1. *Let $\mathcal{L}$ be an orthomodular poset, $\mathcal{S}$ a σ-convex set of probability measures on $\mathcal{L}$, and let $\{\alpha_i\}$ be a sequence of elements of $\mathcal{S}$ such that every α_i has support in $\mathcal{L}$. If the support of every mixture of the sequence $\{\alpha_i\}$ exists in $\mathcal{L}$, then it is given by $\bigvee_i s(\alpha_i)$.*

Proof. Let α be a mixture of the sequence $\{\alpha_i\}$. If $a \in \{b \in \mathcal{L} : b \geqslant s(\alpha)\}$, then $a \in \mathcal{L}_1(\alpha) = \{b \in \mathcal{L} : s(\alpha_i) \leqslant b\}$ for all i. This means that $\{b \in \mathcal{L} : b \leqslant s(\alpha)\} \subseteq \bigcap_i \{b \in \mathcal{L} : s(\alpha_i) \leqslant b\}$. Conversely if $a \in \bigcap_i \{b \in \mathcal{L} : s(\alpha_i) \leqslant b\}$, then $a \in \{b \in \mathcal{L} : s(\alpha_i) \leqslant b\} = \mathcal{L}_1(\alpha_i)$ for all i. Thus $a \in \mathcal{L}_1(\alpha) = \{b \in \mathcal{L} : s(\alpha) \leqslant b\}$ and $\{b \in \mathcal{L} : s(\alpha) \leqslant b\} \supseteq \bigcap_i \{b \in \mathcal{L} : s(\alpha_i) \leqslant b\}$. These facts prove that $\bigcap_i \{b \in \mathcal{L} : s(\alpha_i) \leqslant b\} = \{b \in \mathcal{L} : s(\alpha) \leqslant b\}$. This equality implies $s(\alpha) = \bigvee s(\alpha_i)$, as is easily checked.

Proof of Theorem 11.4.2. Let a_1, a_2 be two elements of $\mathcal{L}$. If one of them (or both) is zero, there is nothing to be proved, so we suppose $a_1, a_2 \neq 0$. By

ENCYCLOPEDIA OF MATHEMATICS and Its Applications, Gian-Carlo Rota (ed.). Vol. 15: E. G. Beltrametti and G. Cassinelli, The Logic of Quantum Mechanics

ISBN 0-201-13514-0

the hypothesis we know that there exist $\alpha_1, \alpha_2 \in \mathbb{S}$ such that $a_1 = s(\alpha_1)$ and $a_2 = s(\alpha_2)$. Then by the convexity of $\mathbb{S}$ there exists $\alpha = t\alpha_1 + (1-t)\alpha_2$, $0 \leqslant t \leqslant 1$, and $s(\alpha) = a_1 \vee a_2$ by Lemma C.1.1. In quite a similar way we can prove the existence of $b = a_1^{\perp} \vee a_2^{\perp}$, whence $b^{\perp} = a_1 \wedge a_2$. The proof of σ-completeness (assuming σ-convexity of $\mathbb{L}$) is identical: it is known[1] that every separable orthomodular lattice where joins of disjoint sequences of elements exist is complete.

Proof of Theorem 11.4.3.   Suppose that $\mathbb{S}$ is sufficient on $\mathbb{L}$ and (11.4.2) holds true. Let us show first that if (11.4.2) holds, then every $\alpha \in \mathbb{S}$ has support in $\mathbb{L}$. If $\mathbb{L}_0(\alpha) = \{0\}$, then $\mathbb{L}_1(\alpha) = \{1\}$ and 1 is the support of α. If $\mathbb{L}_0(\alpha) \neq \{0\}$, then there exists a maximal orthogonal subset of $\mathbb{L}_0(\alpha)$ (by Zorn's lemma) and such a subset is at most countable (by the separability of $\mathbb{L}$); we denote it by $\{a_i\}$. Put $a = +a_i$; then $\alpha(a) = \sum_i \alpha(a_i) = 0$, and $a \leqslant \bigvee (c : c \in \mathbb{L}_0(\alpha)) = b$. If we show that $\alpha(c) = 0 \Rightarrow c \leqslant a$, this will imply $a \geqslant c$ for all $c \in \mathbb{L}_0(\alpha)$; then $a \geqslant b$; hence $a = b$ and $\alpha(b) = 0$. Let $\alpha(c) = 0$, $c \not\leqslant a$; then there exists by hypothesis $d \geqslant c$, a such that $\alpha(d) = 0$, and, by the assumption $c \not\leqslant a$, we have $d \neq a$. We have $d = a + (d - a)$ and $\alpha(d) = \alpha(a) + \alpha(d-a)$; hence $\alpha(d-a) = 0$ and $d - a \in \mathbb{L}_0(\alpha)$. Now $d - a$ is orthogonal to a_i for all i; then $\{a_i\} \cup \{d - a\}$ is an orthogonal set of elements of null α-measure; this contradicts the maximality of $\{a_i\}$. Hence $c \leqslant a$. This shows that $\mathbb{L}_0(\alpha) = \{b \in \mathbb{L} : b \leqslant a\}$ and that $a^{\perp}$ is the support of α. Conversely suppose that if every $\alpha \in \mathbb{S}$ has support in $\mathbb{L}$, then (11.4.2) holds true; notice that if a, b are such that $\alpha(a) = \alpha(b) = 0$, then $a, b \leqslant s(\alpha)^{\perp}$ and $\alpha(s(\alpha)^{\perp}) = 0$; thus we choose $c = s(\alpha)^{\perp}$. Note now that if every nonzero element of $\mathbb{L}$ is the support of some $\alpha \in \mathbb{S}$, then trivially $\mathbb{S}$ is sufficient on $\mathbb{L}$. To prove Theorem 11.4.3 we have to show that if $\mathbb{S}$ is sufficient on $\mathbb{L}$ and (11.4.2) holds, then every $a \in \mathbb{L}$ $(a \neq 0)$ is the support of some $\alpha \in \mathbb{S}$. In fact, let $X = \{b \in \mathbb{L} : b = s(\alpha), \alpha \in \mathbb{S}_1(a)\}$, where a is any nonzero element of $\mathbb{L}$. We have $\mathbb{S}_1(a) \neq \varnothing$ by the sufficiency of $\mathbb{S}$, and of course $X \subseteq \{b \in \mathbb{L} : b \leqslant a\}$. Let $b_1, b_2, \ldots, b_n, \ldots$ be a maximal orthogonal set of elements of $\mathbb{L}$ (countable by the hypothesis of separability). Put $b = +b_i$; of course $b \leqslant a$; if we had $b < a$ strictly, then $b - a \neq 0$ and $b - a \perp b_i$ for all i; then $\{b_i\} \cup \{b - a\}$ would be an orthogonal set of elements of X, contrary to the maximality of $\{b_i\}$. Then $+b_i = a$, and we have $b_i = s(\alpha_i)$ for some sequence $\{\alpha_i\}$ of probability measures (recall that (11.4.2) implies that every $\alpha \in \mathbb{S}$ has support in $\mathbb{L}$). Thus a is the support of any mixture $\alpha = \sum w_i \alpha_i$.

LEMMA C.1.2.   *If $\mathbb{L}$ is a separable poset, $\mathbb{S}$ is a σ-convex set of probability measures on $\mathbb{L}$, and any of the equivalent properties stated in Theorem 11.4.3 holds, then if X is a subset of $\mathbb{L}$ and $\alpha(a) = 0$ for all $a \in X$ we have $\alpha(\bigvee (b : b \in X)) = 0$.*

Proof. By Theorems 11.4.2 and 11.4.3, $\mathcal{L}$ is a complete lattice and the property $\alpha(a)=\alpha(b)=0 \Rightarrow \alpha(a \vee b)=0$ holds. Let $X \subseteq \mathcal{L}$ and $\alpha \in \mathcal{S}_0(X)$. In a separable complete lattice the join of any family of elements is obtainable as the join of a countable subfamily;[1] thus there exists a sequence $\{b_i\} \subseteq X$ such that $\vee(b : b \in X)= \vee_i b_i$. Let $a_n = b_1 \vee \cdots \vee b_n$; then $a_1 \leq a_2 \leq \cdots \leq a_n \leq \cdots$ and $\alpha(a_n)=0$ for all n. We have $\alpha(\vee(b \in X)=\alpha(\vee b_i)= \alpha(\vee a_n)$. Putting $a_0=0$, we have $\vee a_n = +(a_n - a_{n-1})$, and $\alpha(\vee a_n)= \Sigma_n \alpha(a_n - a_{n-1})=\lim_{N \to \infty} \Sigma_{n=1}^N \alpha(a_n - a_{n-1})=\lim_{N \to \infty} \alpha(a_N)=0$.

Proof of Theorem 11.4.4. Assume $\mathcal{S}_1(a) \subseteq \mathcal{S}_1(b)$; then if $\alpha \in \mathcal{S}_1(a)$, also $\alpha \in \mathcal{S}_1(b)$ and $\alpha(a \wedge b)=1$, and hence $\mathcal{S}_1(a) \subseteq \mathcal{S}_1(a \wedge b)$, but also $a \wedge b \leq a$, whence $\mathcal{S}_1(a \wedge b) \subseteq \mathcal{S}_1(a)$. Suppose, if possible, that $a \wedge b < a$ strictly; then $a-(a \wedge b) \neq 0$ and $a-(a \wedge b) \perp a \wedge b$. By the sufficiency of $\mathcal{S}$ there exists $\alpha_0 \in \mathcal{S}$ such that $\alpha_0(a-(a \wedge b))=1$; hence $\alpha_0(a)=1$ and $\alpha_0(a \wedge b)=0$. This shows that $\alpha_0 \in \mathcal{S}_1(a)$ and $\alpha_0 \notin \mathcal{S}_1(a \wedge b)$; then $\mathcal{S}_1(a) \neq \mathcal{S}_1(a \wedge b)$. This is a contradiction. Therefore $a \wedge b=a$, and $a \leq b$.

C.2 Results of Section 11.5

In this section we assume that $\mathcal{L}$ is a complete lattice and $\mathcal{S}$ is a sufficient σ-convex set of probability measures such that every element of $\mathcal{S}$ has support in $\mathcal{L}$. In the proof of Theorem 11.4.3 we have shown that if every element of $\mathcal{S}$ has support in $\mathcal{L}$, then (11.4.1) holds. In Theorem 11.4.4 we have shown that if $\mathcal{S}$ is sufficient and (11.4.1) is satisfied, then $\mathcal{S}$ is strongly ordering on $\mathcal{L}$; moreover, by a straightforward modification of the proof of Lemma C.1.2 we have that in our present hypotheses

$$\text{if} \quad \alpha \in \mathcal{S}_1(X) \ (X \subseteq \mathcal{L}) \quad \text{then} \quad \alpha(\wedge(a \in X))=1.$$

DEFINITION C.2.1. $a \in \mathcal{L}$ is called the *support* of the (nonempty) subset S of $\mathcal{S}$ when a is the smallest element in $\mathcal{L}_1(S)$. We shall denote it by $s(S)$.

Notice that in case S contains a single element this definition agrees with that of the support of a probability measure.

LEMMA C.2.1. $s(S)$ *exists in* $\mathcal{L}$ *and* $s(S)= \wedge(a \in \mathcal{L}: a \in \mathcal{L}_1(S))= \vee(s(\alpha): \alpha \in S)$.

Proof. $\alpha(\wedge(a : a \in \mathcal{L}_1(S)))=1$ if $\alpha \in S$; thus $\wedge(a : a \in \mathcal{L}_1(S))$ is the smallest element in $\mathcal{L}_1(S)$. To show that $\vee(s(\alpha): \alpha \in S)=s(S)$, suppose $a \in \mathcal{L}_1(S)$. Then $a \geq s(\alpha)$ for all $\alpha \in S$, and $a \geq \vee(s(\alpha): \alpha \in S)$. Moreover, $\beta(\vee(s(\alpha): \alpha \in S))=1$ for all $\beta \in S$ implies $\vee(s(\alpha): \alpha \in S) \in \mathcal{L}_1(\beta)$; hence $\vee(s(\alpha): \alpha \in S) \in \mathcal{L}_1(S)$.

LEMMA C.2.2. *If $S \subseteq \mathbb{S}$ and $S \neq \emptyset$ then $\bar{S} = \mathbb{S}_1(s(S))$.*

Proof. To show $\mathbb{S}_1(s(S)) \subseteq \bar{S}$, suppose $\alpha \in \mathbb{S}_1(s(S))$ and $a \in \mathcal{L}_1(S)$. Then obviously $s(S) \leq a$; hence $\alpha(a) \geq \alpha(s(S)) = 1$, which implies $a \in \mathcal{L}_1(\alpha)$, so that $\alpha \in \bar{S}$. To show the reverse inclusion, let $\alpha \in \bar{S}$; then $s(S) \in \mathcal{L}_1(S) \subseteq \mathcal{L}_1(\alpha)$, which implies $\alpha \in \mathbb{S}_1(s(S))$.

Recalling that if $a \in \mathcal{L}$ then $\mathbb{S}_1(a) = \{\alpha \in \mathbb{S} : \alpha(a) = 1\}$, we get the following useful characterization of superposition of a set of probability measures: $\beta \in \bar{S}$ if and only if $s(\beta) \leq \bigvee(s(\alpha) : \alpha \in S)$.

LEMMA C.2.3. *For all $a \in \mathcal{L}$, $a \neq 0$, we have $a = s(\mathbb{S}_1(a))$.*

Proof. Since $a \in \mathcal{L}_1(\mathbb{S}_1(a))$, we have to show that $a \leq b$ for all $b \in \mathcal{L}_1(\mathbb{S}_1(a))$. Now $b \in \mathcal{L}_1(\mathbb{S}_1(a))$ implies $\mathbb{S}_1(a) \subseteq \mathbb{S}_1(b)$; hence $a \leq b$.

THEOREM C.2.1. *$a \mapsto \mathbb{S}_1(a)$ is a bijection between $\mathcal{L}$ and $\mathcal{K}$ that is order-preserving. Its inverse is given by $S \mapsto s(S)$, and it is order-preserving too.*

Proof. $\mathbb{S}_1(a) \in \mathcal{K}$; thus $a \mapsto \mathbb{S}_1(a)$ is well defined. We have to show that this mapping is injective, surjective, and order-preserving. Injectivity: if $\mathbb{S}_1(a) = \mathbb{S}_1(b)$ then $a = b$, because $\mathbb{S}$ is strongly ordering. Surjectivity: by Lemma C.2.2, for all S such that $S = \bar{S}$ we have $S = \mathbb{S}_1(a)$, where $a = s(S)$. Order preserving: trivial. The inverse of the mapping $a \mapsto \mathbb{S}_1(a)$ is given by $S \mapsto s(S)$, by Lemma C.2.3. The mapping $S \mapsto s(S)$ is order-preserving because $\mathbb{S}$ is strongly ordering.

LEMMA C.2.4. *Let $S \subseteq \mathbb{S}$; if $S \mapsto S^\perp$ is the mapping defined in (11.5.5), then $S^\perp = \mathbb{S}_1(s(S)^\perp)$ and $S^{\perp\perp} = \bar{S}$.*

Proof. Suppose first that $S^\perp = \emptyset$: we show that $s(S) = 1$. Indeed, let us suppose that there exists $a \in \mathcal{L}_1(S)$, $a \neq 1$. Since $a^\perp \neq 0$, there exists $\alpha \in \mathbb{S}$ such that $\alpha(a^\perp) = 1$; then $\alpha(a) = 0$ and $\alpha \in S^\perp$, which contradicts the hypothesis. Suppose now $S^\perp \neq \emptyset$. Let $\beta \in S^\perp$; we have $s(\beta) \perp s(\alpha)$ for all $\alpha \in S$; then $s(\beta) \perp \bigvee(s(\alpha) : \alpha \in S) = s(S)$, and hence $\beta(s(S)) = 0$, that is, $\beta \in \mathbb{S}_1(s(S)^\perp)$. Conversely, $\beta \in \mathbb{S}_1(s(S)^\perp)$ implies $\beta(\bigvee(s(\alpha) : \alpha \in S)) = \beta(s(S)) = 0$; then $\beta(s(\alpha)) = 0$ for all $\alpha \in S$; this implies $s(\beta) \perp s(\alpha)$ for all $\alpha \in S$. Then $\beta \perp \alpha$ for all $\alpha \in S$, that is, $\beta \in S^\perp$. Thus the equality $S^\perp = \mathbb{S}_1(s(S)^\perp)$ is proved. To show the second statement of the lemma, we note that, due to the first part of the proof, we have

$$S^{\perp\perp} = \mathbb{S}_1\left((s(S^\perp))^\perp\right) = \mathbb{S}_1\left(s\left(\mathbb{S}_1\left(s(S)^\perp\right)\right)^\perp\right)$$

$$= \mathbb{S}_1\left((s(S)^\perp)^\perp\right) = \mathbb{S}_1(s(S)) = \bar{S}.$$

LEMMA C.2.5. $S \mapsto S^{\perp}$ *is an orthocomplementation in* $\mathcal{K}$.

Proof. By the preceding lemma $S^{\perp}$ is an element of $\mathcal{K}$. Moreover if $S = \bar{S}$, we have $S^{\perp\perp} = S$. For all $S \in \mathcal{K}$ we have $S \wedge S^{\perp} = S \cap S^{\perp} = S \cap \mathcal{S}_1(s(S)^{\perp})$ $= \varnothing$. If $S, T \in \mathcal{K}$ and $S \subseteq T$, then $s(S) \leqslant s(T)$; hence $s(S)^{\perp} \geqslant s(T)^{\perp}$. This implies $\mathcal{S}_1(s(S)^{\perp}) \supseteq \mathcal{S}_1(s(T)^{\perp})$; then $S^{\perp} \supseteq T^{\perp}$.

THEOREM C.2.2. *The mapping* $a \mapsto \mathcal{S}_1(a)$ *and its inverse* $S \mapsto s(S)$ *preserve orthocomplementations.*

Proof. We can assume without loss of generality that $a \neq 0$. By Lemmas C.2.3 and C.2.4 we have

$$\mathcal{S}_1(a^{\perp}) = \mathcal{S}_1\big(s(\mathcal{S}_1(a))^{\perp}\big) = \mathcal{S}_1(a)^{\perp}.$$

Moreover

$$s(S^{\perp}) = s\big(\mathcal{S}_1(s(S)^{\perp})\big) = s(S)^{\perp}.$$

C.3 Results of Section 11.6

Proof of item (i). Assume (11.6.1), let $\beta \in \overline{\{\alpha\}}$ (here by the bar we mean closure with respect to $\mathcal{S}^P$), and let $a \in \mathcal{L}_1(\alpha)$. By $\alpha(a) = 1$ we have $\beta(a) = 1$; then $a \in \mathcal{L}_1(\beta)$. This implies $\mathcal{L}_1(\alpha) \subseteq \mathcal{L}_1(\beta)$; hence $\alpha = \beta$ and $\overline{\{\alpha\}} = \{\alpha\}$. Now assume (11.6.2); let $\mathcal{L}_1(\alpha) \subseteq \mathcal{L}_1(\beta)$ and $\alpha \in \mathcal{S}^P$. Then $\beta \in \overline{\{\alpha\}} = \{\alpha\}$; hence $\alpha = \beta$.

Proof of of item (ii).. Let $\alpha \in \mathcal{S}^P$. If there exists $a \in \mathcal{L}$, $a \neq 0$, such that $a \leqslant s(\alpha)$, then by the sufficiency of $\mathcal{S}^P$, there exists $\beta \in \mathcal{S}^P$ such that $\beta(a) = 1$; hence $\beta(s(\alpha)) = 1$. Therefore $s(\beta) \leqslant s(\alpha)$ and $\mathcal{L}_1(\alpha) \subseteq \mathcal{L}_1(\beta)$. This shows $s(\beta) = s(\alpha)$; since $a \geqslant s(\beta)$, $a \leqslant s(\alpha)$, we get $a = s(\alpha)$. Thus the support of a pure state is an atom. If $a \in \mathcal{L}$, $a \neq 0$, then there exists $\alpha \in \mathcal{S}^P$ such that $\alpha(a) = 1$, and, since $s(\alpha)$ is an atom, $a \geqslant s(\alpha)$. This shows that $\mathcal{L}$ is atomic. Consider now the restriction of s to $\mathcal{S}^P$: it maps $\mathcal{S}^P$ into $\mathcal{C}(\mathcal{L})$, the set of atoms of $\mathcal{L}$. This mapping is injective. In fact, suppose $s(\alpha_1) = s(\alpha_2)$, $\alpha_1, \alpha_2 \in \mathcal{S}^P$; then $\mathcal{L}_1(\alpha_1) \subseteq \mathcal{L}_1(\alpha_2) \Rightarrow \alpha_1 = \alpha_2$. This mapping is also surjective. Let a be an atom; then choose $\alpha \in \mathcal{S}^P$ such that $\alpha(a) = 1$. Since a is an atom and $s(\alpha) \leqslant a$, we have $a = s(\alpha)$. Conversely suppose (A). Then trivially $\mathcal{S}^P$ is sufficient on $\mathcal{L}$. Moreover, if the restriction of s to $\mathcal{S}^P$ is a bijection between $\mathcal{S}^P$ and $\mathcal{C}(\mathcal{L})$, and $\mathcal{L}_1(\alpha) \subseteq \mathcal{L}_1(\beta)$, $\alpha \in \mathcal{S}^P$, then $s(\beta) \leqslant s(\alpha)$. Since $s(\alpha)$ is an atom, we have $s(\beta) = s(\alpha)$, and by the hypothesis of bijection we have $\alpha = \beta$.

Reference

1. N. Zierler, *Pacific J. Math.* 11 (1961) 1151.

Subject Index